T0094026

JOHNS HOPKINS STUDIES IN THE HISTORY OF TECHNOLOGY
Kate McDonald, W. Patrick McCray, and Asif Siddiqi, Series Editors

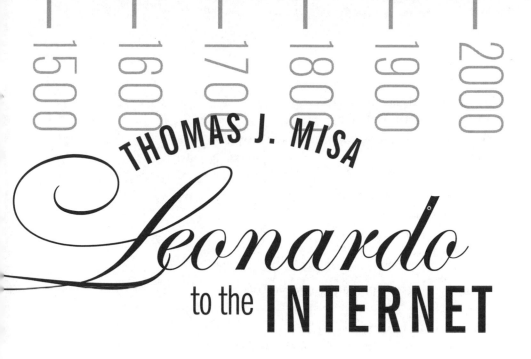

1500 1600 1700 1800 1900 2000

THOMAS J. MISA

Leonardo
to the INTERNET

Technology and Culture from the Renaissance to the Present

THIRD EDITION

Johns Hopkins University Press

Baltimore

© 2022 Johns Hopkins University Press
All rights reserved. Published 2022
Printed in the United States of America on acid-free paper
9 8 7 6 5 4 3 2 1

Johns Hopkins University Press
2715 North Charles Street
Baltimore, Maryland 21218–4363
www.press.jhu.edu

Library of Congress Cataloging-in-Publication Data

Names: Misa, Thomas J., author.
Title: Leonardo to the internet : technology and culture from the
  Renaissance to the present / Thomas J. Misa.
Description: Third edition. | Baltimore : Johns Hopkins University Press,
  2022. | Series: Johns Hopkins studies in the history of technology |
  Includes bibliographical references and index.
Identifiers: LCCN 2021018629 | ISBN 9781421443096 (hardcover) | ISBN
  9781421443102 (paperback) | ISBN 9781421443119 (ebook)
Subjects: LCSH: Technology—History. | Technology and civilization.
Classification: LCC T15 .M575 2021 | DDC 609—dc23
LC record available at https://lccn.loc.gov/2021018629

A catalog record for this book is available from the British Library.

Special discounts are available for bulk purchases of this book.
For more information, please contact Special Sales at specialsales@jh.edu.

CONTENTS

FIGURES AND TABLES

FIGURES

TABLES

PREFACE

This book explores the varied character of technologies over a long period of time, roughly the half-millennium from the Renaissance to the present. It spans the preindustrial past; the age of scientific, political, and industrial revolutions; and more recent topics such as imperialism, modernism, war, global culture, and security. Such a long duration provides a solid empirical base for exploring wide-ranging notions about technology in the final chapter.

This study began years ago in an effort to understand the work of Leonardo da Vinci. The time-honored image of Leonardo as artist and anatomist, who did some nice-looking technical drawings on the side, was not the whole story. Leonardo spent his most active years working as a technologist and an engineer. It is scarcely an exaggeration to say that he did his famous paintings and strikingly realistic anatomy drawings in the periodic lulls between technology projects. I wondered about Leonardo's technical work. Was he really, as some enthusiasts claimed, the "prophet of automation," the inventor of laborsaving machines (and such wonders as helicopters, airplanes, and automobiles) that catapulted Europe from the "dark ages" directly into the modern era? Thinking more deeply, I began to see a distinctive focus in Leonardo and in the numerous engineers with whom he shared notebook drawings and technical treatises. The technological activities of these Renaissance engineers related closely to the concerns of the Renaissance courts and city-states that commissioned their work. Leonardo was not much concerned with laborsaving or "industrial" technologies, and for that matter few of his technological projects generated wealth. Quite the opposite. Leonardo's technologies were typically wealth-*consuming* ones: the technologies of city building, courtly entertainments and dynastic display, and war making.

The Renaissance court system was the conceptual key. While it's well accepted that Johann Gutenberg invented the Western system of moveable type printing, it's not well known that he himself was a court pensioner. Printing shops throughout the late Renaissance depended, to a surprising extent, on court-generated demand. Even printed books on technical subjects were objects of courtly patronage. I began to appreciate that Renaissance courts across Europe, in addition to their well-known support of famous artists, scientists, and philosophers, were at the time the dominant patrons of the most prominent technologists. These included royal courts in Spain and France, ambitious regional rulers in Italy, the papal court in Rome, and courtlike city-states such as Florence. The technical projects they commissioned—from the Florence Cathedral to the mechanical robots for courtly entertainment, as well as the printed works on science, history, philosophy, religion, and technology—created and themselves constituted Renaissance culture. This is chapter 1.

There are good reasons to see the industrial revolution as a watershed in world history, but our common inclination to seize on industrial technologies has confounded a proper understanding of the great commercial expansion that followed the Renaissance. This is chapter 2. Economic historians ask of the great Dutch Golden Age in the seventeenth century: How fast did the Dutch economy grow, and why didn't it industrialize? The Dutch not only had rising per capita incomes and a healthy and diverse national economy; they were also the architects and chief beneficiaries of the first multicentered global economy. Again, on close inspection, I found a distinct character in Dutch technological activities. Commerce, like the courts, fostered distinctive if nonindustrial technologies. Dutch merchants and engineers were highly attuned to generating wealth, and they took active steps to sharpen their focus on making high-quality items with relatively high-paid labor. The typical Dutch cloth was not cheap cotton—the prototypical industrial product—but high-priced woolens, linens, and mohairs. Dutch technologies formed the sinews for their unprecedented international trading system, including shipbuilding, sugar-refining, instrument making, and innovations in finance like joint-stock companies and stock markets. I began not only to think of technologies located historically and spatially in a particular society, and shaped by that society's ideas of what was possible or desirable but also to see how these technologies evolved to shape the society's social and cultural developments. To capture this two-way influence, I took up the notion of distinct "eras" of technology and culture as a way of organizing the material for this book.

This notion of distinct eras provides a kernel for new practical insight into our own social and cultural prospects. This book argues against the common billiard-ball model propounded by many popular writers and journalists who see technologies coming from "outside" society and culture and having "impacts" on them—for good or ill. Indeed, the question whether technologies are "outside" a society or "within" it is a far from trivial matter. If technologies come from outside, the only critical agency open to us is slowing down their inevitable triumph—a rearguard action at best. By contrast, if technologies come from *within* society and are products of ongoing social processes, we can in principle alter them—at least modestly—even as they change us. This book presents an extended empirical evaluation of this question. It shows, in many distinct eras, historical actors actively choosing and changing technologies in an effort to create or sustain their vision of the future (whatever this happened to be).

With these issues in mind, I came to the industrial revolution in Britain with a new interest in comprehending "industrial society" in chapter 3. The older view that the industrial revolution radically transformed Britain in a few decades around 1800, and soon forcibly propelled the world into the modern age, has just not stood the test of time. This version of the billiard-ball model of technology changing society is too simple. In recent years, historians have learned too much about the vast range of historical experiences during industrialization and the dynamics of industrial change and economic growth. For example, surprisingly few industrial workers in industrial Britain actually labored in large-scale industrial factories (only about one in ten), and there were simply too few steam engines to do anything like "transform" the entire economy—until well into the nineteenth century.

In a convulsion of skepticism and debunking, some historians have all but thrown out the concept of an "industrial revolution." But there is a distinct and specific logic that shaped technologies during the early industrial revolution. In industrial-era Britain there were precious few Dutch-style technologists focusing on high-quality materials and high-paid labor. Instead, the predominant focus of British technologists was, let's say, industrial: cutting costs, boosting output, and saving labor. Inventions of the era expressed these socioeconomic goals. Cheap cotton cloth, and lots of it, made by ill-paid factory "hands," was a distinguishing product of industrial-era Britain. If mechanizing industry was not the highest calling in life, as Victorian moralists repeatedly warned, it was nevertheless a central and defining purpose for inventors, engineers, and industrialists of the time. Beyond Britain, commentators and technologists sometimes looked to copy British models of industry but more frequently adapted industrial technologies to their own economic and social contexts. The result was a variety of "paths" through the industrial revolution.

Given these ideas about court, commerce, and industry as defining purposes for technology, I began thinking about what helped define technologies in the next two centuries, closer to home as it were. The task became more difficult. It was impossible to isolate a single distinct "type" of society with a corresponding set of technologies. The legacy of the industrial revolution, it seemed, was not a single "industrial society" with a fixed relationship to technology but rather a multidimensional society with a variety of purposes for technology. Between the middle decades of the nineteenth century and the early decades of the twentieth, we can identify at least three varied purposes—the themes of chapters 4, 5, and 6.

The first of these technology-intensive activities was that of empire, the effort by Europeans and North Americans to extend economic and political control over wide stretches of land abroad or at home. This is chapter 4. Imperialists confronted unprecedented problems in penetrating unfamiliar lands, often in the face of determined resistance by native peoples, and in consolidating military and administrative control over these lands and their peoples. Steamships, telegraphs, and transcontinental railroads were among the technologies that made imperialism effective and affordable. The "gunboat" diplomacy, deployed by the British with great success in China, India, and Africa against lightly armed native peoples, depended on the construction of iron-hulled, steam-driven, shallow-draft vessels that were heavily armed. (Also in the mid-nineteenth century, and in parallel with steamboats, the use of quinine gave Europeans for the first time reasonable odds against endemic malaria in Africa and Asia.) It would be foolish not to recognize the economic dimensions of the imperialist venture, since Britain's factory-made cotton textiles were shipped off to captive markets in India to be exchanged for tea and raw cotton (with a side-trade in opium).

All the same, the disposing of surplus factory goods and the importing of cheap raw materials was not the complete story. No one at the time tried to justify or defend empire in purely economic terms. Feelings of national pride, the accepted imperatives of linking overseas colonies to the homeland, the often-bizarre economics of empire (where tremendously expensive steamboats or railroads "saved" money in transporting military forces or colonial officials), sometimes missionaries' visions of saving the heathen—these were the ways Britain's imperialists coaxed taxpayers' money for such extravagant ventures as the round-the-world telegraph system, easily the most demanding high-technology effort of the nineteenth century. Long-distance repeating telegraphs, efficient coal-burning steamships, and undersea telegraph cables tempted imperial officials to exert oversight and control over far-flung possessions. Imperialists credited the telegraph with "saving" British rule in India during the Mutiny of 1857–58. The same reasoning—national pride, imperial imperatives, and the economics of empire—helps us understand the urgency behind the transcontinental railroads in India, North America, and South Africa. Economics as traditionally understood had little to do with empire or imperial-era technologies.

A second impulse in technology gathering force from the 1870s onward was in the application of science to industry and the building of large systems of technology, the subject of chapter 5. For the first time, in the rise of the science-based chemical and electrical industries, scientific knowledge became as important as land, labor, and capital as a "factor" of production. The new importance of science led to the rise of research-based universities (universities per se were centuries old), government research institutes, and industrial research-and-development laboratories, all of which appeared on a large scale in Germany before 1900 and in the United States a little later. The rise of the chemical, electrical, steel, and petroleum industries, and the associated large corporations that funded and managed them, constituted a "second" industrial revolution. Britain, the first industrial nation, played a surprisingly small role in the movement.

An intensification of interactions between technology, aesthetics, and consumption in the "modern" movement points out that technology in the twentieth century formed the backdrop of our productive apparatus as well as the foreground of our daily lives. The twentieth-century modern movement was built on technological capabilities developed during the science-and-systems era and yet led to new and distinctive cultural results, explored in chapter 6. The achievement around 1900 of mass-produced steel, glass, and other "modern materials" was the material precondition and artistic inspiration for the modern movement in art and architecture between 1900 and 1950. Modernism led not only to avant-garde designs for private houses, public-housing blocks, and office buildings but also to factories, museums, hospitals, and schools. The movement, through its association with German household reformers, the "modernist mafia" known as CIAM, and with the Museum of Modern Art and other self-appointed arbiters of "good taste," shaped public fascination with new modernist designs for domestic and public spaces.

These middle chapters, besides setting down some engaging stories, advance my sustained argument concerning the "question of technology." Highly articulate figures—the artists, architects, and household reformers of the modern movement—self-consciously embraced technology to achieve their positive visions of housing the poor, embracing modern urban life, and even enhancing modern society's cultural and spiritual development. Even if the modernists' enthusiasm for technology seems a bit naïve today, the modern movement encouraged specific technologies as well as shaped cultural developments. And if you are wondering, you will also find in these middle chapters industrial technologies implicated in filthy, disease-ridden cities and imperial technologies implicated in slaughtering native peoples in India and North America. Technology has been and can be a potent agent in disciplining and dominating.

This book began during the waning days of the Cold War. Since then, it has become easier to properly evaluate the superpowers' role in finding and funding military technologies. The "command economies" that took shape during World War II fanned the development of atomic power, microwave radar, and digital computing, as chapter 7 recounts. In these Cold War decades, scientists and engineers learned that the military services had the deepest pockets of all technology patrons. For dreamers and schemers of the most massive technology projects, the military was the main chance. The story is told of the Nazi rocket scientist Wernher von Braun preparing his laboratory's surrender in the chaotic closing days of World War II. "Most of the scientists were frightened of the Russians. They felt the French would treat them like slaves, and the British did not have enough money to afford a rocket program. That left the Americans."[1]

It is worth recalling the social and political changes during the military-dominated era of technology. In all the great powers during the Cold War era, state-imposed secrecy pervaded the weapons, aerospace, nuclear, and intelligence sectors. (With the opening of archives, we have learned just how similar were the Soviet, American, and French nuclear technocrats, for instance, sharing kindred visions of limitless energy, technology-induced social change, and at times contempt of safety concerns.) In the United States the fortunes of research universities, such as the Massachusetts Institute of Technology (MIT) and Stanford, and industrial contractors such as Bell Labs, Boeing, RCA, IBM, and General Electric, depended on the Pentagon. With the end of the Cold War, the sharp contraction of military R&D (research and development) budgets traumatized many technology-based companies, universities, and government institutes. In the West we are comparatively lucky. In the former Soviet Union, frighteningly enough, high-level nuclear technicians were told to find work elsewhere.

We recognize that "globalization" or "global culture" oriented technology and society in the final three decades of the twentieth century, the topic of chapter 8. Think about a world, as recent as 1970, without automatic teller machines and cell phones. Take away ready access to email and the internet and bring back the library's paper-card catalog. Do away with NASDAQ, Microsoft, Dell Computer, and Amazon. com. Make it impossible for a middle-class Western person to invest his or her retirement savings in anything but domestic government bonds. Now ease your way back into the present. Where a troubleshooting phone call to Apple is likely to be answered by a call-center worker in India. Where anyone with an internet connection can risk a stock investment in the Indian software titans Infosys, Tata Consultancy, or Wipro. Where a German media giant owns Penguin and Random House publishers, where

Disney owns a choice piece of Paris, where a private Chinese company owns Volvo, and where McDonald's owns a smaller or larger slice of 117 countries. No wonder that the "promises of global culture" brought both exhilaration and fear.

As the first edition of this book was going to press, I suggested that the pronounced military build-up in the wake of 11 September 2001 had brought an end to the optimistic era of "global culture" and possibly even heralded a new, national security–dominated era for technology.[2] Yet the many billions spent on security did not achieve that goal. Chapter 9 explains how the structure of our technological systems and networks—for energy, information, and global shipping—makes it difficult to build security into them. It also explores how, unintentionally, well-meant decisions about technologies can even increase systemic risk and geopolitical insecurity.

The intellectual framework set out in this book offers a way of thinking about the perennially puzzling "question of technology." Ever since Alvin Toffler's best-selling *Future Shock* (1970), we hear that technology is "quickening" its pace and forcing cultural changes in its wake, that our plunge into the future is driven by technology. Cue Moore's Law, an observation of electronics pioneer Gordon Moore that computer chips double their complexity every year or two. For decades, the increase in complexity meant smaller chips and faster chips. Chapter 10 on dominance of the digital gives the first complete historical assessment of Moore's Law, from its beginnings in 1965, its growth and elaboration with formal roadmapping efforts that synchronized semiconductor R&D, its technical challenges confronting "thermal death" in 2004, and the recent demise in 2016 of the international roadmapping that sustained it worldwide. Chapter 10 also analyzes the transformation of the internet and the unsettling divergence of democratic and authoritarian internets.

My framework demonstrates the character of our society is entailed with our choice of technologies. The "eras" that form the architecture of this book demonstrate that the *adoption* of different technologies encourages the *formation* of distinct types of societies. Sometimes we think there is only one path to the future—a "singularity" seems popular these days—but this book shows there are vastly different paths with consequential differences. The early chapters describe distinct types of societies—court, commerce, industry, and empire—and the different technologies that each selected. No one precisely voted on them, of course, but court patrons during the Italian Renaissance promoted the city-building, military, and entertainment technologies that sustained their vision of court society, and likewise the early-capitalist Dutch commercial era promoted wealth-generating technologies such as economical fluyt ships and high-value material processing that likewise sustained their commercial society. Similar techno-cultural patterns accompanied industry and empire in the nineteenth century as well as modernism and globalism in the twentieth. For a time, these technologies sustained and reinforced the social, cultural, and economic priorities of their respective societies. Eras, even if they do not last forever, are durable.

We do not "choose" technologies from a blank slate of course. Some technologies can be readily undone or bypassed, but many cast a long shadow. These include large-scale transportation and energy infrastructures. The railroads of the imperialist era shaped the political fortunes of southern Africa around 1900, while from the science and systems era of a century ago most of us still experience 120- or 240-volt electricity, depending on where we live. The Fukushima earthquake in 2011 revealed that Japan has two different electricity systems, both unusually at *100* volts but divided by alternating-current frequencies. When the earthquake knocked out the Tokyo Electric Power Company's nuclear plant, it left the grid in the county's northeast (Tokyo, Tohoku, and Hokkaido) operating at its standard 50 hertz, or cycles per second, but severely underpowered. Excess electricity from the southwest (Osaka, Nagoya, Kansai, Kyushu) might have saved the day, but that grid operates at an incompatible 60 hertz and there are only three small-capacity frequency converters in the middle of the country. It turns out that Tokyo and the northeast was wired according to the German model at 50 hertz while Osaka and the southwest followed the US model at 60 hertz. A near 50–50 split between the two incompatible grids bedevils attempts to create a single national grid. In the short term, Japan gutted its environmental aspirations by restarting two giant coal-fired generating plants.[3]

Some technologies can adapt to different types of societies, but often they form deeper "structures" that resist such easy modification or appropriation to distinct cultures. Super-efficient steamships changed the world's economy beginning in the 1880s, so that even countries and cultures that did not directly own steamships still had their economies and society changed profoundly. Low-value jute, wool, mutton, and wheat were sent around the world. Islands in the immense south Pacific—thousands of miles from any sizable port—participated in a global economy of coconut oil production, though today their economic prospects might depend more on tourism than copra. Globalization makes it difficult for societies to form isolated economic areas, whether or not they possess efficient steamships or latter-day container ports. Chapter 9 on systemic risk describes the effects when "flexible" technologies like the internet are extended to the operation of electricity plants, air-traffic control, and global shipping, while chapter 10 examines when and how political values from democratic as well as authoritarian societies are built into the internet.

Finding points of reference to understand the deeper structures of technological society, oddly enough, cuts against current trends in historical reflections on technology. Historians remain skeptical of claims about technology directly shaping history, emphasizing instead the ways that societies and cultures "shape" technologies to reflect societal preferences and values. The "social construction" and "user heuristic" approaches encourage research on the temporally earlier phases of technical change, such as inventive efforts and early emergent technological developments, where substantial user initiatives are often evident and easily documented. In a similar vein, current research in the field documents how different subcultures derive distinct identities and meanings from their intentional adoptions of consumer technology, such as cell phones, retro-style sneakers, alternative music, or customized automobiles. Still, to put it bluntly, all users of internal combustion automobiles—whether Latino "low riders" or Wyoming ranchers or Michigan snowmobilers—are equally dependent on gasoline obtained from a highly integrated global energy system. And, thanks to air pollution and imported fossil fuel, they all contribute to systemic risk of climate change and geopolitical instability (see chapter 9).

Cell phones and social media are apt sites to examine users, but it turns out that what one finds is shaped by what one's looking for. A pioneering study of mobile phones was done by ethnographers Heather Horst and David Miller in 2006, intentionally discounting "generalized observations [that] lend themselves to explanations that invoke technology as causation." Instead, Miller notes, "we simply assumed that you would find a wide range of [individual] appropriations of the phone, depending upon the local cultural context." Users are again at the forefront with his recent *How the World Changed Social Media* (2016). "We showed how the same genre of activity, such as playground banter or the meme, quite happily migrated from BlackBerry phones to Orkut or Facebook or Twitter. If the genre of usage remained largely the same across these very different platforms, then the properties or affordances of the platforms were largely irrelevant to explaining their usage." OK. Focusing on what an *individual* does with Facebook, of course, does not reveal what *Facebook* does with its two billion users. Some alarm bells went off in January 2020 with the announcement of Facebook's Off-Facebook [sic] activity tool that tracks "the information the social network receives about your use of non-Facebook apps and websites." *Forbes* reported that users can "opt out," but they cannot erase or delete the data Facebook is harvesting. Even with our mobile phones switched off, "we're all living in a reality TV program where the cameras are always on."[4]

These currently fashionable celebrations of users, consumption, and the role of subcultures in creating technological identities have come at a price. This emphasis on individual agency has tended to sidestep, downplay, or even ignore the societal effects of large-scale, mature technological systems. These "effects" are never simple, and they are always and everywhere bound up with complex social and cultural processes. Today, our evolving social norms about personal empowerment and instant communication and control over our bodies are expressed through and mediated by cell phones, personal-fitness trackers, and perhaps the internet of things. And yet this is not the whole picture. Technologies can and do structure society and culture. Early computers using the standard ASCII character code—that is, the *American Standard Code for Information Interchange*—easily handled the small set of letters that English makes do with, but they could not manage the accents and "special" characters needed for most European languages, let alone the complex character sets needed for many other languages. With ASCII's wide adoption in the 1960s, computers around the world became "hard wired" for English. In 2010, the Internet Corporation for Assigned Names and Numbers (ICANN) permitted—for the first time—Internet domain names that use character sets other than ASCII, such as Arabic or Chinese. "The coming introduction of non-Latin characters represents the biggest technical change to the Internet since it was created four decades ago," announced the organization's chairman.[5]

Finally, while this book traces a long history from the Renaissance to the very recent past, it in no way argues "the triumph of the present." Our own relationships with technology significantly changed, in scarcely a decade, prompting the second edition of this book, and another decade brought the need for a third edition. Indeed, this work has found its mark if it prompts readers to open up a mental space for thinking both more widely and more deeply about technology. We face supreme challenges in climate change, geopolitical security, and economic disparities, as well as many others, that have not sufficiently engaged the world's technologists. This is a pity. By the end of this book, I hope that readers can understand why this is so. In a nutshell, it goes like this. Societies with distinct goals and aspirations have chosen and sustained certain technologies, and these technologies have powerfully molded the economic, social, and cultural capabilities of the societies that have adopted them. What type of society we wish for the future is at once an open question and a constrained choice. While our societal debate should be an open one, accessible by all members, our options for society tomorrow will be framed by the choices we make about our technologies today.

ACKNOWLEDGMENTS

I am indebted to many people who shared their knowledge and perspectives with me. For comments on the first edition, I would like to thank Henk van den Belt, Mikael Hård, Kevin Harrington, Cheryl Ganz, Thomas Hughes, Richard John, Donna Mehos, Joel Mokyr, Margaret Power, Merritt Roe Smith, and Ed Todd. Arne Kaijser shared with me his enthusiasm about Dutch technology. The chapters also benefited from comments at seminars at University of Pennsylvania, Newberry Library, Technical University of Eindhoven, Science Museum (London), Institute for Architecture and Humanities (Chicago), and Massachusetts Institute of Technology. I tried out ideas about writing large-scale history at the 1999 summer school in history of technology, at Bjerringbro, Denmark, and at a plenary session called "How Do We Write the History of Technology?" at the 1999 annual meeting of the Society for the History of Technology, in Detroit.

For assistance with finding and gathering illustrations, I thank several persons: Hayashi Takeo for permission to use his father's haunting photograph of Nagasaki, Siân Cooksey for the Leonardo drawing held at Windsor Castle, Jenny O'Neill at the MIT Museum, and Johan Schot, Lidwien Kuipers, and Frank Veraart for the images of globalization. Lastly, thanks to Sohair Wastawy for arranging access to Illinois Institute of Technology's special collections.

For the staff at Johns Hopkins University Press, I have the following heartfelt thanks: to Henry Tom for his early enthusiasm for this project (and for his persistence and patience); to Anne Whitmore for her robust editing of the manuscript for the first edition; and to Glen Burris for the design of the book itself.

I dedicated the first edition to the memory of my father, Frank Misa, who was born in the last years of the science-and-systems era, grew up in the era of modernism, lived through the era of war, and grasped the promises of global culture.

I gratefully acknowledge assistance in preparing the second edition of this book. For feedback on the first edition, thanks to my students at the Illinois Institute of Technology (Chicago) and the University of Minnesota (Twin Cities). Thanks also to Dick van Lente (Rotterdam) and Eda Kranakis (Ottawa) for perceptive assessments of the first edition and to Erik van der Vleuten for conceptualizing history, risk, and systems. I benefited from comments and discussion at the Tensions of Europe plenary conference (Budapest), CONTEC seminar (Eindhoven), the Society for Scholarly Publishing (Philadelphia), Carlsberg Academy Conference on the Philosophy of Technology (Copenhagen), University of Michigan (Dearborn), and Max-Planck-Institut für Wissenschaftsgeschichte (Berlin).

My move in 2006 to the University of Minnesota's Charles Babbage Institute decisively shaped the second edition's chapter 9 and its treatment of the history of computing. I have a great many intellectual debts in this new field, starting with daily conversations with Jeffrey Yost and other members of the CBI staff as well as extended conversations with the international community of historians of computing. I also acknowledge feedback and suggestions at Indiana University, the Second Conference on History of Nordic Computing (Turku), the International Congress for the History of Science (Budapest), a summer school at the University of Turku, the European Social Science History Conference (Lisbon and Ghent), and the University of Minnesota. At Johns Hopkins University Press, I thank Michele Callaghan for her perceptive editing. And, not least, thanks to Henry Tom for his well-timed prompt on writing a historical account of the near-present. The news of Henry's death came to me the same week that page proofs for the second edition arrived. Henry's knowledge, editorial judgment, and sense of fit were unsurpassed.

This third edition would scarcely have been possible absent the dozen years I spent at the University of Minnesota's Charles Babbage Institute, with superlative colleagues including Jeffrey Yost, Arvid Nelsen, Katie Charlet, Jim Cortada, and others at CBI as well as enthusiastic faculty and students in the Program for History of Science, Technology and Medicine and supportive colleagues in the Department of Electrical and Computer Engineering. At CBI a research project on Moore's Law directly shapes this edition's new chapter 10. Our CBI research on e-government, women in computing, and computer security informs chapters 9 and 10. And the undergrad course I started on "Digital World" focused chapter 10's treatment of the political internet. Colleagues in the IEEE History Committee and ACM History Committee lent insights and insider perspectives.

I'm grateful also for opportunities to try out these ideas at the University of Minnesota; the Cryptologic History Symposium; a workshop on Computing and Environment, Université Pierre et Marie Curie, France; the Georgia Tech Center for International Strategy, Technology, and Policy; Department of History of Science at the University of Wisconsin; the Faculdade de Ciências e Tecnologia at Universidade

NOVA de Lisboa; Department of Information Science at University of Colorado–Boulder; and Department of History and Philosophy of Science at University of Athens. Colleagues and friends in the Society for the History of Technology kept me grounded during the 2020 coronavirus pandemic. This multi-decade effort was started in Chicago over a dinner conversation with Henry Tom. I learned from his deep experience something about how books work; and I am profoundly indebted to his suggestion of separating a preliminary "journalist's glance" from a more-complete "historian's gaze." At Hopkins I'm again thankful for the support of Matt McAdam, Juliana McCarthy, Kathryn Marguy, Kimberly Johnson, and Andre Barnett. And, for forty years, series editor Merritt Roe Smith has supported my research and writings in the history of technology.

Supplemental materials, including reading questions, images, web resources, and bibliographies can be found at tjmisa.com/L2i/index.html.

CHAPTER 1

1450 – 1600

Technologies of the Court

Even Renaissance men struggled with their careers. Niccolò Machiavelli, celebrated as a political philosopher, and Leonardo da Vinci, hailed as a universal genius, saw tough times during turbulent years in Italy. Machiavelli (1469–1527) grew up in Florence when it was under the control of Lorenzo de' Medici, who ruled the city-state by what we might now call cultural machine politics. Florence was nominally a republic governed by committees elected from its six thousand guild members. But during the long decades of Medici rule, uncooperative committees were disbanded, while the family created and wielded effective power through an elaborate network of cultural patronage, financial preferences, and selective exile. A Medici-inspired vogue for showy urban palaces, lavish rural villas, and prominent churches, convents, and monasteries kept Florence's artists and builders busy—and the Medici coat of arms embedded in the family's numerous buildings kept its citizens ever mindful of the benefits of Medici rule. Machiavelli's two-decade-long public career, serving his beloved but rickety republic, lasted only so long as the Medici dynasty was out of power.[1]

The Medici dynasty in power was more important for Leonardo da Vinci (1452–1519). Leonardo did his apprenticeship in Medici-ruled Florence, worked then and later on projects sponsored by the Medici, and launched his career with the blessing of the Medici. Whether from the Medici family or from his numerous other courtly patrons, Leonardo's career-building commissions were not as a painter, anatomist, or visionary inventor, as he is typically remembered today, but as a military engineer and architect. Courtly patrons, while offering unsurpassed fame and unequaled resources to the artists, philosophers, architects, and engineers they deemed worthy, did not guarantee them lifetime employment. The Medicis' return to power brutally ended Machiavelli's own political career. He was jailed and tortured, then in forced retirement he wrote his famous discourse on political power, *The Prince,* which was dedicated to the new Medici ruler in an unproductive hope of landing a job.

Political turmoil in the court system shaped Leonardo's career as a technologist just as decisively. In 1452, the year of his birth, the city-states of Italy were gaining economic and cultural prominence. Venice dominated Europe's trading with the Near East, Florence flourished as a seat of banking and finance, and Rome doggedly asserted the primacy of the pope. The princely dominions of Milan and Urbino were ruled outright by noble families, who lavishly dispensed artistic and cultural patronage. Leonardo's career was so wide-ranging (he worked at each of these five locations) because the era was politically unstable. And the era was unstable because military conflict was never far away. Rulers up and down the Italian peninsula confronted hostile powers at every turn: to the east the Ottoman Turks, to the north assertive France, and across the peninsula they battled among themselves for valuable land or strategic ports or dynastic advantage. Military engineers such as Leonardo were much in demand. During these decades, secular and religious powers melded uneasily. What else can we say of the cardinals of Rome electing as pope a worldly member of the Borgia family who, at his court, openly showed off his mistress and their four children?

Whether secular or religious, Renaissance courts and city-states had expansive ambitions, and technical activities figured prominently among their objects of patronage. The history of Renaissance technology often seems a story of individual engineers such as Leonardo da Vinci, Francesco di Giorgio, and others, as if they worked as independent free agents; and the history of Renaissance politics and culture is typically related to the shifting fortunes of the varied city-states and noble courts and individuals, like Machiavelli. We will see, however, that technology, politics, and culture were actually never far apart. For a time, Machiavelli and Leonardo even collaborated on a grand engineering plan to improve the commercial and military prospects of Florence by moving its river.

This chapter locates Renaissance technologists squarely within the system of court patronage. We will see that the papal court in Rome sponsored or employed such landmark technological figures as Alberti, Leonardo, and Biringuccio. Leonardo's career as an engineer is inseparable from his work for the Medici family, the Sforza court, and the Borgia clan. The pattern of court-sponsored technologies extended right across Europe and beyond. For "royal factories" and "imperial workshops" in China, Ottoman Turkey, Mughal India, and the Persian Empire, historian Arnold Pacey notes that government control over manufacturing was meant not so much to increase trade but "to supervise production of quality textiles and furniture in support of a magnificent style of court life."[2] Likewise, royal courts in France, Spain, and England supported innovations in silk weaving and even shipbuilding. The well-known history of moveable-type printing looks differently in the light of pervasive court sponsorship of technical books and active court demand for religious publications. Characteristically, Leonardo and his fellow Renaissance-era technologists had surprising little to do with improving industry or making money in the way we typically think of technology today. Instead, Renaissance-era courts commissioned them for demanding technical projects in city-building, courtly entertainment, dynastic display, and the means of war. As one Leonardo scholar observes, "It was within the context of the court that the engineer carried out his many duties, first and foremost of a military nature."[3]

THE CAREER OF A COURT ENGINEER

The practical world of artisans that nurtured Leonardo as a young man was far removed from the educated world of courtly scholars like Machiavelli. In the absence of direct evidence, we must guess at much of Leonardo's early life. Documents recording legal disputes with his stepbrothers make it clear that he was the illegitimate son of a low-level lawyer, and that his paternal grandparents raised him. The clarity of his handwriting points to some formal education, while a desire to hide his inventions from prying eyes is suggested by his famous backward or "mirror" writing. At age fourteen he was apprenticed in the Florence workshop of sculptor and painter Andrea del Verrocchio. While Italy's thriving universities were the sites for educating scholars, workshops such as Verrocchio's principally educated artisans and craftsmen. In his decade with Verrocchio (1466–76), Leonardo learned the basics of architecture, sculpture, bronze casting, painting, and possibly some mathematics. During these years, Verrocchio had several major commissions, including an ostentatious family tomb for the Medici and a large bronze statue commemorating a Venetian military hero.

As a member of Verrocchio's workshop, Leonardo had a small but visually prominent role in building the Florence cathedral. Officially the Cathedral of Santa Maria del Fiore, it was described as an "enormous construction towering above the skies, vast enough to cover the entire Tuscan population with its shadow . . . a feat of engineering that people did not believe feasible . . . equally unknown and unimaginable among the ancients" (fig. 1.1). In 1420, more than a century after the cornerstone was laid, Filippo Brunelleschi (1377–1446) gained the city's approval for completing its impressive dome and began work. Brunelleschi's design followed the official plans for an eight-ribbed dome, more than 100 feet high, which began 170 feet above the cathedral's floor. The dome measured more than 140 feet in diameter—then and now the largest masonry dome in the world. Brunelleschi directed that the dome's lower level be built of solid stone and its upper levels consist of a two-layered shell of stone and brick; at its apex stood a 70-foot-high lantern tower to admit light and air. To construct this novel structure, without wooden bracing from underneath, Brunelleschi devised numerous special cranes and hoisting apparatuses that could lift into place the heavy stones and bricks. Brunelleschi, often with Cosimo de' Medici's support, designed numerous other buildings in Florence, including the Ospedale degli Innocenti, a charitable orphanage commissioned by the silk merchants guild and a notable example of a distinctive Florentine style of architecture. Although the cathedral's dome and lantern had been completed (in 1461) before Leonardo came to Florence, there remained the difficult task of placing an 8-foot-diameter copper sphere at its very top. Verrocchio's workshop built and finished the copper sphere and, using one of Brunelleschi's cranes, placed it to stand a dizzying 350 feet above the city's streets.[4]

The cathedral project occupied Verrocchio's workshop from 1468 to 1472 and made a distinct impression on the young Leonardo. Forty years later he wrote, "Keep in mind how the ball of Santa Maria del Fiore was soldered together."[5] At the time Leonardo made numerous drawings of Brunelleschi's hoisting machines, revolving crane, and ancillary screwjacks and turnbuckles. In addition to his work as an architect and sculptor, Brunelleschi utilized geometrical perspective (discussed later) to capture the three dimensionality of machines in two-dimensional drawings. Leonardo's notebooks make clear that he mastered this vital representational technique. To depict Brunelleschi's heaviest hoisting machine, an ox-powered, three-speed winch with reversing ratchets that allowed workers to raise a stone and then carefully lower it into place, Leonardo sketched a general view of the hoist along with a series of its most important details. These multiple-view drawings, done in vivid geometrical perspective, are a signature feature of his notebooks.

Leonardo's career as an independent engineer began as an assignment from Lorenzo de' Medici. Lorenzo had directed Leonardo to formally present a certain gift to Ludovico Sforza, the new and self-installed ruler of the duchy of Milan. Leonardo talked his way into a full-time job as the Sforza court's engineer. It was a fortunate move, and he spent nearly two decades there (1482–99). Milan was twice the size of Florence, and the Sforza family presided over an active and powerful court. In a famous letter to his new patron, Leonardo proclaimed his engineering talents and highlighted their application in war. "I will make bombards, mortars, and light ordnance of fine and useful forms, out of the common type," he stated. He also had promising designs for "covered chariots, safe and unattackable," "covered ways and ladders and other instruments," a way to "take the water out of trenches," "extremely light, strong bridges," and "many machines most efficient for attacking and defend-

FIG. 1.1. DOME OF THE FLORENCE CATHEDRAL

Renaissance-era building projects frequently mobilized advanced technology to create impressive cultural displays. To complete the Florence cathedral Brunelleschi erected the largest masonry dome in the world, measuring 143 feet in diameter. The octagonal, ribbed dome was formed in two layers, using bricks laid in a spiral herringbone pattern. During his apprenticeship, Leonardo da Vinci helped construct and then set in place the large copper sphere at the top.

William J. Anderson, *The Architecture of the Renaissance in Italy* (London: Batsford, 1909), plate 4. Courtesy of Galvin Library, Illinois Institute of Technology.

ing vessels." After enumerating no fewer than nine classes of weapons ("I can contrive various and endless means of offense and defense"), Leonardo then mentioned civilian engineering. "In time of peace I believe I can give perfect satisfaction and to the equal of any other in architecture and the composition of buildings public and private; and in guiding water from one place to another."[6] Yet even the guiding of water, as we will see, was no innocent task.

Ludovico would be formally recognized as duke of Milan some years later (he had seized power by pushing aside his ten-year-old nephew), and then Leonardo would become known as *ingeniarius ducalis* (duke's engineer). But when Leonardo arrived in the 1480s, Ludovico's claim to rule over Milan was shaky. Leonardo came to the rescue. To strengthen his regime, Ludovico planned a huge bronze statue, more than 20 feet in height, to commemorate his deceased father, the former and legitimate ruler of Milan. In his letter to the duke, Leonardo shrewdly offered to "undertake the work of the bronze horse, which shall be an immortal glory and eternal honour to the auspicious memory of the Prince your father and of the illustrious house of Sforza."[7] Leonardo, it seems, knew exactly what was expected of him. From this time forward, Leonardo fashioned a career as a court engineer with a strong military slant. His notebooks from Milan are filled with drawings of crossbows, cannons, attack chariots, mobile bridges, firearms, and horses.

Leonardo's drawings create such a vivid image in the mind that it is not easy to tell when he was illustrating an original invention, copying from a treatise, recording something he saw firsthand, or, as was often the case, simply exercising his fertile imagination. Among Leonardo's "technological dreams"—imaginative projects that were beyond the realm of technical possibility—are a huge human-powered wheel apparently meant to drive the winding and firing of four oversize crossbows and an immense wheeled crossbow at least 40 feet in breadth. A seemingly impossible horse-drawn attack chariot with rotating knives might actually have seen the battlefield, since a somber note by Leonardo admitted the rotating knives "often did no less injury to friends than to enemies." Pirate raids on the seaport of Genoa prompted Ludovico Sforza to direct Leonardo to design means "for attacking and defending vessels." Leonardo envisioned a submarine, a sealed, watertight vessel temporarily kept afloat by air in leather bags. But he put the lid on this invention to prevent (as he put it) "the evil nature of men" from using such a vessel to mount unseen attacks on enemy ships.[8]

Another "technological dream" with obvious military potential was Leonardo's study of human-powered flight. In the mid 1490s, Leonardo attempted to apply his philosophical stance that nature was mechanically uniform and that humans could imitate the natural "equipment" of flying animals. He devised several devices for transforming a human's arm or leg motions into the birdlike flapping of a mechanical wing. Several of these appear absurdly clumsy. In one, the pilot—in midflight—

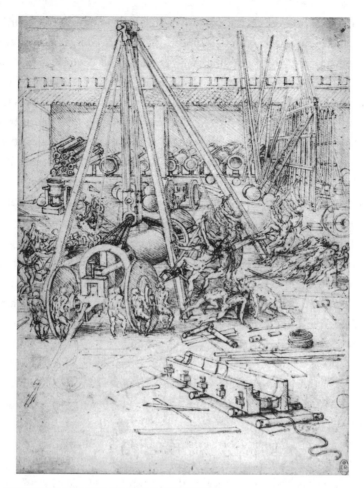

FIG. 1.2. LEONARDO AND THE MILITARY REVOLUTION

Cannons were the expensive high-tech weapons of the court era. In the late thirteenth century the Chinese invented gunpowder weapons capable of shooting projectiles, and gunpowder cannons were in use by Europeans by the early fourteenth century. In this drawing Leonardo captured the bustle of a busy cannon foundry, using geometrical perspective. The prohibitively high cost of high-nitrate gunpowder, however, constrained the wide use of gunpowder weapons for many decades to come.

The Royal Collection, © 2003 Her Majesty Queen Elizabeth II, RL 12647.

continually rewound springs that mechanically flapped the contraption's wings. In another, an especially stout design, there were heavy shock absorbers to cushion crash landings. At least one of these wonders took flight. Leonardo advised, "You will try this machine over a lake, and wear a long wineskin around your waist, so that if you should fall you will not drown."[9]

Leonardo also worked on the expensive, high-technology military hardware of the Renaissance: gunpowder weapons. Indeed, because the Sforzas already employed a court architect for civilian building projects, Leonardo devoted himself to military projects. In his notebooks, one finds full engagement with gunpowder weapons. There are characteristic exploded-view drawings for wheel-lock assemblies (to ignite the gunpowder charge), a water-driven machine for rolling the bars to be welded into gun barrels, and a magnificent drawing of workers guiding a huge cannon barrel through the midst of a bustling foundry (fig. 1.2). Still, as historian Bert Hall emphasized, the "military revolution" we often associate with gunpowder weapons was slow to develop, given the prohibitively high cost of gunpowder, the laughable state of firearm accuracy, and the surprising deadliness of crossbows, pikes, and battle axes.[10] (A classic turning point, considered in chapter 2, is Maurice of Nassau's integration of firearms into battlefield tactics in the early 1600s.) Leonardo also devised numerous means for building and defending fortifications. He illustrated several devices meant to knock down the ladders that an attacking force might place to scale a fortified wall. While perhaps not as exciting as gunpowder weapons, the varied means for attacking or defending fortifications were at the center of Renaissance-era warfare.

Beyond the military projects that occupied Leonardo in Milan, it was entirely characteristic of the court era that laborsaving or industrial technologies were little on his mind. Fully in tune with his courtly patrons, Leonardo focused much of his technological creativity on dynastic displays and courtly entertainments. For the oversize statue commemorating Sforza's father and legitimizing his own rule in Milan, Leonardo built a full-scale clay model of the impressive horse-mounted figure. The story is told that Leonardo would leave "the stupendous Horse of clay" at midday, rush over to where he was painting the Last Supper fresco, "pick up a brush and give one or two brushstrokes to one of the figures, and then go elsewhere." The immense mass of bronze set aside for the monument was in time diverted to make cannons.[11]

Sforza also charged Leonardo with creating court culture directly. Leonardo took a prominent role in devising the lavish celebration of the marriage in 1489 between a Sforza son and Princess Isabella of Aragon. Marriages between noble families were serious affairs, of course. Marriage celebrations offered a prominent occasion for proclaiming and cementing dynastic alliances, and hosts often commissioned allegorical performances that "were at the same time ephemeral court entertainments and serious occasions for political propaganda." Leonardo's contribution to the Sforza event mingled the heavens and the earth. Traditionally, scholars have understood *Il Paradiso*, created and staged for the Sforza fest, as a human-centered dance spectacle that included "seven boys representing the seven planets" circling a throne and that featured the unveiling of paradise. At the finale, the figure of Apollo descended from on high to present Princess Isabella with the spectacle's text. Leonardo also built moving stage platforms and settings—and perhaps even an articulated mechanical robot for these festivities. An eyewitness described the set of planets as a mechanical model built by Leonardo. Leonardo's notebooks have designs for at least two other moving-set theatricals, one featuring the mechanical elevation of an allegorical "star" while another offered a glimpse of hell complete with devils, furies, and colored fires. The mechanisms would have been complex spring- or cable-powered devices, constructed mostly of wood. Leonardo's much-celebrated "automobile" was most likely also for courtly entertainments. These theatrical and courtly "automata" were a culturally apt expression of Leonardo's interest in self-acting mechanisms.[12]

A fascination with self-acting mechanisms is also evident in Leonardo's many sketches of textile machines found around Milan. These self-acting automata have led some breathless admirers to call him the "prophet of automation," yet it seems more likely that he was simply recording the interesting technical devices that captured his imagination. (This is certainly the case in his drawings of Milan's canal system.) Quite possibly, he was sketching possible solutions to problems he had spotted or heard about. One of these drawings that is likely Leonardo's own creation (ca. 1493–95) was a rotary machine to make sequins, used for ornamenting fine gowns or theatrical costumes. Again, Leonardo expressed his mechanical talents in culturally resonant ways, just as inventors in later eras would focus on self-acting mechanisms for industry and not for courts.

Leonardo's life at the Sforza court ended in 1499 when France invaded the region and ousted Sforza. "The duke lost his state, his property, and his freedom, and none of his works was completed for him," recounted Leonardo. In the chaotic eight years that followed, Leonardo acted essentially as a mercenary military engineer. He traveled widely, with the application of military technologies never far from his mind, and worked all sides of the conflicts rending Italy. (Even while designing a warm-water bath for the duchess of Milan, Leonardo recorded in his notebooks "a way of flooding the castle" complete with drawings.) In Venice, which was reeling under attack by the Ottoman Turks, he suggested improvements to that city's all-important water defenses; while several years later, armed with this special knowledge, he would advise the new ruler of Milan on how to flood Venice. In Florence once again during 1500–1501, he failed to gain any technical commissions (the city had its own engineers), and so he devoted himself to painting and to the study of mathematics.[13]

In the summer of 1502, the ill-famed Cesare Borgia tapped Leonardo to be military engineer for his campaign in central Italy. It is impossible to make a simple reckoning of the shifting alliances that shaped Leonardo's career during these turbulent years. The Sforza family had originally gained control of Milan with the support of the Florentine Medici. The French invasion of Milan, which cost Leonardo his job at the Sforza court, was aided by Alexander VI (1492–1503), the so-called Borgia pope, none other than Cesare Borgia's father, who in effect traded Milan to the French in exchange for their backing of his rule as pope and their support for his son's military venture. We can be certain that while on Cesare Borgia's military campaign Leonardo witnessed the sack of Urbino, where the Montefeltro family had collected a famous library and an illustrious court (discussed later). In addition to its political consequences, the Borgia campaign furnished the occasion when Leonardo first met Machiavelli and when Machiavelli found his "model" prince in Cesare Borgia.

Returning to war-torn Florence in 1503 Leonardo gained from that city two characteristic commissions. With the Medici family temporarily exiled, the struggling republican government leaned on the fine arts and public works. The city kept a squad of historians busy chronicling its glories. "Florence is of such a nature that a more distinguished or splendid city cannot be found on the entire earth," gushed one account. In the same celebratory vein, the republic commissioned Leonardo to paint a fresco commemorating the Florentines' victory over Milan in the battle of Anghiari in 1440. The second commission, a giant hydraulic-engineering scheme hatched with Machiavelli's behind-the-scenes assistance, took aim at the enemy town of Pisa. The two men planned to move the Arno River—at once devastating the downstream town of Pisa and improving Florence's access to the sea. Two thousand workers began digging a diversion canal, but when the results were unsatisfactory, the scheme was halted. During this unsettled period, Leonardo also served as a consultant on military fortifications for Florence's ally, the lord of Piombino, and on painting and architecture for French-held Milan. Leonardo's frequent travels between Florence and Milan brought work on the *Battle of Anghiari* fresco to a crawl. Perhaps even Leonardo experienced some discord when simultaneously working for Milan while memorializing Milan's defeat.[14]

Leonardo's famous anatomical studies began during a lull in the military action. After 1507 Leonardo's notebooks are crowded with detailed sketches of the muscles, bones, and tendons of the human body; his access to corpses in the Florence city hospital sharpened his empirical investigations. After Leonardo witnessed one old man's calm passing, he did a dissection "to see the cause of so sweet a death" traced to the man's heart and liver.[15] Leonardo also conceived of a series of elaborate analogies between geometry and mechanics, anatomy and geology, and the human body and the cosmos. He also outlined a theoretical treatise on water, but keeping in mind the Leonardo-Machiavelli scheme to destroy Pisa by diverting its river, it may be unwise to see this treatise as a "pure" theoretical tract.

In French-held Milan once again from 1508 to 1513, Leonardo gained the patronage of Louis XII, king of France, who occasionally directed Leonardo toward a specific technical project but mostly provided him general patronage and time for theoretical investigations. Yet Leonardo's second Milan period ended when the Sforza family *recaptured* the city and drove out his newly acquired French patron. Leonardo fled to Florence, where the Medici family had returned to power, then went on to Rome in the service of Giuliano de' Medici, the brother of Pope Leo X (1513–21). In the next two years, Leonardo worked on several specific commissions from Giuliano, who flourished as his brother sent patronage his way. While in Rome Leonardo mapped the malaria-ridden marshes on the nearby coast, with the aim of draining them, a public health project given to Giuliano by his pope-brother. A Leonardo drawing of a rope-making machine is even stamped with the Medici's diamond-ring symbol. But when Giuliano died in 1516, Leonardo once again found himself without work or patron. At least Leonardo saw the situation plainly, writing, "The Medici made me and ruined me."[16]

Leonardo spent the last three years of his life at the royal court of France. The new French king, François I, gave him the grand title of "the King's foremost painter, engineer and architect." Officially, Leonardo was to design a new royal palace complex, but these plans were never realized. Instead, Leonardo became in true fashion a distinguished figure at court. The king, wrote one courtier, "was extremely taken with his great virtues, [and] took so much pleasure in hearing him speak, that he was separated from him only for a few days out of the year. [The king] believed that there had never been another man born in the world who knew as much as Leonardo, not so much about sculpture, painting, and architecture, as that he was a very great philosopher."[17]

In addition to his duties as a great philosopher, Leonardo created more courtly entertainments. His friend Francesco Melzi described one automaton that Leonardo built to honor the French king. A lion with a bristling mane, it was led by a hermit. On its entrance, we are told, women in the audience drew back in terror; but when the king touched the lion three times with a magic wand handed to him by the hermit, the lion-automaton broke open and spilled at the king's feet a mound of fleur-de-lys.

Such events were famously packed with symbols to test the audience's savvy. Everyone understood the flower as a classic symbol of the French royal house; the split-open lion was a prominent heraldic symbol of a rival court.[18] To the end, Leonardo understood how to adapt his considerable talents to court desires. He died, in France, on 2 May 1519.

The special character of technological creativity in the Renaissance, as we have seen, resulted from one central fact: the city-states and courts that employed Leonardo and his fellow engineers were scarcely interested in the technologies of industry or commerce. Their dreams and desires focused the era's technologists on warfare, city building, courtly entertainments, and dynastic display. A glance at Leonardo's technical notebooks confirms that he was an outstanding engineer, architect, and artist. But a closer examination of them reveals that he was not the solitary genius imagined by some authors. For instance, the Renaissance engineers who sketched Brunelleschi's hoisting machinery for the Florence cathedral include, besides Leonardo, Francesco di Giorgio, Buonaccorso Ghiberti, and Giuliano da Sangallo. And while we have a contemporary's quip that Leonardo "was as lonely as a hangman,"[19] we also know that he met personally with, borrowed ideas from, and likely gave inspiration to a number of fellow court engineers. The intellectual resources and social dynamics of this technological community drew on and helped create Renaissance court culture.

Foremost among these intellectual resources was the distinctive three-dimensionality and depth of Renaissance art and engineering. This owed much to Leon Battista Alberti, renowned as the "father of perspective." Alberti (1404–72) was born into a prominent Florentine merchant-banking family, studied law at the University of Bologna, and worked for three decades in the administrative offices of the pope's court. As historian Anthony Grafton made clear, Alberti was a courtier of courtiers. Even more than Leonardo's, his career was bound up with the court system. In his latter years, to secure a place at the Ferrara court, he wrote what amounted to a how-to manual for succeeding in the court system—with apt examples from his own life! He was experienced at the game. His patrons, besides the papal curia, included the princely courts at Ferrara, Rimini, Mantua, and Urbino, as well as the great Florentine builder Giovanni Rucellai. In his astonishing career, he was variously a humanist writer and rhetorician, architectural designer and consultant, literary critic, sculptor, mapmaker, and a leading theorist of painting. "He is the sort of man who easily and quickly becomes better than anyone else at whatever pursuit he undertakes," thought one contemporary.[20] If he had done nothing else, Alberti's treatises on architecture and on practical mathematics might prompt latter-day readers to label him an engineer.

Linear perspective, his most far-reaching theoretical achievement, was one thing that Alberti did *not* invent. Leading Florentine artists such as Masaccio were already practicing something like linear perspective a decade or more before Alberti's famous treatise *On Painting* (1436). During the same time, Brunelleschi, at work on the Florence cathedral, staged dramatic public events that popularized the new technique. Positioned on the steps of the cathedral, Brunelleschi painted precise "show boxes" of the prominent Baptistry across the square, and dazzled passersby when they had difficulty telling the difference between his painting of the scene and the real thing. He did the same for the Piazza della Signoria, site of the city's government. Alberti, too, later painted show boxes in which (as he wrote) the viewer could see "huge mountains, vast provinces, the immense round of the sea surrounding them, and regions so distant from the eye that sight was dimmed . . . they were such that expert and layman alike would insist they saw, not painted things, but real ones in nature."[21]

With his treatise on painting, Alberti turned the practice of perspective into a structured theory. In what is now a commonplace of drawing, he directed artists to treat the two-dimensional picture plane on which they worked, whether it was a wall, a panel, or a canvas, as if it were a window in which a three-dimensional scene appeared (fig. 1.3). The classic exercise, illustrated and popularized by Albrecht Dürer (1471–1528), is to view an object through a nearby pane of glass ruled into squares, and then to transfer what you see, square by square, onto a piece of paper or canvas that is similarly ruled. Dürer's most famous "object," illustrating his 1525 treatise on geometry and perspective and reproduced widely ever since, was a naked woman on her back, suggesting that perspective was not merely about accurately representing the world but about giving the (male) artist power over it. Whatever the object, vertical lines will remain vertical, while receding horizontal lines will converge toward a vanishing point at the drawing's horizon. Parallel horizontal lines, such as on a tiled floor, must be placed at certain decreasing intervals from front to back. Finally, for maximum effect, the viewer's eye must be precisely positioned. In fact, the show boxes directed the observers' point of view by locating the peephole at the picture's vanishing point, from which they looked at the image reflected in a mirror on the box's far side. Alberti showed how the exact use of perspective can trick the mind's eye into seeing a three-dimensional original.

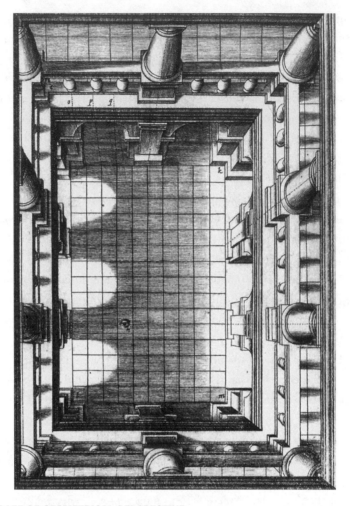

FIG. 1.3. THE GAZE OF GEOMETRICAL PERSPECTIVE

Painters, architects, and engineers used geometrical perspective to depict their ideas with precision and persuasion, gaining commissions from the courts and in the process creating culture. Geometrical perspective not merely represented the world but also changed people's ways of thinking. The penetrating eyeball at the symbolic center of this geometrical-perspective drawing hints at the power of the artist's gaze.

Joseph Moxon, *Practical Perspective* (London, 1670), plate 36. Courtesy of Illinois Institute of Technology Special Collections.

The geometrical ideas and devices described in Alberti's writings were widely adopted by artists, mathematicians, cartographers, and engineers. And the entertaining show boxes did not harm his flourishing court career. Leonardo, in addition to studying Alberti's technical innovations and expanding on his empirical investigations, included passages of Alberti's writings on painting into his own treatise on the subject. Alberti's distinctive phrases passed into Leonardo's corpus without attribution. It is Alberti's ideas we are reading when Leonardo writes that the perspective picture should look as though it were drawn on a glass through which the objects are seen; or that the *velo*, or square-marked net, locates the correct positions of objects in relation to one another; or that one should judge the accuracy of a picture by its reflection in a mirror. Leonardo even echoed Alberti's maxim that the only "modern" painters were Giotto and Masaccio.[22]

Leonardo's membership in a community of engineers has only been recently appreciated. In the absence of any sort of "engineer's guild" listing its members, and owing to the press of military secrecy, we will never know all the engineering colleagues with whom Leonardo schemed and dreamed. We do know that topmost among Leonardo's contemporaries was Francesco di Giorgio (1439–1501) and that the two men met and in time shared a great deal of technical material. Born in Sienna, Francesco studied painting, sculpture, and architecture. He began his career as a sculptor, then worked as an engineer on his native city's water supply (1469). Francesco became the era's leading military engineer. His career, like Leonardo's and Alberti's, was inseparable from the court system.

Beginning in 1477, Francesco joined the Montefeltro family's fortress-court at Urbino. Court artists there, such as Piero della Francesca, painted the noble family (including one of the Sforza sisters) and authored treatises on painting that were dedicated to the duke and placed in his library. That famous library held some 1,100 richly bound books reportedly costing the whopping sum of 30,000 ducats, about half a year's worth of the payment the Duke of Urbino, a *condottiere* (soldier of fortune) like Sforza, received for his military services. Francesco became the chief designer for the duke's ambitious palace expansion (its scale might be estimated by its stables for 300 horses), and he built a series of military fortresses throughout the state of Urbino. His first treatise on technology—with drawings of artillery, pumps, water-powered saws, methods for attacking fortified walls, catapults, and siege equipment, as well as tips on defending fortifications in an era of gunpowder—was dedicated to the duke (and with its ascribed date of ca. 1475 likely led to his employment by the duke two years later).

Francesco stayed at the Montefeltro court some years after the duke's death, until 1487, when he once again took up work as chief city engineer in Sienna. Also during his post-Urbino years, he traveled frequently to Naples and there advised the Duke of Calabria on military matters. Not by accident, the second edition of the abovementioned treatise on technology was dedicated to this duke while its third edition was illustrated by the duke's court painter. When he was called in to advise on the construction of the cathedrals of Pavia and Milan in 1490, Francesco met Leonardo in person. Francesco had long left the Montefeltro court by the time Leonardo assisted Cesare Borgia with its capture in 1502.

Close study of the two men's notebooks has revealed that Francesco was one source of technical designs previously attributed to Leonardo alone. For instance, the devices that Francesco described in his treatise on military architecture are in many cases more sophisticated than similar devices in Leonardo's notebooks. We know that Leonardo read and studied Francesco's treatise, since there is a manuscript copy of it clearly annotated in Leonardo's distinctive handwriting. As with Alberti, it seems Leonardo here simply borrowed from a well-regarded author.

In a curious way, the presence of Leonardo's voluminous notebooks has worked to obscure the breadth and depth of the Renaissance technical community, because researchers overzealously attributed all the designs in them to him. At his death in 1519, Leonardo left his notebooks and papers to his "faithful pupil" Francesco Melzi, who kept them at his villa outside Milan for fifty years. Eventually, Melzi's son peddled the papers far and wide. Scholars have found caches of Leonardo's papers close to his areas of activities in Milan, Turin, and Paris and as far afield as Windsor Castle, in England. In Madrid a major Leonardo codex was found as recently as 1966. Scholars believe that about one-third (6,000 pages) of Leonardo's original corpus has been recovered; these papers constitute the most detailed documentation we have on Renaissance technology. Fortunately, scholars have abandoned the naïve view that each and every one of Leonardo's notebook entries represents the master's original thinking. In his varied career, Leonardo traveled widely and often sketched things that caught his attention. His notebooks record several distinct types of technical projects: his specific commissions from courtly patrons; his own technological "dreams," or devices, that were then impossible to build; his empirical and theoretical studies; and devices he had seen while traveling or had heard about from fellow engineers; as well as "quotations" from earlier authors, including Vitruvius.

Scholars have intensely debated what should count as Leonardo's own authentic creations. These probably include a flying machine that was a completely impractical mechanical imitation of a bat, some textile machines that remain tantalizing but obscure, and a machine for polishing mirrors. Characteristic of the court culture of Renaissance technology, the textile machine that made sequins to decorate fancy garments is most certainly Leonardo's own. Leonardo's "inventions" that should be attributed to others include an assault chariot, endless-screw pumps, several lifting apparatuses, and other pump designs that clearly were known in ancient Rome. Leonardo also copied drawings for a perpetual motion machine.

Perhaps the most distinctive aspect of Leonardo's career was his systematic experimentation, evident in his notebooks especially after 1500. He wrote:

> In dealing with a scientific problem, I first arrange several experiments, since my purpose is to determine the problem in accordance with experience, and then to show why the bodies are compelled so to act. That is the method which must be followed in all researches upon the phenomenon of Nature. . . . We must consult experience in the variety of cases and circumstances until we can draw from them a general rule that [is] contained in them. And for what purposes are these general rules good? They lead us to further investigations of Nature and to creations of art. They prevent us from deceiving ourselves or others, by promising results to ourselves which are not to be obtained.[23]

Some objects of Leonardo's systematic investigations were gears, statics, and fluid flow. He made a survey of different types of gears, designing conical and helical gears of wood. He also investigated the causes of crumbling in walls and found that the direction of cracks in walls indicated the source of strain. In studying the resistance of beams under pressure, he arrived at the general principle that for square horizontal beams supported at each end, their resistance varied with the square of their side and inversely with their length—a fair approximation. Fluids were more challenging. To study the flow of water, he built small-scale models involving colored water. He also sought general guidelines for placing dams in rivers. Indeed, it may have been with the failed Arno River scheme in mind that Leonardo reminded himself about "promising results to ourselves which are not to be obtained."

The influence of Renaissance engineers on Europe was substantial. The medieval historian Lynn White wrote, "Italian engineers scattered over Europe, from Madrid to Moscow and back to Britain, monopolizing the best jobs, erecting wonderful new machines, building palaces and fortifications, and helping to bankrupt every government which hired them. To tax-paying natives they were a plague of locusts, but rulers in the sixteenth century considered them indispensable. Their impact upon the general culture of Europe was as great as that of the contemporary Italian humanists, artists, and musicians."[24]

GUTENBERG'S UNIVERSE

The invention of moveable type for printing led to an information explosion that profoundly altered scholarship, religious practices, and the character of technology. Much about the printing revolution conjures up images of enterprising capitalist printers breaking free of tradition-bound institutions. All the same, courts across Europe created large-scale markets for printed works and shaped the patronage networks for writings about technology. The first several generations of printers as well as the best-known early technical authors were, to a surprising extent, dependent on and participants in late-Renaissance court culture.

Printing was a composite invention that combined the elements of moveable metal-type characters, suitable paper, oil-based ink, and a wooden press. Inks and presses were commonly available in Europe, while paper came to Europe by way of China. Paper was being made in China by the third century AD from the fibers of silk, bamboo, flax, rice, and wheat straw. Papermaking spread to the Middle East by Chinese papermakers taken prisoner in 751. From Samarkand, the great Arabic center of astronomy and learning, one "paper route" passed through Baghdad (793), Cairo (900), Fez (1000), Palermo (1109), Játiva (1150), Fabriano (1276), and Nuremberg (1390). Another route passed along the coast of Northern Africa, through Spain (beginning of thirteenth century), and into France.

Moveable type was also "first" developed in the Far East, centuries before Gutenberg. Printing from carved stone or wood plates was well advanced in China at the turn of the first millennium; 130 volumes of Chinese classics were printed in 953, and a key Buddhist text was printed using 130,000 plates in 982. The first truly moveable type is credited to Pi Sheng (1041–48), who engraved individual characters in clay, fired them, and then assembled them on a frame for printing. This method was reinvented and improved around 1314 by Wang Cheng, a government magistrate and prolific compiler of agricultural treatises. He commissioned artisans to carve a set of 60,000 specially designed characters and used them to publish up to 100 copies of a monthly gazette. During the Ming Dynasty (1368–1644), moveable wooden characters were used to publish the official court gazette. In Korea characters were first cast from metal (lead and copper) in 1403 and used to print numerous works during that century. The Korean characters spread to China (end of the fifteenth century) and to Japan (1596), and yet metal type did not entirely displace woodblock printing. Traditional woodblock printing persisted, since metal type could not print on two sides of the typically thin paper. The further difficulty of moveable type printing in Chinese is evident in Wang Cheng's extraordinarily large set of carved characters.

While it was a vector for the "paper route," the Arab world halted the spread of Asian printing to the West. Islam encouraged handwriting the words of Allah on paper but for many years forbade its mechanical printing. The first Arabic-language book printed in Cairo, Egypt, did not appear until 1825.[25]

"The admirable art of typography was invented by the ingenious Johann Gutenberg in 1450 at Mainz," or so stated the son of Gutenberg's partner in 1505. Remarkably enough, the only direct evidence to assess this claim are two legal documents from 1439 and 1455. We know surprisingly little about Gutenberg himself. For almost ten years of his life (1429–34 and 1444–48), not even his city of residence is known with certainty. And while, for instance, tax records tell us that Gutenberg in 1439 had a wine cellar storing 2,000 liters, no one can be sure if he was married to the legendary Ennelin—who may have been his wife, his lover, or a Beguine nun. It is clear that Gutenberg was born into a wealthy and established family of Mainz, sometime around 1400—by printers' lore it was St. John the Baptist's Day, 24 June 1400. His father and family were well connected with the church mint at Mainz, and from them, it seems, Johann gained his knowledge of metal casting, punch-cutting, and goldsmithing.

In 1434 Gutenberg moved to Strasbourg, near the Rhine River in northeast France, and by 1436 he was engaged in experiments on printing. In Strasbourg he also practiced and taught goldsmithing, gem cutting, and mirror making. In 1438 he agreed to convey "the adventure and art" of printing to two of his mirror-making associates; in effect they paid Gutenberg 250 guilders to create a five-year partnership. However, one of the two died, and his brothers sued Gutenberg in 1439 to be taken into the partnership. The court records indicate little more than that Gutenberg owned a press, had bought lead and other metals, and from them had cast what were called "forms" (a word used to describe the molding or casting of metal). A goldsmith testified that Gutenberg had paid him 100 guilders in 1436 for "that which pertains to the use of a press." Secrecy was a prime concern. When his partner was dying, Gutenberg had all the existing "forms" melted down in his presence. He also directed a helper to take apart an object with "four pieces" held together by two screws, quite likely an adjustable mold for casting type. More than this we will probably never know. To finance these experiments, Gutenberg borrowed large sums of money.[26]

Gutenberg left Strasbourg in 1444, but it is unclear what he did and where he went until his return to his native Mainz four years later. Money surely mattered. Shortly after his reappearance in Mainz, he secured a 150-guilder loan guaranteed by a relative. In 1450 he borrowed 800 guilders from a merchant named Johann

Fust, and soon enough an additional 800 guilders with the proviso that Fust was to become a profit-sharing partner in "the works of the books" (namely, the printing of the Latin Bible). These "works" comprised perhaps six presses and twenty printers and assistants. In 1455, generating the second court dossier, Fust sued Gutenberg for the repayment of the two loans including interest. It is often assumed that this court action wiped out Gutenberg's finances and deprived him of his livelihood, but in fact the court required Gutenberg only to repay the first 800 guilders.

Gutenberg retained his half-share in the Bible sales and revenue from other printing jobs, and a recent researcher concludes that he not only repaid Fust but also regained possession of the printing equipment held as collateral. Fust set up a new and highly successful partnership with Peter Schöffer, who had also worked with Gutenberg; between Fust's death in 1466 and his own in 1503, Schöffer published at least 228 books and broadsides. For his part, Gutenberg continued in the printing trade and employed at least two workmen, who became successful printers in their own right after his retirement in 1465. In that year, as the Archbishop of Mainz formally declared, "by special dispensation have we admitted and received him as our servant and courtier." The honor gave courtier Gutenberg financial security and, until his death in 1468, the privileges of a nobleman.[27]

Gutenberg's principal inventions were the adjustable mold for casting type and a suitable metal alloy for the type. Understandably, the two lawsuits reveal few technical details, but a Latin grammar and dictionary printed in Mainz in 1460 (by whom is not clear) points out that its successful printing resulted from the "marvellous consistency in size and proportion between patterns and moulds." Such consistency would be important for type to be assembled in lines for printing and locked into a frame when placed on the printing press. The letters "M" and "W" are roughly the same width, but "i" and "W" are not. Making different size letters of exactly the same depth and height-on-paper required an adjustable-width type mold. Gutenberg also experimented to find a type metal that would be hard enough to be durable yet easy to melt and mold. Again, while the exact composition of Gutenberg's own alloy is not known, the first printed book that describes typecasting, Biringuccio's *Pirotechnia* (1540), of which more will be discussed later, states, "the letters for printing books are made of a composition of three parts fine tin, an eighth part of lead, and another eighth part of fused marcasite of antimony." Samples from the printshop of Christopher Plantin from 1580 are 82 percent lead, 9 percent tin, and 6 percent antimony, with a trace of copper.[28]

Printing traveled speedily. By 1471 printing shops had sprung up in a dozen

European cities as far away as Venice, Rome, and Seville. By 1480 there were two dozen printing cities in northern Italy alone, while printers had moved east to Cracow and Budapest and north to London and Oxford. By 1500, there were printers as far north as Stockholm and as far west as Lisbon.

The printing press made a little-known German theology professor named Martin Luther into a best-selling author and helped usher in the Protestant Reformation. The Reformation is officially dated from 31 October 1517, the day Luther tacked his Ninety-Five Theses onto the church door at Wittenberg, Germany. His theses were written in Latin, not the language of the common people, and church doors had been "the customary place for medieval publicity." Yet, printers sensed a huge market for his work and quickly made bootleg copies in Latin, German, and other languages to fill it. It was said that Luther's theses were known across Germany in two weeks and across Europe in a month. Luther wrote to Pope Leo X, another Medici pope and brother of Leonardo's onetime patron: "It is a mystery to me how my theses . . . were spread to so many places. They were meant exclusively for our academic circle here." Within three years (1517–20) enterprising printers had sold over 300,000 copies of Luther's writings. "Lutheranism was from the first the child of the printed book, and through this vehicle Luther was able to make exact, standardized and ineradicable impressions on the minds of Europe," writes historian Elizabeth Eisenstein. Eventually, Luther himself hailed printing as "God's highest and extremest act of grace, whereby the business of the Gospel is driven forward."[29]

The Catholic Church responded to Luther's theological arguments but could not halt the spread of printing. The Protestant movement's new emphasis on individuals' reading the Bible required a massive printing effort. Whatever their personal beliefs, printers thus had material reasons to support Protestantism. Even the Catholic Church unwittingly helped the cause. Beginning in 1559, the church issued its infamous *Index of Prohibited Books*, a terror to freethinking authors and publishers in Catholic countries. But for Protestant-leaning authors it amounted to free publicity, and for printers in Protestant countries it pointed to potentially best-selling titles.[30] Machiavelli and Galileo were among the many well-known authors targeted by the *Index*.

Owing to the repression of printing in the Catholic countries of southern Europe, naturally printing flourished in northern Europe. During the forty years from 1500 to 1540, no fewer than 133 printers produced more than 4,000 books in the Netherlands, centered in the city of Antwerp. Only Paris surpassed Antwerp in population and commercial activity, and it was in Antwerp that Christopher Plantin (ca. 1520–89) transformed the artisan craft of printing into an industry. Born in France and appren-

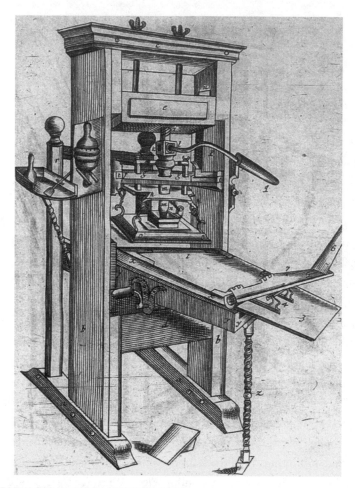

FIG. 1.4. EARLY DUTCH-STYLE PRINTING PRESS

Christopher Plantin's secret weapon? The pioneering authority on early printing, Joseph Moxon, identified this as a "new fashioned" printing press invented in Amsterdam and widely used in the Low Countries. It is similar to the seventeenth-century printing presses on exhibit today at the Plantin-Moretus Museum (Antwerp).

Joseph Moxon, *Mechanick Exercises* (London, 1683), plate 4. Courtesy of Illinois Institute of Technology Special Collections.

ticed in Paris as a bookbinder, Plantin moved to Antwerp and by 1550 was listed in that city as "boeckprinter." He established his own independent press in 1555. Ten years later, when other leading printers had between two and four presses, Plantin had seven; at the height of his career in 1576 he had a total of twenty-two presses and a substantial workforce of typesetters, printers, and proofreaders (fig. 1.4).

Although it is tempting to see printers as proto-capitalists—owing to their strong market orientation and substantial capital needs—their business owed much to the patronage and politics of the court system. Plantin's first book, printed in 1555, was an advice manual for the upbringing of young ladies of the nobility, printed in Italian and French. Plantin benefited from the goodwill and patronage of Cardinal Granvelle and of Gabriel de Zayas, secretary to King Philip II of Spain. With Philip II's patronage, including a healthy direct subsidy, Plantin in the 1560s conceived and printed a massive five-language Bible in eight volumes. During 1571–76, "his period of greatest prosperity," Plantin kept his many presses busy printing Bibles and liturgical books for the king of Spain. In these years his firm sent to Spain no fewer than 18,370 breviaries, 16,755 missals, 9,120 books of hours, and 3,200 hymnals. Botanical works, maps, dictionaries, and literary and scholarly works rounded out his catalogue. "Plantin's firm in Antwerp thus gleaned a lion's share of rewards from the expansion of an overseas book trade in Spain's golden age," writes Eisenstein. "Plantin himself became not only Royal Typographer but chief censor of all his competitors' output while gaining a monopoly of all the liturgical books needed by priests throughout the far-flung Habsburg realms."[31]

When the political winds shifted, Plantin quickly changed his tack. In 1578, when the Spanish were temporarily driven from Antwerp, Plantin became official printer to the new authorities, the States General, and soon printed many anti-Spanish tracts. But when in 1585 the Spanish threatened to take Antwerp again, and this time for good, Plantin opened a second office in Leiden and for two years served as printer to the newly founded University of Leiden. As Eisenstein observes, "Plantin's vast publishing empire, which was the largest in Europe at the time, owed much to his capacity to hedge all bets by winning rich and powerful friends in different regions who belonged to diverse confessions."[32]

Plantin's massive output suggests the huge scale of book production at the time. In the first fifty years of printing (1450s–1500), eight million books were produced in Europe. The economics of the new technology were impressive. For instance, in 1483 a certain Florence printer charged about three times more than a scribe would have for reproducing Plato's *Dialogues*, but the printer made more than a thousand copies while the scribe would have made just one. Moreover, handwritten scribal copies often contained idiosyncratic alterations from the original, while the printer's copies were identical to each other, an additional dimension of quality. This economy of scale sharply reduced the cost of books, which meant that one scholar could have at hand multiple copies from several scholarly traditions, inviting comparison and evaluation. Eisenstein writes, "Not only was confidence in old theories weakened, but an enriched reading matter also encouraged the development of new intellectual combinations and permutations."[33] In this way, the availability of vastly more and radically cheaper information led to fundamental changes in scholarship and learning. Printing lent a permanent and cumulative character to the fifteenth-century Renaissance. The same was true for many other technologies.

TECHNOLOGY AND TRADITION

We tend to think of technology as cumulative and irreversible, permanent and for all time, but it has not always been so. Technologies have ebbed and flowed with the cultures that they were intrinsically a part of. The Middle Ages became known as the "dark ages" because Western Europeans had "lost" contact with the classical civilizations of Greece and Rome (whose knowledge of science, medicine, and technology resided in the libraries of the Islamic world). Not until the incremental technical changes of the late Middle Ages were combined with the innovation of three-dimensional perspective, courtesy of Alberti, Leonardo, and others, and with the possibility of exact reproduction through printing did technology become a cumulative and permanent cultural element. This cultural characteristic of the "modern" world deserves to be better understood.

Transfer of technology before the Renaissance could be hit-or-miss. Machines invented in one time, or place, might well need to be rediscovered or even reinvented. Indeed, the great technological advances of Song China (960–1279) were practically lost for many years. The list of technologies invented in China during these years includes not only paper and printing, as noted above, but also gunpowder weapons, magnetic compasses, all manner of canals, locks, and hydraulic engineering, as well as the industrial-scale production and consumption of iron. China's ironmasters in the north-central Heibei and Henan regions were using coke to smelt iron ore fully five centuries before the English industrial revolution. (The Henan region, despite its rich ancient history, languished for subsequent centuries as a relatively poor region of China until the 1990s, and it recently gained worldwide attention as the possible origin for the coronavirus pandemic as noted in chapter 10.) Chinese silk-reeling mills, farm implements (made of iron), and bridges are among the numerous other advanced technical fields where, as historian Arnold Pacey writes, "techniques in use in eleventh-century China . . . had no parallel in Europe until around 1700."[34]

Yet these groundbreaking Chinese technologies were not reliably recorded with the rigorous geometrical perspective that allowed Renaissance engineers to set down their ideas about the crucial workings of machines. Chinese drawings of silk-reeling mills, for example, are often so distorted that to Western eyes, accustomed to geometrical perspective, it is very difficult to tell how they might be built. The definitive *Science and Civilisation in China* series offers numerous instances of inaccurate and/ or incomplete illustrations of textile machines. In the volume on textile spinning, it is clear that hand scrolls (themselves works of fine art) invariably have the "best" illustrations of textile technology while the post-Song-era encyclopedias and technical books are plagued by imperfect and misleading illustrations. In fact, the step-wise corruption in images of silk-reeling machinery in successive encyclopedias of 1313, 1530, and 1774 rather clinches the point. In the wake of political disruptions after 1279, Song China's technical brilliance was lost not only to the Chinese themselves but also to the West, whose residents formed an inaccurate and incomplete view of China's accomplishments.[35]

Eugene Ferguson, a leading engineer-historian, demonstrated how quickly technical drawings might be corrupted, even in the West. He compares a series of drawings made by Francesco di Giorgio around 1470 with copies made in the 1540s by artist-scribes who had been specially trained to copy mechanical drawings.[36] The results are startling. Francesco's original perspective drawings of hoisting cranes and automobiles are workable, if not always brilliant. But the scribes' copies are something else; many mechanical details are missing or distorted. For instance, a hoisting crane's multiple block-and-tackle system, useful for gaining mechanical advantage, was condensed in the copy to a single pulley that provides no mechanical advantage. Similarly, Francesco's original versions of a bar spreader and bar puller resembled modern turnbuckles in that the *right*-handed threads on one side of a square nut and the *left*-handed threads on the other served to push apart or pull together the *ends*. In the copyists' version, however, left-handed threads are shown throughout, creating a device that would simply move the *nut* toward one end or the other. While Francesco's originals rigorously observe the conventions of three-dimensional perspective (so that parallel lines running right-to-left are really parallel while lines running back-to-front converge to a vanishing point), the copyists' versions often have no vanishing point.

For some of these devices, Ferguson thinks a knowledgeable technical worker might have been able to "see through" the damaged view of the copied drawing and "fix" the errors. But now imagine the copyist's version was again copied by hand with yet-additional distortions and errors. In time, recreating Francesco's original devices would become hopeless. His designs, like those of the Chinese technologists, would effectively be "lost." The point is that before the combination of printing and geometrical perspective, inventions made in one generation might not be available to successive generations or for that matter beyond the close circle of colleagues sharing notebooks, apprenticeships, or craft practices. In these circumstances, a disruption to the social order, like the fall of a ruler and destruction of his library, would disrupt a technical tradition. Technological change could not be permanent and cumulative.

In these terms a permanent and cumulative tradition in technology, enabled by the inventions of printing and perspective, appeared first in central Europe's mining industry. Vannoccio Biringuccio's *Pirotechnia* (1540), Georgius Agricola's *De re metallica* (1556), and Lazarus Ercker's *Treatise on Ores and Assaying* (1580) are brilliantly illustrated printed books that even today vividly convey distant technologies. These three volumes have surprisingly close connections to late-Renaissance court culture, even though these authors came from diverse social backgrounds. Biringuccio was a supervisor of silver and iron mines with extensive practical experience, Agricola was a university-trained humanist, and Ercker was an upwardly mobile mining supervisor who married into a prominent family. As Pamela Long has shown, each author was associated with the mining industry of central Europe at a time when technology-won efficiencies were urgently needed. Costs were rising because an earlier mining boom (of 1460 to 1530) had already tapped the ready veins of silver, gold, and copper, while prices were falling due to the influx of Spanish gold and silver stripped from the New World. Thus, wealthy investors and the holders of royal mining rights eagerly sought and avidly consumed information about mining technology that promised substantial economic returns. As Long writes, "the great majority of 16th-century mining books were written by Germans in the regions of the empire where the capitalist transformations of mining were most pronounced—the Harz Mountains near Goslar, the Erzgebirge Mountains in Saxony and Bohemia, and the Tyrolian Alps to the south."[37] Biringuccio, an Italian, had visited mines in this region and wrote to urge his countrymen to make comparable investments in technological changes in their own copper mines.

Each of these three authors flourished by gaining the courtly favor of "prince-practitioners," as Long calls them, who had a special interest in mining or metal technology. Biringuccio (1480–1539) worked for Italian princes, including the Farnese of Parma and Alphonso I d'Este of Ferrara, for the ruling Petrucci family of his native Sienna, and for the Florentine and Venetian republics, and at the time of his death was director of the papal foundry and munitions plant in Rome. Agricola (1494–1555) was born in Saxony during a great expansion of that region's silver mines, graduated from Leipzig University, studied medicine in Italy, and served as town physician in the Erzgebirge region. He turned his close observations of mining and careful investments to his personal enrichment (by 1542 he was among the twelve wealthiest inhabitants of Chemnitz, Saxony) and to courtly recognition by the Saxon prince Maurice, who gave him a house and land in 1543 and three years later made him burgomaster and later councilor in the court of Saxony. His famous *De re metallica* is dedicated to the Saxon princes Augustus and Maurice.

The same Saxon princes figured in the career of Lazarus Ercker (ca. 1530–94). After attending the University of Wittenberg and marrying into a prominent family, he was named assayer at Dresden by Augustus. Within a year, he dedicated a practical metallurgical handbook to Augustus, who named him general assay master for three districts. Then, Ercker aligned himself with a new prince-practitioner, Prince Henry of Braunschweig, who named him assay warden of the Goslar mint. After dedicating his second book, on minting, to Henry's son, Julius, duke of Braunschweig-Wolfenbüttel, Ercker was named master of the Goslar mint. Moving to Bohemia in the mid 1560s, Ercker was made control assayer in Kutná Hora (Kuttenberg) through the influence of his second wife's brother. His wife, Susanne, served alongside Ercker as "manager-mistress" of the mint for many years. His *Treatise on Ores and Assaying*, first published in 1574, was dedicated to the Emperor Maximilian II, from whom he received a courtly position. Ercker was named chief inspector of mines by Maximilian's successor, Rudolf II, who also gave him a knighthood in 1586.[38]

Each of these three authors praised the values of complete disclosure, precise description, and openness often associated with the "scientific revolution." Their books detailed the processes of mining, smelting, refining, founding, and assaying. Biringuccio and Agricola used extensive illustrations to convey the best technical practices of their time (fig. 1.5). Biringuccio's *Pirotechnia* features 85 woodblock engravings while Agricola's *De re metallica* has more than 250. Agricola hired illustrators to make detailed drawings of "veins, tools, vessels, sluices, machines, and furnaces . . . lest descriptions which are conveyed by words should either not be understood by the men of our times or should cause difficulty to posterity."[39]

Even the illustrated volumes known as "theaters of machines" published by Jacques Besson (1578) and Agostino Ramelli (1588) reflected the support of aristocratic and royal patrons. These books were something like published versions of Leonardo's technological dreams, recording highly imaginative designs that no one

FIG. 1.5. AGRICOLA'S LIFTING CRANE

Agricola hired illustrators to make detailed drawings, like this vivid illustration of a two-directional lifting crane powered by a 36-foot-high waterwheel, "lest descriptions . . . conveyed by words . . . not be understood."

Georgius Agricola, *De re metallica* (1556; reprint, Berlin: VDI Verlag, 1928), 170. Courtesy of Illinois Institute of Technology Special Collections.

had built. Besson was engineer at the court of France's King Charles IX. Ramelli, an Italian military engineer, was also in the service of the king of France when his book—*Diverse and Ingenious Machines*—was published. While Besson's book had 60 engraved plates, Ramelli's book featured 195 full-page copperplate engravings of grain mills, sawmills, cranes, water-raising machines, military bridges, and cat-apults. Ramelli devised more than 100 different kinds of water-raising machines. Eugene Ferguson observes that Ramelli "was answering questions that had never been asked and solving problems that nobody . . . [had] posed. There is no sug-gestion that economic forces induced these inventions. The machines were clearly ends, not means." Yet many of Ramelli's machines, including a sliding vane water pump that is original to him, were eventually realized, as the problems of friction and materials were overcome. Ferguson concludes, "The seeds of the explosive expansion of technology in the West lie in books such as these."[40]

The scientific revolution itself was also surprisingly dependent on printing technology and courtly patronage. Tycho Brahe, a young Danish nobleman who taught himself astronomy through printed books in the 1560s, had access not only to accurate printed versions of all of Ptolemy's work (including a new translation of the *Almagest* from the Greek), but also to tables that had been recomputed on the basis of Copernicus's work, to printed sine tables, trigonometry texts, and star cat-alogues. Tycho could directly compare star tables computed from Copernicus and Ptolemy. On the island of Hveen, a gift from the Danish king, Tycho's observatory, Uraniborg, had no telescope; but it did have a well-stocked library, fifty assistants, and a busy printing press.

Galileo was not only a first-rate scientist but also a prolific popular author and imaginative scientist-courtier. His career began with a landmark study of mechan-ics that owed much to his observations of large hoisting machines at the Arsenal of Venice, and he secured his fame with another technological breakthrough—the telescope. Among the astronomical discoveries he reported in *Starry Messenger* (1610), something of a best seller across Europe, were the four moons of Jupiter. Galileo named them the "Medicean stars" in a frank bid for the favor of Cosimo de' Medici, the namesake and latter-day heir of the great Florentine powerbroker. In 1616 the Catholic Church placed Copernicus's work on its *Index* of prohibited works, while Galileo's own *Dialogue on Two World Systems* (1633), which left little doubt about his sympathies for the Copernican system, was put on the *Index*. Because of his open support of the sun-centered cosmology, Galileo was placed under house arrest, where he worked on the uncontroversial topic of classical mechanics. Yet even this treatise had to be smuggled out of Italy by a Dutch printer, Louis Elsevier, and printed in Leiden as *Discourses on Two New Sciences* (1638). "I have not been able," wrote Galileo in 1639, "to obtain a single copy of my new dialogue. . . . Yet I know that they circulated through all the northern countries. The copies lost must be those which, as soon as they arrived in Prague were immediately bought by the Jesuit fathers so that not even the Emperor was able to get one."[41]

The desires and dreams of Renaissance courts and city-states defined the character of the era's technology and much of its culture. Leonardo, Francesco, Alberti, and other engineers of the Renaissance era worked on war making, city building, courtly entertainments, and dynastic displays because that is what courtly patrons valued and that is what they paid for. We can easily broaden our view to include Italy's famous luxury glass and fancy silk industries and the impressive state-directed Arsenal of Venice without significantly altering this basic picture of court dominance. In early modern Europe, mechanical clocks were sought by "the numerous courts—royal, princely, ducal, and Episcopal" as well as by urban merchants, according to economic historian David Landes. For the silk industry, Luca Molà writes, "While in the fourteenth and in the first half of the fifteenth century merchant oligarchies ruling over Italian cities dedicated considerable energy to the development of silk manufacturing, later on champions of the new industry were to include the major princely families of the peninsula (such as the Visconti, the Sforza, the Gonzaga, the Este, the Medici, the Savoia, the della Rovere), more than one pope, and the monarchs of Spain, France, and England."[42] Courts in China, Turkey, India, the Persian Empire, and Japan also during these years, in varied ways, channeled technologies toward court-relevant ends.[43]

The patrons of Renaissance technologies, especially when compared with those of eras discussed in later chapters, were not much concerned with laborsaving industrial technologies or with profit-spinning commercial ones. Similarly, the early generation of moveable-type printing was deeply dependent on European courts for large-scale printing jobs, while authors of the books that made technology into a cumulative and progressive tradition depended on courtly patronage networks. The pervasiveness of the court system in the Renaissance should not really surprise us, since it was the dominant cultural and political actor at the time, fully analogous to the commercial and industrial institutions as well as the nation-states, corporations, and government agencies that followed with different imperatives and visions for technologies.

CHAPTER 2

1588 – 1740

Techniques of Commerce

The noble courts, city-states, and prince-practitioners who employed Renaissance technologists to build cities, wage war, entertain courts, and display dynasties were not using technologies primarily to create wealth or improve industries. Rather, they were using their wealth—from land rents, banking, and mercenary activities—to pay for the creation and deployment of technologies.[1] This practice shaped the character of that era's technology and—through the resulting cathedrals and sculptures, urban palaces and rural villas, court automata and printed books—the character of Renaissance society and culture as well. We have seen how this occurred in the growth and expansion of the mining industry in central Europe around 1600 and in the profusion of court-sponsored technology books on mining.

The imperatives of creating wealth reshaped the content and purpose of technology during the great expansion of commerce across Europe. In Venice, Florence, and other Italian city-states, commercial activities began expanding even before the Renaissance, of course, but the commercial era was fully realized a bit later in Antwerp, Amsterdam, and London. Each of these three cities was the node for far-flung trading networks, constructed when commercial traders, following up on the Europeans' "voyages of discovery," created maritime trading routes to Asia, Africa, and the New World. Even though no single year marks a shift from one era to another, the influence of Renaissance-era courts was on the wane by around 1600 while the influence of commerce was distinctly rising.[2] It is important to recognize the historical distinctiveness of the commercial era and to avoid reducing it to an "early" but incomplete version of industrial capitalism. As we shall see, the era of commerce was thoroughly capitalistic but not industrial in character. The imperatives of commerce included carrying goods cheaply, processing them profitably, and creating the means for moving cargoes and money. Technologies such as innovative ship designs, import-processing techniques, and a host of financial innovations expressed these commercial impulses, just as attack chariots, court automata, and princely palaces expressed the court impulse of Renaissance technologies.

The age of commerce, anticipated in Spain and Portugal as well as in China and

India, found its fullest expression during the seventeenth-century Golden Age of the Dutch Republic. The Dutch Republic at this time was reminiscent of Renaissance Italy in that all manner of cultural activities flourished to "the fear of some, the envy of others and the wonder of all their neighbors." Indeed, the many parallels between the older centers in the south of Europe and the rising commercial centers in the north led Fernand Braudel to posit long-term "secular trends" in which each city-state had, as it were, its historical moment at the center. Yet such a view, however appealing as an overarching world-historical theme, suggests a troubling inevitability about historical change and undervalues how the Dutch developed new technologies to capture the leading economic role in Europe and to construct a trading empire of unprecedented global scope.[3]

Venice had dominated trade in the Mediterranean, while Spain captured the riches of the New World; but the Dutch Republic became the center of a worldwide trading and processing network that linked enslaved people from Africa, copper from Scandinavia, sugar from Brazil, and tea and spices from Asia, with maritime, processing, and trading technologies at home. It was an improbable success for a small country that lacked surpluses of labor, raw materials, and energy sources. "It seems a wonder to the world," puzzled one English writer of the time, "that such a small country, not fully so big as two of our best shires [counties], having little natural wealth, victuals, timber or other necessary ammunitions, either for war or peace, should notwithstanding possess them all in such extraordinary plenty that besides their own wants (which are very great) they can and do likewise serve and sell to other Princes, ships, ordnance, cordage, corn, powder, shot and what not, which by their industrious trading they gather from all the quarters of the world."[4] This chapter examines how citizens of the Dutch Republic shaped technologies in the pursuit of commerce and how commercial technologies shaped their culture.

TECHNOLOGY AND TRADE

The Rhine River is the central geographical feature that propelled the so-called Low Countries to economic preeminence even before the seventeenth century and continues to shape commercial prospects there today. The Rhine begins with melting ice in the mountains of central Switzerland and runs 760 miles through the heart of Western Europe. The river passes in turn through the cities of Basel, Strasbourg, Cologne, and Düsseldorf, before exiting to the North Sea. Its several tributaries connect to a score of commercial centers, including Frankfurt, Stuttgart, and Mulhouse. Even today, the Rhine's vessel-tonnage nearly exceeds the tonnage carried on the Mississippi River and St. Lawrence Seaway combined. Handling this flow of cargo occupies two of the largest ports in the world, Rotterdam and Antwerp, which organize the multifarious businesses of offloading, processing, and reexporting the commerce of all Europe. Yet, these basic port and processing functions have changed surprisingly little since the Dutch Golden Age, allowing, of course, for the immense scale of today's port complexes (see fig. 9.6). A high level of trading activity occurred early on; by 1550 imports for the Low Countries were already four times those for England and France in per capita terms.

The Dutch Republic as a political and cultural entity took form in the wake of Luther's Bible-printing Reformation. Beginning in the 1570s, Catholic Spain and the Protestant core of the northern Low Countries, the United Provinces, fought for control over the region, then part of the Spanish empire. Spain's dominance over the southern provinces of Flanders and Brabant, the campaign for which included its recapture of Antwerp in 1585, prompted many merchants and craftsmen, including the theologically flexible printer Christopher Plantin (see chapter 1) to set up shop elsewhere. The tide turned just three years later, however, when the English Navy defeated Spain's "Invincible" Armada (1588) and thereby eliminated one threat to the Dutch maritime economy. During the next decade, the Dutch blockaded Spanish-occupied Antwerp, cutting off that rival port from North Sea shipping and striking a blow to its leading role in commerce. The Dutch then forcibly reclaimed from Spain the inland provinces surrounding the Zuider Zee.

During these battles Maurice of Nassau, the Dutch military commander, turned handheld firearms into winning weapons. He transformed the matchlock muskets of the time into formidable battlefield weapons through systematic drilling of his soldiers. It went like this. Maurice divided his troops into small units and separated the loading, aiming, and firing of a musket into forty-two individual steps, each one activated on the battlefield by a particular shouted command. For instance, steps eleven to sixteen directed each soldier: "hold up your musket and present; give fire; take down your musket and carry it with your rest [pole that supported musket during firing]; uncock your match; and put it again betwixt your fingers; blow your [firing] pan." Maurice also drilled his soldiers to form revolving ranks in battle: in the protected rear rank, reloading their muskets; in the middle, preparing to fire; in the front, firing a coordinated volley at the enemy; and then falling back in orderly fashion to reload. Drilled in this methodical procedure, the disciplined Dutch soldiers, when directed, let loose deadly barrages of musket balls. The Spaniards fell in droves.[5]

As a result of Maurice's military victories, the newly formed Dutch Republic in 1609 secured an advantageous truce with Spain. The new independent political confederation comprised the maritime provinces (Zeeland, Holland, and Friesland, which had federated thirty years earlier) and the inland provinces to the north. They were assertively Protestant. The southern provinces (more or less today's Belgium) remained under the control of Catholic Spain. "This had disastrous effects on the once wealthy economies of Brabant, Flanders, and the Walloon provinces," writes economic historian Joel Mokyr. "Spanish mercenaries devastated the land, a huge tax burden crippled the economy, and the relentless persecution of Protestants prompted thousands of highly skilled artisans to flee abroad, where they strengthened their homeland's competitors."[6]

Those fleeing the Spanish-controlled lands included craftsmen and sailors as well as merchants and financiers. Walloon exiles introduced new mills for the "fulling," or finishing, of woolen cloth at Leiden and Rotterdam, in 1585 and 1591, respectively. Many merchants fleeing the Spanish-dominated lands eventually settled in Amsterdam. In Amsterdam, Walloon exiles were among the leading financial shareholders in the Dutch East India Company (discussed later). Walloons and other groups who had suffered religious persecution welcomed the climate of religious tolerance in the Dutch Republic, exceptional in Europe at the time, which extended across the spectrum of Protestant sects and included even Catholics and Jews. Jews became leading figures in Amsterdam's tobacco and diamond trades.

Louis de Geer (1587–1652) numbered among the emigrant merchants and technologists whose influence was felt from Portugal to Prague, especially in the fields of fortification, urban drainage, mining, and harbor engineering. De Geer, originally from the Walloon province city of Liège, began his career in the Dutch city of Dordrecht trading a mix of goods, then moved to Sweden. De Geer's wide-ranging activity in Sweden exemplifies the substantial Dutch technical influence in northern Europe as well as a subtle but pronounced shift from court to commerce. Sweden at the time was seeking to develop its rich mineral holdings in iron and copper, much in demand for coinage and cannons. The king of Sweden chartered the Royal Company (in 1619) to win back control of the copper trade, which had slipped to Amsterdam. Since Sweden's state finances were dependent on copper, the trade was a matter of some urgency. This court-sanctioned venture did not succeed, however, and in the next decade control of the Swedish copper industry passed firmly into Dutch hands. Dutch investors before and after the Royal Company's failure provided investment capital to Sweden that required repayment in copper, which further tilted the field against the crown and toward commerce.

Louis de Geer played a leading role not only in routing Swedish copper exports to Amsterdam but also in transferring valuable mining and smelting technologies to Sweden. In Stockholm, de Geer created the largest industrial complex in Sweden, making iron and brass, ships and ropes, cannon and cannonballs. The *garmakeriet* method for refining copper was an import from the southern Netherlands, as was the *vallonsmidet* (literally, "Walloon smithy") process for ironmaking. Another Dutchman, with the backing of the Swedish king, set up a famous ironworks at Eskilstuna that grew into another of Sweden's major industrial districts. At the peak of its copper industry, around 1650, Sweden accounted for fully half of Europe's total copper production. The only other nation with substantial copper production at the time was Japan, and Dutch merchants coordinated its copper trade, too.[7]

The emergence of specialized ship designs in the Netherlands signaled that the Dutch understood how to bring technology and trade together in the pursuit of commerce. Most seafaring nations either possessed the raw materials needed for wooden shipbuilding or had ready access to overseas colonies that did, while the Dutch had neither. Among the necessary raw materials for wooden shipbuilders were, obviously, timber for masts and planking; resin, pitch, and turpentine (so-called naval stores) for waterproofing; and rope for rigging. In a sense Dutch shipbuilding began with wheels of cheese, which along with other export goods, including salt, wine, and herring, were shipped off to Norway and the Baltic countries in exchange for timber and naval stores. The Dutch economy relied on savvy trading, such as that practiced by merchants in Dordrecht and Zeeland, the leading exporters of German and French wines, respectively. The real key, though, was in spotting where a low value import such as salt from Spain, Portugal, or France could be processed into a high value export. In this respect the salt-refining industry in Zeeland was an important precursor to the distinctive and wide-ranging "traffics" system (discussed later).

An innovative, purpose-built ship secured to the Dutch thorough dominance in the North Sea herring fishery. The full-rigged herring "buss" was a sizeable fishing vessel—with a crew of fifteen or more—and was effectively a self-contained factory, designed to stay out in open seas for up to eight weeks and through the roughest weather while carrying the salt and barrels and manpower needed to gut, salt, and pack the freshly caught herring right on board (fig. 2.1). By the 1560s the northern provinces already had around 500 herring busses; the largest packed 140 tons of salted fish. Yet the real distinction of the Dutch herring fishery was not so much its volume of production but rather the consistently *high quality* of the packed herring and its correspondingly high price—signature characteristics in the Dutch commercial era

FIG. 2.1. DUTCH HERRING BUSS

Something like a fishing vessel married to a factory, the herring buss carried all the salt, wooden barrels, and human labor needed to transform freshly caught North Sea herring into a marketable commodity.

Courtesy of The Kendall Whaling Museum, Sharon, Mass.

that sharply distinguish it from the high-volume but lower quality early industrial era that followed.

Use of the factorylike herring busses was just one of the distinctive Dutch maritime activities. The province of Holland at that time had 1,800 seagoing ships. (Amsterdam's 500 ships easily surpassed Venice's fleet of seagoing ships, estimated at 300 at its peak around 1450.) The labor force required just for this shipping fleet was perhaps 30,000 people, not counting the thousands more needed to build the ships, rigging, and sails and to transport the off-loaded goods. This shift of labor out of the agricultural sector, a defining feature of "modern" economies, relied on the large imports of Baltic grain that allowed Dutch rural workers to specialize in exportable goods like cheese and butter. For a small population—the Dutch Republic had two million at its height in the seventeenth century—such an export-oriented agriculture was indispensable.[8]

The Dutch effected their most brilliant departure from traditional ship designs in the field of commercial trading vessels. "English shipping in the late sixteenth century was, generally speaking, multi-purpose and, being often used for voyages to the Mediterranean, tended to be strongly constructed, well manned, and well armed," writes Jonathan Israel. "Dutch shipbuilders, by contrast, concentrated on low-cost hull forms that maximized cargo space, discarded armaments, and used only the simplest rigging. Dutch shipping was designed for minimal crews and maximum economy."[9] Venice's ships were, even more than Britain's, designed for the pirate-infested Mediterranean (and given the recurrent European wars and the long-running contest with the Ottoman Turks, one country's loathsome "pirates" might be another country's heroic privateers). Spanish ships were oriented to the imperial lifeline of New World silver, in which heavily armed sailing ships loaded down with bullion plied the Atlantic. Between silver runs, the same Spanish ship might be ordered to shuttle goods or ferry soldiers to a European port. The higher costs of moving goods with armed ships, where space for soldiers and weaponry crowded out cargo capacity, hampered the commercial prospects of both Venice and Spain.

The famous Dutch *fluytschip* was in several ways the precise embodiment of commerce. The fluyt, a distinctive cargo ship design which dominated northern European shipping for a century or more, emerged from the city of Hoorn in the mid-1590s (fig. 2.2). This was no graceful clipper ship with sleek lines and tall masts. The fluyt was instead stubby and squat: its sail area was small and its full-rigged masts short compared to its capacious cargo hold. The vessel was ploddingly slow, yet with its excellent handling qualities and the extensive use of pulleys and blocks, a small crew of around a dozen could easily control it. Another economy was achieved by the use of cheap light pine in the vessel, except for the heavy oak hull to resist rot.

The basic fluyt design set aside the typical reinforced construction, which every armed ship needed to support the weight and recoil of cannons, in exchange for increased cargo space and simplicity of handling. The resulting ship—an artifact shaped by commerce—was exquisitely adapted to the peaceful trade of northern Europe, if not for the more dangerous waters of the Mediterranean or the Atlantic. Since transit taxes in the Baltic were levied according to a ship's deck area (rather than its cargo capacity), Dutch fluyt builders logically enough reduced the topdeck's area while retaining a large cargo hold underneath. In fact, surviving models of Dutch fluyts from this era are so narrow decked, wide bodied, and round shaped at the bow and stern that they appear to be whimsically "inflated" versions of real ships; nevertheless, these models agree closely with contemporaneous documentation of the ships' actual shapes and sizes.

Following on these innovations, Dutch shipbuilders practiced a high degree of design specialization. In contrast with most other European countries, where military ships and commercial ships were used interchangeably, Dutch shipbuilders separated military and commercial shipbuilding while segmenting the basic fluyt design. "The output of private shipbuilding yards was so much greater than that of naval yards that the republican Admiralties found it almost impossible to meddle in the affairs of commercial shipbuilders," writes Richard Unger. In the Netherlands, "the design of warships was rather influenced by successful experiments with merchant vessels, the inverse of what happened in the rest of Europe."[10]

The Dutch East India Company (discussed later) had its own shipbuilding yard, where it evolved a distinctive broad-hulled design for the famous East Indiaman. Also known as a *retourschip*, this hybrid vessel was a larger and more heavily armed version of the pinnance, itself an armed version of the fluyt. But while most fluyts were no larger than 500 tons displacement, the largest *retourschip* displaced 1,200 tons and might be an impressive 45 meters long. Meanwhile, Dutch commercial shipbuilders extensively customized the basic fluyt design. So whaling fluyts had double-thick bows to protect against the ice in far northern waters; timber fluyts had special hatches to permit the loading of entire trees; and the so-called *Oostvaarder* fluyts were specially designed for cheating the Danish king out of transit taxes. For a time, until the Danish saw through the scheme, the *Oostvaarders* had bizarre hourglass-shaped topdecks that were extra narrow precisely at the high midpoint where the all-important tax measurement was taken.

If it appears the Dutch experienced only maritime successes, there is at least one cautionary tale from these years. In the mid-1620s the Swedish crown and Dutch shipbuilders were entangled with the ill-fated *Vasa*. It was an oversize warship commissioned by the Swedish king for his military adventures in the Baltic region, and although it was built in Stockholm, the designer and shipbuilder was a Dutchman, shipwright Henrik Hybertsson, together with a Dutch business partner. They purchased timbers and planking from Sweden, ropes and rigging from Latvia, tar from Finland, and sailcloth from France and Holland. The king demanded 4 dozen heavy 24-pounder bronze cannons, which required two full upper decks of armament. Today, visitors to the Vasa Museum in Stockholm's harbor can gaze up at the impossibly tall vessel and wonder how could it possibly have floated? Indeed. On its maiden voyage on 10 August 1628, the vessel barely cleared the inner harbor when it

FIG. 2.2. DUTCH CARGO FLUYT

Line drawing of a fluyt; the original vessel was 100 feet long.

Courtesy of The Kendall Whaling Museum, Sharon, Mass.

started listing to the port side under a light breeze then with a stronger puff of wind it simply toppled over and sank, taking thirty of its crew to the bottom.

An official inquest uncovered several liabilities of the *Vasa*. The vessel was impressively but dangerously tall. Its design was decidedly "top-heavy" with a hull-length of 47 meters (equal to the largest *retourschip*), width (beam) of 11.7 meters, and a top mainmast height of 52.5 meters with just 4.8 meters draft below the water-line. Another problem was that the two sides of the vessel were measured and constructed by different building teams, one using the Swedish foot (of 12 inches) and the other the Amsterdam foot (of 11 inches). As built, the vessel was not evenly balanced. Some hurry attended the trial voyage (with the king away on campaign), and so the vessel's captain did not secure the cannon-deck hatches. When the vessel heeled over, water poured in catastrophically. *Vasa* sat in shallow water in Stockholm harbor for three centuries until 1961 when it was raised up, carefully preserved, and then turned into a major technical and cultural attraction for the city.[11]

The commerce-inspired designs for herring busses and cargo-carrying fluyts are impressive evidence of the Dutch embrace of commerce. Yet their real distinction was to make these innovations into broad society-wide developments that shaped Dutch culture not only at the top of wealthy merchants and investors but also right through the merchant and artisan classes. Even rural workers milking cows for cheese exports participated in the international trading economy. The distinctive *trekvaarten* network of horse-towed barges provided scheduled passenger service throughout the western region of the Netherlands, enabling merchants and traders to travel to and from a business meeting within the day. The Dutch impulse to enlarge the commercial economy manifested itself in uniquely broad ownership of ships and, as we will soon see, of much else.[12]

At other European shipping centers of the time, owning a single ship had been divided among three or four investors. The Dutch took the practice much further. For the bulk-carrying fluyts, as well as for fishing, whaling, and timber-carrying ships, ownership of sixteenth, thirty-second, or even sixty-fourth shares became common. At his death in 1610, one Amsterdam shipowner left fractional shares in no fewer than twenty-two ships. Middle-level merchants connected with the bulk trades were typically the collective owners of the fluyts. Technology investments also took collective forms, as when twelve Amsterdam brewers shared the cost for mechanizing malt grinding or in the extensive multishare ownership of linseed oil mills in the Zaan district (see later). Although history remembers the wealthiest merchants who patronized the famous painters and built the finest townhouses, the Dutch economy, in shipping and in many other sectors, got much of its vitality from an unprecedentedly broad base of activity. In short order, the Dutch led all of Europe in shipbuilding. A contemporary estimate had it that in the late 1660s fully three-quarters of the 20,000 ships used in the European maritime trade were Dutch, with England and France falling into distant second and third places.[13]

CREATING GLOBAL CAPITALISM

While the British East India Company had its heyday somewhat later, presiding over nineteenth-century British imperialism in India (chapter 4), its Dutch counterpart (1602–1798) was quick off the mark. By the middle of the seventeenth century, the Dutch East India Company's economic reach and political influence spread from its singular port at Nagasaki, Japan, clear across the Far East to Yemen in the Middle East. The company's trading patterns—a web of intra-Asian routes in addition to bilateral Asian-European routes—differed fundamentally from the simple bilateral trading practiced by Europeans since the days of the renowned Silk Route. By comparison, the "global" trading empire of which Spain boasted was mostly an Atlantic one; from the 1530s onward, it centered on extracting silver from Mexico and Peru. Spanish trading in the Pacific focused on the Philippines, and goods from the Philippines reached Spain only after they were transshipped across land in Mexico and then put on a second vessel across the Atlantic. In the 1620s the Spanish lost their dominance over the Chinese silk trade and were soon expelled from Japan. At the British East India Company, significant intra-Asian trade emerged only in the 1830s. At the height of Dutch activities in Asia during the mid-seventeenth century:

> Bengal sugar and Taiwan sugar were sold in Persia and in Europe, while silk from Persia, Bengal, Tongking, and China was sold in Japan and in Europe. One key set of exchanges could be traced as follows: Pepper, sandalwood, and other Southeast Asian goods were sold in China . . . , and gold, tutenag [a semiprecious alloy of copper, zinc, and nickel], raw silk, and silk fabrics were bought. Much of the raw silk and silk fabrics was sold in Japan, and the silver earned, along with the gold and tutenag from China, was taken to India to buy cotton fabrics, which were the principal goods that could be exchanged for spices in Indonesia.

The Dutch—through their East India Company in the Pacific and West India Company in the Atlantic, coupled with the extensive trading in Europe and Africa— in effect created the first truly global economy.[14]

This global expansion depended on Dutch innovations in the basic institutions of commercial capitalism, including commodity exchanges, a public exchange bank, and a stock exchange. If the Dutch did not exactly invent capitalism, they created the first society where the principles of commerce and capitalism pervaded the culture. Capitalism is often traced back to the merchants of Venice, and it is true that at least six other European cities had commodity exchanges during the sixteenth century. But while these exchanges had usually set only regional prices, Amsterdam's exchanges became the world centers for commodity pricing and trade flows. The commodity traders' guild began publishing weekly lists of prices in 1585. Within a few years, the Amsterdam commodity exchanges—for grain, salt, silks, sugar, and more—had surpassed their regional rivals and become global exchanges. By 1613 the Amsterdam Exchange was lodged in a new building, where a well-organized guild ensured honest trades by tightly regulating its 300 licensed brokers. Grain merchants soon built a separate grain exchange. By 1634 the commodities exchange's weekly

FIG. 2.3. DELIVERY OF A FUTURES CONTRACT

Amsterdam's financial markets dealt extensively with futures contracts in grain, wool, silks, and even the "buying of herrings before they be catched."

Courtesy of The Kendall Whaling Museum, Sharon, Mass.

bulletins set authoritative prices across Europe for 359 commodities; the exchange added nearly 200 more in the next half-century (fig. 2.3).

The Amsterdam exchanges can be measured not only by their size, scope, and specialization but also in their financial sophistication. For example, futures contracts emerged, speculating on grain that had not yet been delivered and the "buying of herrings before they be catched." In time, Amsterdam merchants were purchasing such varied goods as Spanish and German wools and Italian silks up to twenty-four months in advance of their physical arrival. Maritime insurance became yet another financial activity linked to global trade. At least until London in the 1700s (see chapter 3), there was simply no rival to Amsterdam in the breadth, depth, and refinement of its financial markets.[15]

The rising volume of trading in physical commodities depended on robust means of trading money. The Amsterdam Wisselbank, or exchange bank, founded in 1609, was the first public bank outside Italy and was without peer in northern Europe until similar banks were organized in Hamburg and London in 1619 and 1694, respectively. Before this, a merchant needing to pay for a delivery of foreign goods might visit a money changer to purchase sufficient foreign currency. Where possible, merchants preferred so-called bills of exchange, which were in effect a merchant's paper promise to pay the bearer a certain sum of money on demand. However, bills of exchange issued by merchants circulated at a profit-sapping discount that varied with the trustworthiness of the merchant and the trading distance.

The Amsterdam Wisselbank slashed these burdensome transaction costs. In essence, it provided an institutional means for merchants to quickly and efficiently pay bills, backed by its huge pool of capital. By 1650 a merchant's deposit into the Wisselbank, because of the certainty of getting payment in settling an account, commanded a premium (not discount) of 5 percent. Secrecy surrounded the Wisselbank's accounts; one contemporary estimated its holdings in hard currency at an astounding 300 million guilders, although modern scholars believe the bank more likely held one-tenth this amount. This deep pool of exchange capital meant that Dutch interest rates were half those in England and an even smaller fraction of those in France and Germany. A Dutch merchant borrowing money to purchase goods paid substantially lower interest charges and pocketed the difference. As one Englishman put the matter in 1672, low Dutch interest rates "hath robbed us totally of all trade not inseparably annexed to this Kingdom."[16]

It would be a mistake, even in the face of this determined financial innovation, to see the Dutch solely as sober paragons of economic calculation. After all, in the 1630s the Dutch fostered the great tulip mania, certainly the world's most colorful economic "bubble." Whatever the actual value of a flower bulb, crazed Dutch traders bid tulip prices to remarkable heights. Already by 1633 an entire house in the shipbuilding town of Hoorn was traded for three rare tulips. More to the point, tulip trading embodied several of the classic Dutch financial techniques, including futures contracts, commodity pricing, and multiple-share ownership. Traditionally tulip bulbs had changed hands only during the months from June to September when the bulbs were "lifted" from the soil after flowering and were available for inspection. The first step toward a futures market in tulips was when traders, sensibly enough, began accepting a future delivery of a particular tulip bulb; a slip of paper promised delivery when the bulb was lifted the following June.

The circulation of these valuable slips of paper seems quickly to have gotten out of hand. It was difficult to know, in the absence of any regulatory scheme, whether a given slip of paper represented a "real" tulip in the soil or merely the hope of a seller's obtaining one in the future. An increasing number of traders cared not at all to possess a physical bulb but simply to buy and sell these slips of paper advantageously. Other tulip trading techniques modeled on the commodity exchanges were the elaboration of weight-based pricing and the distinction between "piece" and "pound" goods. The most valuable bulbs were always sold as a "piece" good, after 1635 invariably priced according to their weight in "aces" (around one-twentieth of a gram), while aggregate lots of the less valuable bulbs were traded as "pound" goods similar to commodified grain. Finally, in a move that made tulips more like ships than flowers, at least one Amsterdam grower sold a half share in three expensive bulbs to a confident customer.

These financial practices, fanned by greed and fear, pushed tulip prices to a sharp peak in February 1637. In the first week of that month, an exceptionally fine harvest of tulip bulbs, owned outright by the surviving children of an Alkmaar tulip grower, was auctioned with great fanfare. The prize in the collection was a mother bulb, including a valuable offset, of the variety Violetten Admirael van Enkhuisen, for which a private buyer paid 5,200 guilders, while the most valuable bulbs auctioned publicly were two Viceroys that fetched 4,200 and 3,000 guilders. Dozens of other bulbs, both piece and pound goods, commanded record prices. (At the time, 5,000 guilders would purchase a large townhouse in Amsterdam while the lesser sum of 3,000 guilders, as one contemporary pamphleteer enumerated it, would purchase 2 lasts of wheat [around 24 tons], 4 lasts of rye, 4 well-fed oxen, 8 well-fed pigs, 12 well-fed sheep, 2 oxheads of wine, 4 tons of eight-guilder beer, 2 tons of butter, 1,000 pounds of cheese, 1 bed with accessories, 1 stack of clothes, and 1 silver chalice. And a boat to load the goods into.) Altogether, this prime lot of tulips brought in the breathtaking sum of 90,000 guilders.[17]

One might guess that tulip traders during the mania, while calculating and profit minded, were far from sober. While tulip trading in earlier years had been the province of wealthy connoisseurs, the run-up in tulip prices in the 1630s drew in a large number of fortune seekers. Instead of the tightly regulated commodity exchanges or the Amsterdam Stock Exchange, which opened in 1610 and oversaw a set two-hour trading day, the trading floor for tulips consisted of the back rooms of inns, where trading might extend from ten o'clock in the morning until the early hours of the next day. With wine for the wealthy, and beer and cheap spirits for everyone else, Dutch taverns had age-old traditions of heavy social drinking. "All these gentlemen of the Netherlands," complained one Frenchman, "have so many rules and ceremonies for getting drunk that I am repelled as much by the discipline as by the excess." Traditionally a center of beer brewing, Haarlem emerged as a center of tulip trading as well; its taverns served up 40,000 pints of beer each day, one-third of the city's output—not bad for a population of just 30,000 men, women, and children. The combination of tulip trading and taverns is no accident. "This trade," offered one contemporary, "must be done with an intoxicated head, and the bolder one is the better."[18]

In the same month that the Alkmaar bulbs fetched astronomical prices, the price of tulips collapsed, wiping out many down-market tulip speculators. Very shortly, the buzz about priceless tulip bulbs decayed into a squabble over worthless tulip futures. The crash had little effect on the mainstream commodity exchanges, the stock exchange, or Wisselbank. During the same years as the tulip mania, trading was lively in shares of the United East India Company (Verenigde Oostindische Compagnie, or VOC). The VOC represented a characteristically Dutch solution to the problems of importing spices from the East Indies and contesting the dominance of Portugal and Spain. Beginning in the 1590s, Dutch merchants had flocked into the lucrative East Indian trade. The states of Holland and Zeeland offered them not only free import duties but also free firearms for the venture. By 1601 there were fourteen separate Dutch fleets comprising sixty-five ships, all plying East Indian waters. The economic consequences of this free-for-all, however, were ominous. The prices Dutch traders paid for pepper in the East Indies had doubled over the previous six years, the prices they charged to Dutch consumers back home had fallen, and their profits had plunged.

This ruinous competition came to an end when, after the merchants' appeals, the Dutch federal government assembled the directors of the numerous trading companies at The Hague. In 1602 this conference created the VOC. It was a state-supported, joint-stock trading monopoly, with a decentralized structure. The VOC's four "chambers" raised separate capital, kept separate books, and chose separate directors. Initially, Amsterdam claimed half the total votes of the VOC's board of directors, because of its leading role in financing the East Indies trade. In fact, Amsterdam got just eight of seventeen voting seats. Four seats went to Zeeland, two each to the North Quarter (Hoorn and Enkhuizen) and South Holland (Delft and Rotterdam), with the last seat rotating among the three *non*-Amsterdam partners. The VOC's founding capitalization of 6.4 million guilders was comparable to the value of 1,000 large townhouses. Over half of this total capitalization (57%) came from the Amsterdam chamber though it controlled only 47 percent of the directors' votes. The richest of Amsterdam's native merchant elite invested sums of 12,000 to 30,000 guilders, and Walloon exiles made some of the largest individual investments, up to 85,000 guilders.[19]

The VOC was much more than an innocuous Dutch spice-trading cartel. It was in fact intended to break Spain's and Portugal's political and trading dominance in

Asia, and the VOC from its founding was granted wide-ranging commercial, military, and political duties. The Dutch federal government gave it warships and an annual subsidy, raised to 200,000 guilders in the mid-1610s. The warships worked wonders. "The Dutch ships were very large and tall, and each carried more than a thousand guns and small arms; they blocked the whole strait," wrote one besieged Chinese official. Within three years of its founding, the VOC snatched control of the legendary Spice Islands from Portugal and thereby cornered the world trade in cloves, mace, and nutmeg. Further gains in Asia—through alliances with local rulers in southern India and the Spice Islands—pushed VOC share prices on the Amsterdam Exchange above 200, or double their face value. However, the independence-making truce between the Dutch Republic and Spain in 1609, because it implied a resurgence of trade and influence by Spain and Portugal, sent VOC shares down sharply; they did not recover for years.[20]

The key to the lucrative spice trade, as the European rivals understood, was in Asia itself. India in the preindustrial era had a thriving cotton industry, and its high-quality and brightly colored cloth could be favorably traded for spices. On the southeast coast of India and on the islands of what is now Indonesia, each of the trading countries sought to establish trading alliances; and when these alliances were betrayed, they tried unarmed trading "factories" (warehouse-like buildings where "factors"— traders—did business). When these were attacked, they built heavily fortified factories garrisoned with soldiers. The Dutch, backed by its fleet of forty-odd warships, erected a massive stone fortress at Pulicat, in the southeast of India, that became the leading European factory in the region and the centerpiece of Dutch control of the Indian cotton trade. By 1617 there were twenty Dutch fortress-factories across Asia. The Portuguese too built fortress-factories in India, at Goa and Ceylon; the Spaniards, in the Philippines; while the English went after the Spice Islands. These "factories" sometimes finished cotton cloth imported from the countryside and as such can be considered the first cotton-textile factories in the world.[21]

Beginning in the 1620s, the VOC created a vast intra-Asian trading network. Not gold or silver bullion alone but traded goods were the medium of exchange. "Shipping fine spices, Chinese silks and porcelain, and Japanese copper to India, the Company purchased cotton textiles [in India] . . . which it then sold in [Indonesia] for pepper and fine spices. In the same way, spices, Chinese silks, Japanese copper, and also coffee from Mocha helped pay for the VOC's purchases of silk and drugs [opium] in Persia. Pepper and spices also supplemented silver in the Company's purchases of Chinese wares, on Taiwan." When in the 1630s the Japanese government restricted its merchants' conduct of foreign trade (one of several harsh measures to preserve its court-based culture), the Dutch took up Japan's active silk and copper trading. As of 1640, the Dutch trading factory at Nagasaki was the sole remaining European trade link with all of Japan. VOC share prices, scarcely registering the collapse of the tulip mania, climbed all the way to 500.[22]

The Tokugawa era in Japanese history (1603–1867) suggests the extreme measures needed to sustain court-based culture in an age of commerce. The arrival of firearms, carried by shipwrecked Portuguese sailors in 1543, into a society dominated by competing warlords had made for a volatile situation. Japanese artisans quickly learned to make immense numbers of firearms, and battles between rival warlords soon involved tens of thousands of firearms (far more than anywhere in Europe). As Noel Perrin's *Giving Up the Gun* makes clear, the proliferation of firearms directly threatened the status of the Japanese *samurai,* or warrior class. Low-class soldiers could simply shoot dead a high-class *samurai* regardless of his formidable sword. In 1603, the warlord Ieyasu Tokugawa asserted central control on the country and, remarkably enough, achieved it across the next two decades. The warring rivals who had torn the country apart with firearms were turned into subservient *daimyo* and given feudal lands to administer. On religious pretext, the troublesome firearms were confiscated and melted down. Japanese merchants were forbidden to conduct trade with or travel to foreign countries. Europeans, having brought firearms, as well as Christianity, which further disrupted Japanese society and culture, also fell under a cloud of suspicion. The Spanish and Portuguese missionaries were especially offensive. By contrast, the Dutch kept their religion to themselves and were permitted to stay (and trade) even after the forcible expulsion of the other Europeans in the 1620s and 1630s. These strict measures to preserve Japan's court-centered culture persisted through the mid-nineteenth century.[23]

It was also in the go-ahead decade of the 1620s that a second Dutch overseas company was organized. The purpose of the West India Company, created in 1621–24, was to secure Atlantic trading routes between West Africa and the West Indies (roughly, from Brazil to Cuba). Like the VOC it was a state-sanctioned trade monopoly federated among five regional "chambers." Even more than the VOC, the West India Company drew investors from outside Amsterdam to raise its 7.1 million guilders. Such inland cities as Leiden, Utrecht, Dordrecht, and Groningen together invested substantially more than the city of Amsterdam. Amsterdam once again had eight votes, but this time it was out of an expanded total of nineteen directors. While the VOC dealt with spices and cotton, the West India Company traded in enslaved people and sugar.[24]

The West India Company focused initially on West Africa and the exchange of Swedish copperware for Guinea gold. In this way copper became part of a far-flung trading network. Whereas in the early 1620s 2 ounces of African gold traded for 70 or 80 pounds of copperware, the West India Company's competition-damping monopoly allowed it to exchange just 35 pounds of copper for the same amount of gold. (From an African perspective, gold was thus worth half as much.) In its first thirteen years, the West India Company imported 12 million guilders in Guinea gold, handsomely exceeding its founding capital; in addition, it raided more than 500 Spanish and Portuguese vessels. The company in one stroke netted a whopping 11 million guilders when in 1628 its ships captured an entire Spanish silver fleet off Cuba, paralyzing Spain's silver-dependent imperial economy. In response to these gains, the company's share price around 1630 topped even the VOC's. For a time it looked as if the company's plunder would actually pay for its heavy outlays for 220 large warships, vast store of weapons, and heavy burden of wages and salaries.

The West India Company planted a profitable sugar colony on the northeast coast of Brazil, on land captured from Portugal in 1635. Cultivating sugar demanded a large labor force, which the Dutch West African trade supplied in the form of enslaved people. "The Dutch now controlled the international sugar trade for the first time and, as a direct consequence, also the Atlantic slave trade. Plantations and slaves went together," writes Israel. During the decade 1636–45, the West India Company auctioned nearly 25,000 enslaved Africans in its Brazilian territory, while the price of sugar in Amsterdam skidded downward 20 percent because of the expanded supply. But in 1645 the Portuguese swarmed up from southern Brazil and staged a series of raids that effectively ended Dutch sugar planting on the mainland. When Dutch shipments of Brazilian sugar ended in 1646, sugar prices in Amsterdam climbed 40 percent in one year. As a sugar colony, Netherlands Brazil never recovered; and West India Company shares collapsed. Dutch sugar and slave traders shifted their focus to the islands of the Caribbean. From 1676 to 1689, the Dutch shipped 20,000 enslaved black people from West Africa to the West Indies; most of them were deposited at Curaçao (off the coast of Venezuela), whence these hapless souls were parceled out across the sugar-cultivating region.[25]

"THE GREAT TRAFFIC"

The material plenty of the Dutch Golden Age was created from a heady mix of shipping, financing, trading, warring, and slaving. Yet, these trading activities were

FIG. 2.4. SECRETS OF DUTCH SHIPBUILDING REVEALED

Title page from a book published in Amsterdam in 1697.

Courtesy of Kendall Whaling Museum, Sharon, Mass.

not the mainstay of the Dutch commercial economy. Dutch preeminence came through the targeted processing and selective reexporting of the traded raw or semi-processed materials. "A more or less unique sector emerged, often referred to as *trafieken* (traffics) as opposed to *fabrieken* (manufactures)," writes Joel Mokyr. Among the "traffics" with links to the maritime sector were sugar refining, papermaking, brewing, distilling, soap boiling, cotton printing, and tobacco processing, as well as the complex of activities related to shipbuilding (fig. 2.4). Other highly specialized activities in which the Dutch gained global dominance include processing dyes and glazes, cutting and polishing diamonds, grinding glass lenses, refining whale oil, bleaching linens, and dyeing and finishing broadcloth. The making of highly precise nautical compasses, maps, and chronometers reinforced Dutch maritime dominance. Dutch control over the Swedish copper trade led at least four Dutch cities to set up copper mills, often with skilled artisans or equipment from Germany. For

each of these "traffics," mastering special techniques and attaining superior quality were more important than achieving high levels of output. Indeed, high wages,
modest volumes, and high-quality production typified the traffics, in sharp contrast
with early industrial technologies, which emphasized low wages, high volumes, and
relatively low-quality production (see chapter 3).[26]

The "great traffic" was surprisingly wide reaching. Amsterdam was the largest
city in the Dutch Republic but not by much. Its population just exceeded Haarlem
and Leiden together, and these two towns often schemed with others to outvote
Amsterdam on foreign affairs and trade matters. For instance, although Amsterdam
lobbied for a peace settlement with Spain from the 1620s on—because the war
harmed its European trade—the country as a whole continued the war, following the
lead of Haarlem and Leiden backed by several inland provinces. The textile towns of
Haarlem and Leiden believed that prolonging war helped them compete with their
chief rival, Spanish-occupied Flanders, while the inland towns favored war owing
to their heavy investments in the two colonial trading companies. Amsterdam,
writes Israel, "was merely the hub of a large clustering of thriving towns, all of which
directly participated in the process of Dutch penetration of foreign markets."[27]

Dutch efforts in textiles often aimed for high-quality production and the resulting high-priced products. Textile towns such as Haarlem and Leiden, no less than the
better-known commercial centers, gained new skills with the arrival of exiled Spanish
Netherlanders. In the years after 1585 Haarlem prospered with the arrival of Flemish
linen-bleachers, who found a ready supply of clear water and soon began bleaching "to the highest standards of whiteness then known in Europe." Initially Leiden
focused on lower grades of cloth (says and fustians) produced from cheap imports
of Baltic wool, with expansion doubling its population in two decades to 1600. Later
in the 1630s Leiden dramatically expanded its output of high-value lakens, using
high-quality wool from Spain (see fig. 2.5). "It seems likely that this was in part due
to the important technical innovations which were soon to make Leiden fine cloth
famous throughout the globe and which gave it its characteristic smoothness of tex-
ture," writes Israel. Another high-grade Leiden cloth that experienced rapid growth
was camlet, made from expensive Turkish mohair. By the mid-1650s two-thirds of
Leiden's sizable 36,000 textile workers were employed in making lakens and camlets;
the value of the city's textile output was 9 million guilders. The Dutch at that time took
in at least three-quarters of Spain's wool exports and the majority of Turkey's mohair
yarn. Significantly, the principal export markets for Dutch fine cloth were in just the
countries that had produced the raw materials—France, Spain, and Turkey.[28]

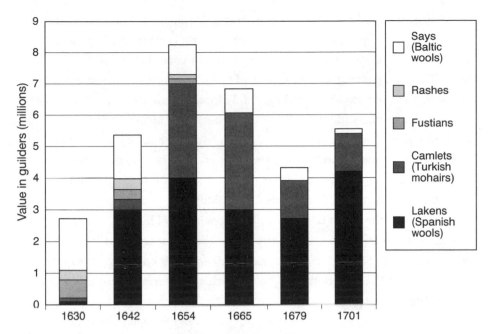

FIG. 2.5. LEIDEN'S HIGH-GRADE TEXTILES, 1630–1701

In the early seventeenth century, Leiden's textile industry made middle-quality fabrics, such as says (from Baltic wools). By midcentury Leiden focused on the upmarket camlets (from Turkish mohairs) and lakens (from superfine Spanish wools).

Data from Jonathan Israel, *Dutch Primacy in World Trade, 1585–1740* (Oxford: Clarendon Press, 1989), 195, 261.

In 1670, when Venice tried to reestablish its fine cloth industry, importing Dutch methods was needed to achieve the Leiden qualities demanded by the principal consumers. French observers of the time noted that the Dutch fine-cloth methods used a third less wool and less labor. Even lakens from Brabant, once the chief rival, lacked the characteristic smooth finish of the Dutch product. "The technical factor was crucial to both the continuance of Dutch dominance of the Spanish wool trade and the dominance of fine-cloth production itself," states Israel. Other Dutch textile centers included the silk industry of Amsterdam and Haarlem (which gained the skills of exiled French Protestants in the 1680s), the linen-weaving industry in the Twente and Helmond regions, and the sail-canvas industry in Haarlem, Enkhuizen, and the Zaan district.[29]

It must have been especially galling for the Spanish wool manufacturers to see all that wool slip through their grasp. From the late Middle Ages on, a massive "herding economy" active through much of central Spain tended the famous Merino sheep that grew superfine wool. In the late fifteenth century, the Spanish crown granted merchants in the northern town of Burgos control over the export trade in wool. Using ships from the north coast of Spain, they sent large sacks of washed, but otherwise unfinished wool to England, France, and the southern Netherlands. The Burgos merchants adopted many of the classic Italian financial techniques, such as double-entry bookkeeping, insurance underwriting, and advance purchasing. Domestic wool manufacturers in Segovia, Cordoba, Toledo, Cuenca, and other centers organized regional "putting-out" networks to spin and weave the wool, mostly for domestic markets. The Spanish wool trade flourished until around 1610. At the very time when the Dutch economy took off, Spain's wool industry was hampered by high labor costs, creeping inflation, and the absence of technical innovation. A French traveler to Segovia, noting the city's past wealth from "the great commerce in wool and the beautiful cloth that was made there," observed in 1660, "the city is almost deserted and poor."[30]

The rise and fall of Dutch sugar refining indicates the dependence of the great traffic on foreign trade. As noted above, Dutch sugar refining expanded in the 1630s when the Dutch briefly controlled northeast Brazil, but the loss of that colony in 1645 forced the Dutch sugar traders to fall back on the Caribbean export trade. By 1662 sugar refining was firmly centered in Amsterdam; the city had fifty or more sugar refineries, around half of Europe's total. In addition, other Dutch cities had another dozen sugar refineries. But sugar was too important to be monopolized for long. The Dutch lost their largest export market for processed sugar when France imposed restrictive trade policies during the 1660s. By the end of the century, the Dutch had lost control of the Caribbean sugar trade to England, with its growing commercial activities, naval power, and insatiable domestic hunger for sugar. Even so, the English frequently reexported sugar for processing in Amsterdam. (They did the same with tobacco.)[31]

After 1650 the Dutch traffics expanded in scope. Delft made pottery that simulated the finest Chinese blue porcelain. Gouda specialized in white clay tobacco pipes. And the Zaan, a vast inland estuary stretching from Amsterdam to Haarlem (now filled in), evolved into a major protoindustrial district. Already a shipbuilding center, the site of numerous timber-sawing mills, and the world's largest timber depot, the Zaan in the latter part of the seventeenth century became the Dutch center of high-quality papermaking and one of two centers for refining whale blubber into soap, lighting fuel, and other products. It also featured rope- and sail-makers and refineries for animal fats and oils. This district alone had 128 industrial windmills in 1630 and perhaps 400 by the end of the century.[32]

Surveying the improbable commercial success of the Dutch Republic, one Englishman observed in 1672 that the Dutch had achieved the "exact making of all their native commodities," and that they gave "great encouragement and immunities to the inventors of new manufacturers." Not only had Dutch traders captured commercial control over many key raw materials, including Spanish wool, Turkish mohair yarns, Swedish copper, and South American dyestuffs; the "traffic" system had also erected a superstructure of value-added processing industries. The Dutch conditions of high wages and labor scarcity put a premium on mechanical innovation, the fruits of which were protected by patents. Another economic role taken on by the Dutch state (at the federal, state, and municipal levels) was the close regulation of industry in the form of setting standards for quality and for the packaging of goods.[33]

Given the commanding heights attained by the Dutch economy during the seventeenth century, many have asked why did it not become the first industrial nation, when that age arrived in the eighteenth century. Three lines of inquiry seem most pertinent: raw materials and energy, international trade, and the traffic industries.[34] Economic historians under the sway of British industrialization's coal paradigm often remark that the Dutch Republic lacked coal. While it is true that the Dutch Republic needed to import many raw materials, it did have numerous sources of industrial energy other than coal, including peat, water power, and wind power. Peat—a biological precursor to coal—was extensively used for heating homes and in the manufacturing of bricks, glass, and beer; while imported coal was increasingly used in industrial activities such as sugar refining, soap boiling, distilling, and copper working. Water mills were powered by inland streams and by tides at the shore. The most famous Dutch mills were of course the windmills. It is estimated that by 1630 there were 222 industrial windmills in the country; they were used to drive sawmills, oil mills, and paper mills. Counting those used for draining land and milling grain, the Dutch had a total of perhaps 3,000 windmills.[35]

International trade overall proved remarkably resilient during these turbulent times (see table 2.1). In 1672 the English declared war on the Dutch Republic, and the French actually invaded the northern and inland provinces, which brought substantial chaos. For two years the French bottled up Dutch merchant ships in port. Yet trade bounced back as soon as Dutch ships were able to leave port. Dutch trade with the East Indies actually grew right across the seventeenth century, despite the

TABLE 2.1. DUTCH INTERNATIONAL TRADE AND TRAFFIC INDUSTRIES, 1661–1702

	1661–1665	1666–1670	1671–1675
Leiden camlets (1,000 pieces)	44	54	54
Leiden lakens (1,000 pieces)	19	18	19
Dutch voyages to Baltic	589	774	496
Dutch whaling voyages			148
Guinea gold imports (1,000 guilders)			
Bengali raw silk imports (1,000 pounds)	79	77	121

Source: Jonathan I. Israel, *Dutch Primacy in World Trade, 1585–1740* (Oxford: Clarendon Press, 1989), 257, 263, 301, 307, 329, 337, 354, 366, 387.

Note: Figures are annual averages for each five-year period

recurrent wars with France and England. Dutch shipments of raw silk from Bengal remained surprisingly strong, too. But by the 1690s, several other sectors of Dutch shipping reached a breaking point. Whaling voyages decreased, while shipments of West African gold fell. Closer to home, Dutch trade with the Baltic withered under stiff competition from the English and Scandinavian fleets.[36]

These reverses, fanned by the sharp protectionist policies of France and England, put great pressure on the "traffics" system. Remarkably, in the years after the 1672 French invasion, there was expansion in several export-oriented industries, including silk, sailcloth, high-quality papermaking, linen weaving, and gin distilling. Silk flourished with the arrival of French Protestant Huguenots in the 1680s. But the traffics system as a whole never quite recovered from the invasion. Leiden, for example, experienced declines in its fine-cloth production between 1667 and 1702. The output of camlets, from fine Turkish mohair, fell steadily (see table 2.1). Growth did not resume in the early eighteenth century, even as the English woolen-textile industry leapt forward with cheaper-grade production (see chapter 3). At Leiden the production of lakens, from fine Spanish wool, dwindled from 25,000 pieces around 1700 to a mere 6,700 pieces in 1750.

The wealth-creating imperatives of traders and merchants, boat-builders and shipowners, sugar refiners and textile makers, and many others—a far more diverse cast than the Renaissance court patrons—altered the character of technology during the era of commerce. While choosing, developing, and using technologies with the aim of creating wealth had been an undercurrent before, the Dutch commercial

1676–1680	1681–1685	1686–1690	1691–1695	1696–1700	1701–1702
34		28	24	31	18
				25	24
694	1,027	885	573	591	484
136	212	175	56	135	
551	525	619	503	317	222
105			160	185	

era saw the flourishing of an international (if nonindustrial) capitalism as a central purpose for technology. It was really a *complex* of wealth-creating technologies and techniques: no other age and place combined cargo-carrying fluyts and factorylike herring busses, large port complexes coupled to buzzing inland cities, the array of added-value "traffic" industries, and the elaboration of world-spanning financial institutions, including exchanges for the trading of stocks and commodities, multi-share ownership of ships, and futures markets for herrings, woolens, and for a time tulips.

These technologies not only set the stage for a Dutch commercial hegemony that lasted roughly a century; they also shaped the character of Dutch society and culture. While fine-arts museums today display the wealthy merchants who owned the Amsterdam townhouses and commissioned portraits from the famous painters, such as Rembrandt van Rijn and Johannes Vermeer, this elite society was only the tip of the iceberg. Just as commerce permeated Dutch society and culture, so too did remarkably broad forms of cultural consumption. A famous generation of oil painters flourished when newly wealthy merchants stood ready to have portraits of themselves and their families (and tropical dyes from overseas gave painters inexpensive access to brilliant coloring agents). Scenes from everyday life were much prized. No less a figure than Rembrandt, in *The Mill* (1645–48), painted the signature Dutch windmills. The consumption of oil paintings was, in comparative terms, exceptionally widespread. "The most modest shopkeeper had his collection of pictures, and hung them in every room," writes Paul Zumthor. A now-collectable Ruisdael landscape or Steen genre picture then cost about a quarter of the weekly wage of a Leiden textile worker.[37]

Wherever one looks—at the diverse stockholders of the two great overseas trading companies, the extensive *trekvaarten* network, the numerous owners of the trading ships, and even for a few years the distinctly down-market traders of tulips—Dutch commerce engaged the talents and wealth of an exceptionally wide swath of society. The depth and breadth of the changes that these activities represent lent a distinctly modern character to Dutch society, not only in the details of "modern" financial institutions and economic growth patterns, but in the pervasiveness of the effect that commerce and technology had on the society as a whole. By contrast, in England, as we will see in the following chapter, a distinctively industrial society would emerge.

CHAPTER *3*

1740 – 1851

Geographies of Industry

Historians once treated the "industrial revolution" with gravity and respect. A generation ago it sat at the center of history. The industrial revolution in Britain, according to Eric Hobsbawm's classic account, "initiated the characteristic modern phase of history, self-sustained economic growth by means of perpetual technological revolution and social transformation." It "transformed in the span of scarce two lifetimes the life of Western man, the nature of his society, and his relationship to the other peoples of the world," stated another influential account. For these historians, the industrial revolution, with its new ways of living and working, created an entirely new industrial society. So-called leading sectors were in the vanguard. Unprecedented growth in the cotton, iron, and coal industries during the decades surrounding 1800, culminating in the steam-powered factory system, fueled a self-sustaining "take-off" in the British economy. The cotton industry was the paradigm, experiencing a flurry of output-boosting inventions in spinning and weaving that made the inventors Hargreaves, Arkwright, Crompton, and Cartwright famous (and Arkwright, at least, vastly wealthy). These inventions required factory-scale application, so the argument went, and factories required James Watt's steam engines. Increases in iron and coal output, for steam engines and cotton factories, made the factory synthesis into a juggernaut that changed the world. In this interpretation, Britain blazed a pioneering path to industrial society while the rest of the world followed behind in its wake.[1]

But in the decades since, the industrial revolution no longer seems such a neat and tidy package. Even if the changes in Britain's core industrial districts were dramatic and well documented, other regions within Britain and even other countries contributed to and participated in these developments. On closer inspection the British economy grew slowly and gradually, with no singular take-off. Early factories in many countries were powered by water, animals, or humans—not steam—and even in coal-rich Britain water power outpaced steam power well into the nineteenth century. Through the 1830s the largest British steam-powered cotton-spinning mills, which employed 1,000 or more workers in eight-story factory buildings, depended on home-based handloom weavers to make the machine-spun yarn into useful cloth.

Even in Britain only one industrial worker in ten ever saw the inside of a factory; most labored in smaller shops or in the so-called traditional sectors, such as construction, wool, and leather. These other activities, not the steam-powered cotton factories, contributed most of the country's manufacturing value, and experienced substantial growth themselves. The new historical view is of "a more long-run, varied and complicated picture of the British path to Industrial Revolution."[2]

All the same, something vital occurred in Britain around 1800. The key industrial-era images—and such central concepts as industry, class, and culture—snapped into focus. The term *industry* was earlier a human attribute in the sense of skill, perseverance, or diligence. Adam Smith, in his *Wealth of Nations* (1776), was perhaps the first to use *industry* as a collective word for manufacturing and productive bodies. In the 1830s *industrialism* was coined to name the new social system. (The term *industrial revolution* passed into common English only in the 1880s, although French authors earlier in the century believed that France was undergoing a *révolution industrielle*, while a little-known German theorized about the *industriellen Umwälzung*.) *Class* also was an existing term, referring to school or college classes in traditional subjects. In reference to ranks of people, *lower class* came first, followed in the 1790s by *higher class* and *middle class*; *working class* emerged around 1815, *upper class* soon after. Even *culture*, earlier "the tending of natural growth" as in *horticulture*, came in the early nineteenth century to mean "a general state or habit of the mind" or "the state of intellectual development in a society as a whole," and, toward the end of the century, "a whole way of life, material, intellectual and spiritual."[3] Writing during the 1850s, Karl Marx argued (originally in German) that the culture of the working class was a creation of industry; a generation earlier no one could have expressed this thought in English. This chapter examines the dynamics of industrial society in Britain, long hailed as the first industrial nation, through extended case studies of London, Manchester, and Sheffield. It then evaluates the notion of "industrial society" through comparisons with other industrial countries and regions.

THE FIRST INDUSTRIAL CITY: LONDON

In the older view of the industrial revolution, there was no need to look at London (fig. 3.1). "The capital cities would be present at the forthcoming industrial revolution, but in the role of spectators," wrote Fernand Braudel in his acclaimed *Capitalism and Material Life*. "Not London but Manchester, Birmingham, Leeds, Glasgow and innumerable small proletarian towns launched the new era." The industrial revolution

FIG. 3.1. CITY OF LONDON

The fashionable center of London, with the dome of St. Paul's Cathedral prominent beyond Southwark Bridge, seen from the gritty south bank of the Thames River. Cargoes arrived in the steamboats and larger vessels, then were offloaded onto shore by smaller boats called lighters, and finally moved on land with the aid of horse and human labor.

Samuel Smiles, *Lives of the Engineers* (London, 1862), 2:100. Courtesy of Illinois Institute of Technology Special Collections.

was, in another influential image, "a storm that passed over London and broke elsewhere." London, supposedly, was stuck in a preindustrial age, with its "gentlemanly capitalists" little concerned with building up new wealth and striving only to enter the established landholding aristocracy.[4] But the notion of an industrial London is worth a more careful look. Around 1800, when manufacturing employed one in three of London's workers, the city had more steam engines than any of the factory towns. In 1850 London had more manufacturing workers than the four largest factory towns in England put together. Chemical, furniture, brewing, printing, shoemaking, textile-finishing, precision-manufacturing, and heavy-engineering industries sprang up to the south and east of London's fashionable center, often in compact specialty districts, while just downstream on the Thames River shipbuilding, provisioning, and processing industries surrounded the Port of London.[5]

Not only was it the country's largest site of industry, London's insatiable hungers and unquenchable thirsts helped transform England from a rural-agricultural economy to an urban-industrial one. London's growth still astounds. In 1700 London, with a half-million residents, was the largest city in Europe (surpassing Paris) and ten times more populous than the next largest British town; of all the world's cities only Tokyo, perhaps, was larger. From 1800 to 1850 London *added* more residents (1.4 million) than the *total* 1850 populations of the country's dozen largest textile-factory towns, even after their industry-driven growth. In 1850 London numbered 2.4 million residents. As Daniel Defoe had earlier observed, "All the people, and all the lands in England seem to be at work for, or employed by, or on account of this overgrown city."[6]

Already by 1700 London's outlying market gardens could not meet its dizzying demand for food. Soon enough, eggs, geese, sheep, cattle, grain, malt, herrings, turkeys, apples, and more were flowing into the city. "Georgian London was said to consume each year 2,957,000 bushels of flour, 100,000 oxen, 700,000 sheep and lambs, 238,000 pigs, 115,000 bushels of oysters, 14,000,000 mackerel, 160,000 pounds of butter and 21,000 pounds of cheese."[7] The need for moving this mountain of food into London helped create a far-flung transport system of roads, rivers, and canals. By 1805 north London was linked by the Grand Junction Canal to the farming centers in southern England and to the industrializing towns, such as Manchester, Birmingham, and Leeds.

Large ships sailed right into the center of London, bearing goods from the world's markets, just as in the Dutch port cities (see chapter 2). Around 1700 London's crowded street-side wharves along the Thames River handled three-quarters of Britain's overseas trade. Crammed into the 488 yards between London Bridge and the Tower of London were sixteen special wharves (the so-called legal quays) plus the Customs House and Billingsgate fish market. (Fresh fish came from the Thames until pollution killed them off in the 1820s.) Additional wharves stood upstream from London Bridge, across the river on the south bank, and downstream from the Tower. But the collection of customs duties required all shipments of goods, wherever they were first offloaded, to pass through one of those sixteen legal quays.

In response to the extreme congestion and the pervasive street-side pilfering of cargo, the city's merchants launched a £5 million dock-building campaign (fig. 3.2). Beginning near the Tower, the new dockyard district eventually stretched twenty-five miles downstream. Built on artificial lakes, and connected to the Thames River by short canals, these dock–warehouse complexes began with the 600-ship West India Docks (opened in 1802) on the Isle of Dogs. Next came the London Docks at

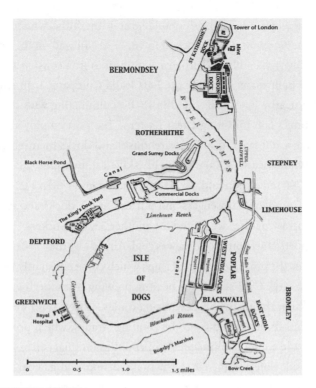

FIG. 3.2. PORT OF LONDON

Before this nineteenth-century dockyard complex was built, all goods entering London passed through the so-called legal quays, located immediately upstream from the Tower (*top of map*). The downriver dockyards became a major site of shipbuilding, provisioning, and processing. Eventually the dock complexes extended far downstream beyond Bow Creek (*bottom of map*).

Charles Knight, *London* (London: Knight, 1842), 3:65. Courtesy of Illinois Institute of Technology Special Collections.

Wapping (1805), the East India Docks at Blackwall (1806), and the Surrey Docks in Rotherhithe (1809). Several years later came the St. Katherine's Docks (1828) next to the centrally located Tower. Downstream the Royal Docks opened in three segments (1855–1921), where today flights operate from the London City Airport.

The sprawling dock complex and flow of cargoes created a vast commercial and shipping infrastructure staffed by a multitude of "agents, factors, brokers, insurers, bankers, negotiators, discounters, subscribers, contractors, remitters, ticket-mongers, stock-jobbers, and a great variety of other dealers in money." Through the mid-1860s London was also Britain's leading shipbuilding district. Sails, ropes, barrels, and pulleys were needed to outfit Britain's overseas fleet. On the south bank were the

Royal Navy Yard at Deptford and the shipbuilders at Rotherhithe, who made the first iron steamship, the *Aaron Manby*, launched in 1822. At Millwall on the Isle of Dogs the prominent millwright William Fairbairn, a pioneer in iron construction, opened shop in 1833 and built many ships for the East India Company. Iron shipbuilding also flourished at nearby Blackwall from the 1840s, culminating with the launch in 1857 of I. K. Brunel's 600-foot-long *Great Eastern*. The port employed thousands. In 1851 there were 4,100 workers at the warehouses and docks themselves; 5,000 in shipbuilding, of whom nearly 3,300 were highly skilled shipwrights; 18,000 in ocean navigation and 5,700 more in inland navigation; and no fewer than 1,214 employees at the East India Company. Messengers and porters, at the docks and elsewhere in the commercial infrastructure, accounted for another 33,000 workers.[8]

Among the port's chief commodities was coal. Already by the 1760s Defoe noted "at London [we] see the prodigious fleet of ships which come constantly in with coals for this increasing city." Coal served for baking, brewing, iron forging, pottery firing, and glassmaking, among the heat-requiring trades, as well as for heating homes and offices. Noxious fumes issued from the "boiling" trades, which produced soap, candles, and glue and were the ultimate fate of countless animal carcasses. Travelers from the countryside entering the city perpetually denounced its coal-fouled environment. By 1830 the volume of coal shipped into London had reached 2 million tons, a 250 percent increase since Defoe's time; in the 1830s coal also began arriving by rail. The Coal Exchange opened in 1849, near the Billingsgate fish market. Two years later the coal sector in London alone employed more than 8,000 persons (1,702 in coal and gas manufacturing, 2,598 as coal merchants and dealers, and 4,023 as coal heavers, including three rugged women).[9]

Beer brewing affords a revealing window into industrial London while illustrating the links between industry and sanitation, consumption, and agriculture. In an age of filthy water, beer was virtually the only safe beverage, and Londoners drank it in huge quantities. In 1767 one dirt-poor clerk, said to be "even worse" off than a day laborer, nonetheless drank beer at each meal: "*Breakfast*—bread and cheese and small [weak] beer from the chandler's shop, *Dinner*—Chuck beef or scrag of mutton or sheep's trotters or pig's ear soused, cabbage or potatoes or parsnips, bread, and small beer with half a pint of porter; *Supper*—bread and cheese with radishes or cucumbers or onions, small beer and half a pint of porter."[10] For decades, British farmers harvested more bushels of barley (mostly for brewing and distilling) than wheat for bread. Taxes on beer making generated one-quarter of Britain's tax revenue before 1750 and one-seventh as late as 1805. During these same decades, London's porter-brewing industry dramatically embodied the hallmarks of industrialization: vast increases in scale and capitalization, labor- and cost-saving technology, and market concentration, as well as a profusion of specialized by-product and supply industries.[11] Reducing costs and increasing output—rather than enhancing quality, as in Dutch commerce—was the focus of technology in the industrial era.

The beer known as porter merits recognition as a prototypical industrial-age product alongside cotton, iron, and coal. Ales, as traditionally brewed, required light malts made from the finest barley, and ale brewers halted yeast fermentation when a little malt sugar remained. The best ales were light and clear, slightly sweet, and drunk soon after brewing. By contrast, porter, first brewed in the 1720s, was dark, thick, and bitter. (Its name derives from the dockworkers who were its earliest consumers.) Porter malts were dark roasted from cheap grades of barley, and no malt sugars remained when the brewers were done. Most important, porter was susceptible to industrial-scale production.

Brewers found that aging porter for a year or more mellowed its rough taste. At first they simply stored porter in 36-gallon wooden barrels, but the practice eventually tied up a quarter or more of a firm's entire capital resources. In the 1730s, the larger brewers began building on-site vats capable of storing up to 1,500 barrels. Larger vats followed soon (fig. 3.3). In 1790 Richard Meux built a 10,000-barrel vat that was 60 feet in diameter and 23 feet high. "They claimed that 200 people had dined within it and a further 200, also inside, drank to its success." Five years later Meux erected a 20,000-barrel behemoth. The competition between brewers to build ever-larger vats waned after 1814, however, when a 7,600-barrel vat at the Horse Shoe Brewery burst open and flooded the neighborhood, killing eight persons "by drowning, injury, poisoning by the porter fumes or drunkenness."[12]

The immense size of these storage vats points to the porter brewers' industrial scale. The annual production of the city's leading porter brewer was 50,000 barrels in 1748, 200,000 barrels in 1796, and 300,000 barrels in 1815; the brewing companies were Calvert, Whitbread, and Barclay, respectively. By 1815 a London porter brewer shipped more in a year than the largest London ale brewer did in a decade. Porter brewers had industrialized, while ale brewers had not. The twelve largest London brewers made virtually nothing but porter and the closely related stout. Between 1748 and 1815 they built factories that more than tripled their production of these "strong" beers (from 0.4 to 1.4 million barrels) expanding their share of the London market from 42 percent to 78 percent. During these decades of expanding output, the city's total number of brewers actually *fell* from 158 to 98.

FIG. 3.3. LONDON BREWERY VAT

The scale of industrial enterprise made visible in this vat for aging porter beer. It held more than 100,000 gallons, saving the use of 3,000 smaller wooden barrels.

Charles Knight, *London* (London: Knight, 1843), 4:13. Courtesy of Illinois Institute of Technology Special Collections.

Beer consumption figured as a significant item in urban household budgets, especially for the poor. If beer consumption is taken as a proxy for workers' standards of living, the early industrial economy did not lift real wage levels until after the tough times of the 1840s. National figures for per capita beer consumption (in England and Wales) stood at 33.9 gallons (1800–1804), increased temporarily around 1830, fell to 31.6 gallons (1860–64), then rose to a much-discussed 40.5 gallons (1875–79). In their study of the British brewing industry Gourvish and Wilson calculate that during that last span of years the average British man each week put away 16 pints of beer.[13]

The porter brewers pioneered industrial scale and led the country in the capitalization of their enterprises. In 1790 the net capital valuation of Whitbread's White Hart Brewery stood at £271,000 while Truman's Black Eagle Brewery and Barclay-Perkins' Anchor Brewery were not far behind, at £225,000 and £135,000, respectively. These are extraordinary figures: at this time £20,000 was the capitalization

FIG. 3.4. PORTER BREWERY

Inside of large brewery showing five men "mashing" water and malt with "oars" (one is held aloft at right). Boiling hot water, from the "coppers" at back, flowed to the "mash tuns" for mixing with malt. The resulting "wort" drained down to the "under-backs" beneath the floor level. Not visible is a horse-driven pump that lifted the wort, collected in the "jack-back" at far left front, back up to the coppers for a boiling with hops. The master brewer is center.

Temple Henry Croker, *The Complete Dictionary of Arts and Sciences* (London: Wilson & Fell, 1764), vol. 1, plate 24.

of Birmingham's very largest manufacturers and, two decades later, of Manchester's largest spinning factory. And the biggest brewers got bigger. In 1828 Whitbread's nine partners had invested capital of £440,000 while Barclay-Perkins' total capital was £756,000.[14]

These outstanding capitalization figures were a sign that porter brewers were early and enthusiastic converts to the industrial-era logic of laborsaving mechanization. Early on, porter brewers purchased horse-driven pumps to move beer throughout the brewery. The big London brewers, despite the heavy cost for maintaining workhorses in the city, mechanized malt grinding also using horsepower. Barclay-Perkins employed just 100 workers to make 138,000 barrels of beer in 1796. In addition to its dozen well-paid brewers and technical managers, the company employed 4 stokers, 7 coopers, 9 yeastmen, 4 miller's men, 18 draymen, 10 stage men, 12 horse keepers, 30 spare men, and 7 drawers-off.[15] (See fig. 3.4.) In 1851 the census found just 2,616 London brewery workers.

While pumps and grinding mills replaced human labor, rotating steam engines retired workhorses from their interminable turnstile rounds. As James Watt's business partner observed in 1781, "the people in London, Manchester, and Birmingham are Steam Mill Mad." In 1784 brewers installed the first of London's new rotating Watt engines, a 4-horsepower model at Goodwyn's and a 10-horsepower one at Whitbread's. These installations came just two years after ironmaster John Wilkinson installed Watt's very first rotating engine and a full two years before mechanization at the Albion Flour Mill, typically remembered as the site where Watt's rotating engine proved itself. At Whitbread's the steam engine was not only reliable and fuel efficient but money saving, retiring twenty-four horses. With the savings from just one year's care and feeding of its horses, at £40 per head, Whitbread's paid for the engine's entire installation cost (£1,000). Brewers indirectly fixed a key measure of the industrial era, since Watt had the "strong drayhorses of London breweries" in mind when he defined "horsepower" at 33,000 foot-pounds per minute. In 1790, Watt recorded, his firm made engines principally for brewers, distillers, cotton spinners, and iron men. Five years later only the cotton and coal industries had more steam engines than the brewing industry (which used twelve engines totaling 95 horsepower).[16]

The porter brewers' unprecedented scale, market domination, and immense capitalization, in addition to their quest for money-saving technical innovation, are signs of their *industrial* character. So is the rise of specialized by-product processing and supply industries, which created a radiating web around the centers of iron, coal, cotton, building—and beer. These ancillary industries have not received the attention they deserve, for they are crucial to understanding how and why industrial changes became self-sustaining and cumulative. Brewing again provides an apt example. Industrial-scale porter brewing fostered several ancillary industries. Spent grains from brewing fattened-up cattle awaiting slaughter, while the rich dregs from beer barrels became fertilizer. Brewers' yeast spawned an entire specialized economy: London's gin and whiskey distillers usually bought their yeast directly from the brewers, while bakers dealt with specialized brewers' yeast dealers. One of these enterprising middlemen devised a patented process for drying and preserving yeast and, after contracting with six large brewers, in 1796 built a large factory to process yeast freshly skimmed from brewing vessels. Specialized trades also sprang up for making beer barrels, copper brewing vessels, and clarifying agents. Beer engines, to pump beer up from a public-house cellar for dispensing, were manufactured by at least three firms located near Blackfriars Road.[17]

Having industrialized production, porter brewers sought influence over beer consumption. (Other competitive means were not readily available, since for decades prices had been fixed by law or custom and the export trade was dominated by rival British brewers in Burton-on-Trent.) Porter brewers' immediate object was the principal site of mass consumption: London's 5,000 public houses, or pubs. Traditionally, brewers had sold barrels of beer to the publican, making monthly deliveries and taking payment on the spot, but the larger porter brewers began retaining ownership right until the beer splashed into the consumer's quart pot. The publican then paid for only the beer he actually sold, while the brewer insisted that the publican take beer deliveries from no one else. Thus began the practice of "tying" a pub's trade. If left unpaid, a publican's monthly balance might become an interest-bearing loan. If the loan balance grew too large, the brewer would ask for the publican's building lease as collateral and if necessary take possession of a defaulted publican's establishment. Some brewers outright bought a public-house lease and installed a dependent tenant. By the early nineteenth century, perhaps half of all London pubs were tied to brewers through exclusive deliveries, financing, or leasing. Tying typified the porter industry. In 1810 Barclay's had tied relationships with 58 percent of the 477 pubs it served, while Truman's had tied 78 percent of 481 pubs, and Whitbread's 82 percent of 308 pubs.[18]

Porter brewers were big supporters of the Beerhouse Act of 1830, which scrapped the centuries-old regulations on the selling of beer. The act, in an effort to save the poor from poisoning themselves with cheap distilled gin, permitted virtually any property-tax payer, upon paying a modest fee, to set up a "beer shop." Across Britain 40,000 willing taxpayers signed up and per capita beer consumption shot up 25 percent. Not surprisingly the big porter brewers of London (the twelve largest now brewed 85 percent of the city's "strong" beer) cheered the new law. Among them, Charles Barclay best captured the industrial logic of porter brewing and mass consumption. "Who are to supply these beer shops? The persons who can sell the cheapest and the best, and we say we can sell cheaper and better than others," he told his fellow Members of Parliament. "We are power-loom brewers, if I may so speak."[19]

We can complete our portrait of industrial London by stepping back to survey the city's multitude of trades and its building and engineering industries. From the first reliable figures, in the 1851 census, London's 333,000 *manufacturing workers* outnumbered Manchester's 303,000 *total residents*. Manufacturing accounted for a near-identical percentage of employment for women and men (33 and 34 percent, respectively), although London's women and men did not do the same jobs. Of the top ten sectors of manufacturing, only in boot- and shoemaking were there roughly equal numbers of men and women. Women worked mostly as boot- and shoemakers, milliners, and seamstresses (making fancy hats and dresses). By contrast, men dominated the trades of tailoring, furniture making, baking, woodworking, printing and bookbinding, iron and steel, and nonferrous metals. Male-dominated shop culture and prevailing conceptions of women's "proper" place in the workforce go some way in accounting for where women and men worked. Tailors, however, used stronger tactics to exclude women (according to labor reformer Francis Place): "where men have opposed the employment of women and children by not permitting their own family to work, or where work is such that women and children cannot perform it, their own wages are kept up to a point equal to the maintenance of a family. Tailors of London have not only *kept* up, but *forced* up their wages in this way, though theirs is an occupation better adapted to women than weaving."[20]

The building trades, although not enumerated as manufacturing, certainly became industrialized during these years. The incessant increase in London's population necessitated continual construction. Bricks were in such short supply in the 1760s "that the makers are tempted to mix the slop of the streets, ashes, scavenger's dirt and everything that will make the brick earth or clay go as far as possible." Bricks fresh out of the kiln came down the street piping hot. "The floor of a cart loaded with bricks took fire in Golden Lane Old Street and was consumed before the bricks could be unloaded," noted one account. By the 1820s such industrial-scale "merchant builders" as the Cubitt brothers emerged. In that decade Thomas and Lewis Cubitt employed more than 1,000 workers in building vast blocks in the fashionable Bloomsbury and Belgravia districts, while their brother William employed an additional 700. The Cubitts not only directly employed all the various building trades but also made their own bricks and operated a steam-powered sawmill. Another of the city's merchant builders, John Johnson, who described himself variously as a paver, stonemason, bricklayer, and carpenter, owned a granite quarry in Devon, a wharf along the Thames, and brickyards, gravel pits, and workshops around London. In the 1851 census London's building sector employed a total of 70,000 bricklayers, carpenters, masons, plasterers, plumbers, painters, and glaziers (including 129 female house decorators).[21]

A stock image of London has its streets filled with small shops, yet in *Days at the Factories* (1843) George Dodd described twenty-two establishments that "were conducted on a scale sufficiently large to involve something like 'factory' arrangements." At midcentury, six of the country's eleven largest printing works were in London, as were twenty-two of the fifty-two largest builders. London's twelve biggest brewers, as noted earlier, towered over the rest. Official figures miscounted many larger firms and simply ignored the city's railways and gasworks, further obscuring a complete picture of industrial London. Official statistics also overlooked the 20,000 construction workers laboring on the London-Birmingham railroad between 1833 and 1838 as well as the untold thousands who in the next decade connected London by rail to the four points of the compass and erected five major railroad stations. The 1851 census found just half of the 334 engineering firms listed in the following year's *London Post Office Directory*. And while the census recorded just six London engineering firms employing 100 or more men, the *Northern Star* newspaper itemized nine such firms' locking out their workers during a strike in 1852. These were Maudslay & Field (800 workers), John Penn (700), Miller & Ravenhill (600), J. & A. Blyth (280), and five additional firms with between 100 and 200 workers.[22]

Despite the faulty census, we can establish London's rightful place as England's first center of engineering by reviewing two impressive mechanical genealogies. Boulton & Watt, as we noted, installed two of their earliest rotating engines in London at prominent breweries while a third went to Albion Mills in the city's Southwark industrial district. Scottish engineer John Rennie came to London in their employ to supervise the construction of the flour mill (1784–88). Albion Mills pioneered the use of steam for grinding flour as well as the use of iron for gearing and shafting. When the mill burned down in 1791—its abandoned charred ruins was poet William Blake's likely model for "dark satanic mills"—Rennie had already made his mark. He then set up his own engineering works nearby. Rennie built the Waterloo and Southwark bridges, and after his death his two sons completed the New London Bridge (1831). His elder son George specialized in building marine steam engines for the Admiralty and in 1840 designed the first propeller-driven naval vessel. At the time, Rennie's engineering works employed 400. In 1856 George Rennie, then heading the mechanical sciences branch of the prestigious British Association for the Advancement of Science, helped a young London inventor, Henry Bessemer, launch his revolutionary steelmaking process (see chapter 6).

Impressive as the Watt-Rennie-Bessemer lineage is, perhaps the most significant London engineering genealogy began with Joseph Bramah. Originally a cabinetmaker from Yorkshire, Bramah patented an "engine" for pumping beer up from a publican's cellar, an improved water closet, and a hydraulic press, and is best known for his precision locks. Henry Maudslay trained at the Woolwich Arsenal, a leading site for mechanical engineering, then worked with Bramah in London for eight years. Maudslay helped realize the grandiose plans of Samuel Bentham, named inspector-general of naval works in 1795. Bentham was determined to alter the prevailing shipbuilding practices, in which, as a result of shipwrights' traditional craft privileges of taking payment-in-kind, "only a sixth of the timber entering Deptford Yard left it afloat." Maudslay's forty-three special machines completely mechanized the manufacture of ships' pulleys, essential in the day of sail, and were installed by Marc Brunel at the Portsmouth navy yard in 1810.[23]

By 1825 Maudslay and Bramah were among the London engineers hailed for their use of specialized machine tools to replace skilled hand-craftsmanship. These "eminent engineers," observed a parliamentary report, "affirm that men and boys . . . may be readily instructed in the making of Machines."[24] Maudslay invented screw cutters and planers and perfected the industrial lathe. He also taught the next generation of leading machine builders. Among them, Joseph Clement, Richard Roberts, Joseph Whitworth, and James Nasmyth created many of the basic machine tools used in shops today, including the horizontal shaper, steam hammer, two-directional planer, and a standard screw thread. (French and American mechanics were also pioneers in precision manufacturing, mechanization, and standardization, as discussed later.) Yet by the 1830s, while Clement struggled to build an early mechanical computer for Charles Babbage, a core group of Maudslay-trained engineers (Whitworth, Nasmyth, Roberts, and Galloway) had departed London. Each of them set up machine-building shops in Manchester.

SHOCK CITY: MANCHESTER

If immense size exemplified London, swift growth provides the conceptual key to Manchester. For many years Manchester served the region as a market town. In 1770 its population numbered 30,000. Then Manchester started doubling each generation, by 1851 reaching the figure of 300,000. In 1786 "only one chimney, that of Arkwright's spinning mill, was seen to rise above the town. Fifteen years later Manchester had about fifty spinning mills, most of them worked by steam." By 1841 Manchester and the surrounding Lancashire region had twenty-five huge cotton firms *each* employing more than 1,000 workers, 270 firms employing between 200 and 900 workers, and 680 smaller firms (fig. 3.5). "There are mighty energies slumbering in those masses," wrote one observer in the 1840s. "Had our ancestors witnessed the assemblage of such a multitude as is poured forth every evening from the mills of Union Street, magistrates would have assembled, special constables would have been sworn, the riot act read, the military called out, and most probably some fatal collision would have taken place."[25]

Legend has it that weavers fleeing the Spanish-occupied southern Netherlands

FIG. 3.5. MANCHESTER'S UNION STREET, 1829

"There are mighty energies slumbering in those massess," wrote one middle-class observer, somewhat nervously, of the thousands of workers who poured out of the massive eight-story factory buildings each evening.

Period engraving from W. Cooke Taylor, *Notes of a Tour in the Manufacturing Districts of Lancashire* (London: Frank Cass, 1842).

(see chapter 2) founded the region's cotton trade. Daniel Defoe as early as 1727 wrote of Manchester, "the grand manufacture which has so much raised this town is that of cotton in all its varieties." At this time families working at home did the necessary cleaning, carding, spinning, and weaving. "Soon after I was able to walk I was employed in the cotton manufacture," wrote the eldest son of Samuel Crompton, the spinning-machine inventor. The young Crompton did duty by stepping into a tub of soapy water and tramping on lumps of cotton. "My mother and my grandmother carded the cotton wool by hand, taking one of the dollops at a time, on the simple hand cards. When carded they were put aside in separate parcels for spinning."[26] Spinning by women and weaving (largely by men) were also done at home; many weavers built their own wooden looms. Manchester merchants imported raw cotton from the American South and distributed it to the region's families, then sent finished cloth to wholesale markets in London, Bristol, and Liverpool. By the 1790s Britain was exporting the majority of its cotton textile output.

Its proximity to the port city of Liverpool constituted only one of the region's natural advantages. The ocean bathed the Lancashire region with damp air that made cotton fibers sticky, making them easier to spin into a continuous thread. (For the same reason, mills later vented steam into the spinning room to humidify the air.) Coal was readily at hand. Indeed, the Duke of Bridgewater built a canal, opened in 1761, to connect his coal mines the roughly six miles to Manchester. Bridgewater cut underground canals directly to the coalface, at once draining the mines and halving the cost of coal in Manchester. The earliest steam engines burned so much coal that only coal mines could use them economically. Even the early Watt rotating engines were employed mainly where the cost of animal power was excessive (as in London) or where coal was cheap (as in Manchester). The Bridgewater Canal's financial success led to additional connections to Liverpool, Leeds, and other industrial cities, eventually sparking a "canal mania" that resulted in a canal network across Britain, including London, well prior to the railroads.

One of the central puzzles concerning the industrial "revolution" is why it took so long to unfold. One clue is the need for ancillary industries that link innovative sectors together and ramify change; another clue is to be found in how and why the paradigmatic steam-driven cotton factory took form. Even though the basic mechanical inventions in carding, spinning, and weaving of cotton, along with Watt's rotating engine, had all been made before 1790, it was not until around 1830 that Manchester had a unified cotton factory system and a recognizable modern factory sector. To comprehend this forty-year "delay," we must consider the gender shift that attended incremental innovations in spinning and weaving as well as structural shifts in Manchester's business community.

Spinning begins with a long cord of loose cotton fibers ("roving") and results in thread suitable for weaving or knitting into cloth. For centuries, spinning had been the province of women. A woman might spin thread her entire life, supporting herself and remaining unmarried, hence the term *spinster*. Many women welcomed the spinning jenny, invented by James Hargreaves in 1764. The jenny was a small-scale hand-operated device that separated spinning into two distinct mechanical motions: an outward motion pulled and twisted the cotton roving, while an inward motion wound the twisted thread onto spindles. It was cheap to buy or build one's own jenny, and a woman might spin twenty or thirty times more thread on a home jenny than she could with a traditional single-spindle spinning wheel.[27]

Spinning took a factory-dominated industrial form owing to the work of Richard Arkwright, a barber and wigmaker from a Lancashire village. Arkwright's "water frame" resembled the Lewis-Paul spinning machines of the 1740s in that both operated continuously with rotating rollers to impart twist to the roving and a winding mechanism to take up the completed thread. But while the Lewis-Paul machines used a single set of rollers, Arkwright used two sets of rollers, covered them with leather, and spaced them about the same distance apart as the length of individual cotton fibers. The second set spun slightly faster than the first, at once twisting and pulling the cotton fibers into thread.

Arkwright patented his "water frame" in 1769 and built several mills at Nottingham, Cromford, and elsewhere in the region using horsepower and water-power before opening Manchester's first steam mill in 1783. At his large five-story Manchester mill, which soon employed 600 workers, an atmospheric steam engine pumped water over a waterwheel to power the machinery.[28] Arkwright's extraordinary profits led many writers ever since to assume that a technical imperative brought about the factory system: that machinery and the factory were one. Yet this view withers on close inspection. Indeed, the term "water frame," which suggests that it required waterpower and hence factory scale, is entirely misleading. Like jennies, water frames were initially built in smaller sizes adaptable to home spinning. Early Arkwright machines were small, hand-cranked devices with just four spindles. The death blow to home spinning came when Arkwright restricted licenses for his water frame patent to mills with 1,000 or more spindles. (Having a smaller number of large-scale water frame mills made it easier for Arkwright and his business partners to keep an eye on patent infringements.) Practically speaking, these large mills needed water or steam power. Water frames could no longer be set up in a kitchen. Arkwright's mills spun only the coarsest grades of thread, but he and his partners employed poor women and paid them dismal salaries that left handsome profits for the owners. It was said that none of the jobs took longer than three weeks to learn. Arkwright's mills—with their low wages and skills, their high-volume production of lower-grade goods, and their extensive mechanization—embodied core features of the industrial era.

The cotton-spinning industry, despite Arkwright's great wealth, came from an inelegant and unpatented hybrid technology. In 1779 Samuel Crompton's first spinning "mule" grafted the twin rotating rollers of Arkwright's machine onto the back-and-forth movements of Hargreaves' jenny. Thereafter, a host of anonymous mechanics tinkered with the mule to make it larger (with 300 or more spindles). Whereas small-scale jennies supported the home industry of female spinners, as did the first mules, the larger mules, still hand-powered, required the greater physical strength of male spinners. The shift to heavier mules, dominated by skilled male spinners, doomed the independent female spinner. By 1811 a nationwide count revealed that spindles on Crompton's mules totaled 4.6 million, as compared with Arkwright's water frame total of 310,000 spindles, and Hargreaves' jennies at just 156,000 spindles. "Within the space of one generation," wrote Ivy Pinchbeck, "what had been women's hereditary occupation was radically changed, and the only class of women spinners left were the unskilled workers in the new factories to house Arkwright's frames."[29]

This wave of spinning inventions did not create a factory system so much as split Manchester into two rival cotton industries, the "factory" (spinning) firms and the "warehouse" (weaving) firms. The town's sixty factory-spinning firms were led by two giants, McConnel & Kennedy and Adam & George Murray. Each was valued at around £20,000 in 1812, and each employed more than 1,000 factory workers. Few spinning firms were this large. Through 1815 it was surprisingly common for as many as ten separate spinning firms to share one factory building; at this time only one-quarter of the town's factory buildings were occupied by a single factory firm. Compared with the factory firms, the "warehouse" firms were even more numerous and smaller (the largest warehouse firm in 1812 was half the largest spinning company in valuation). Weaving firms bought spun yarn from the spinning firms, and "put out" the yarn to a vast network of handloom weavers, male and female, across the region. Weaving firms then collected the woven cloth and marketed it, often shipping to overseas markets via Liverpool.[30]

Through the 1820s substantial conflict and hostility existed in Manchester between these two rival branches of the cotton industry. Factory and warehouse firms clashed in strident debates on exports, tariffs, child labor, and factory inspections. The perennial yarn-export question provoked sharp conflict: whereas the spinning firms hoped to export their expanding output, the warehouse firms naturally feared that exports would hike domestic prices for yarn and in turn favor foreign weavers. Warehouse firms committed to home-based weaving blasted factory-based spinning. In 1829 one warehouseman slammed spinning factories as "seminaries of vice" and, hinting at the unsettling gender shifts, contrasted spinning with the more beneficial occupations of weaving, bleaching, and dyeing "which require manly exertions and are productive of health and independence."[31]

Something like an integrated factory system took shape in Manchester only around 1825. By then Manchester had thirty-eight integrated spinning-and-weaving firms. These were mostly "new" firms, established since 1815, or warehouse-weaving firms that had bought or built spinning firms (thirteen of each). Also by 1825, in marked contrast from just a decade earlier, most factory firms (70 percent) wholly occupied a factory building. The emergence, around 1830, of powered "self-acting" mules and smooth-working power looms thus fit into and reinforced an existing factory-based industrial structure. A gender division became locked in place, too. "Mule spinning" was firmly men's work, as was all supervisory work, while spinning on the smaller "throstles" as well as factory weaving was generally women's work. (Powered self-acting mules, no longer needing a man's strength, might have been tended by women but the strong male-dominated spinners union kept women out of this skilled and better-paid trade.)

In the 1841 governmental report Factory Returns from Lancashire, integrated firms had more employees and more installed horsepower than the set of nonintegrated fine spinning, coarse spinning, and power weaving firms *together.* Yet during the years from 1815 to 1841, the very largest firms (above 500 workers) fell from 44 percent to 32 percent of total factory employment in the region, while medium-sized firms (150–500 workers) more than filled the gap, growing to 56 percent of total factory employment. The largest firms of the earlier period were specialized behemoths—McConnel & Kennedy sent most of their huge yarn output to Glasgow weavers—and were unable to move quickly in rapidly changing markets. The *Manchester Gazette* in 1829 foretold Manchester's prominent role in overseas imperialism (discussed in chapter 4): "You can sell neither cloth or twist [yarn] at any price that will cover your expenses. . . . Your eyes are wandering over the map of the world for new markets."[32]

Ironically, just as Manchester gained notoriety as "Cottonopolis" the city's industrial base was changing. For decades, in fact, Manchester was not merely the region's largest textile-factory town but also its commercial center. The Manchester Exchange, a meeting place for merchants and manufacturers, opened in the center of town in 1809. Manchester's cotton-textile sector grew rapidly during the early decades of the nineteenth century; in 1841, textile workers outnumbered workers in all Manchester's other manufacturing trades together. The size of the city's cotton industry peaked in 1851 at over 56,000 workers. Yet that year, for the first time, more Manchester workers were employed in finishing cloth, making machines, building houses, and miscellaneous manufacturing (together) than in making cotton cloth. From this time on, growth would depend not on the city's but the region's cotton mills.[33] In Manchester, the making of dyeing and bleaching agents grew up as specialized auxiliaries to the textile industry.

Yet, far and away the most important auxiliaries were Manchester's machine builders. While the first generation of them had built textile machines and managed textile factories, the midcentury machine builders—the generation of London transplants—focused on designing, building, and selling machine tools. Richard Roberts (1789–1864) came from London to Manchester in 1816 and specialized in large industrial planers, along with gear-cutting, screw-cutting, and slotting machines. Roberts' first notable textile invention was a machine for making weavers' reeds, manufactured by the productive Roberts, Sharp & Company partnership. Roberts made his most famous invention, an automatic, self-acting version of Crompton's mule spinner, in the midst of a cotton spinners' strike. During the strike a delegation from the region's cotton manufacturers came to Roberts and proposed, as the industrial writer Samuel Smiles related the story, that he "make the spinning-mules run out and in at the proper speed by means of self-acting machinery, and thus render them in some measure independent of the more refractory class of their workmen."[34] After a four-month effort, he hit upon the self-acting mule, patented in 1825 and improved in 1832 with a radial arm for "winding on" spun thread. Two years later Roberts, Sharp & Company began building locomotives in large numbers; and in 1848, once again in the midst of labor strife, Roberts devised a novel rivet-punching machine that made possible the Britannia Bridge over the Menai Straits.

Joseph Whitworth (1803–87) is the best known of the Maudslay protégés trained in London. In 1833 he came to Manchester to manufacture a series of accuracy-enhancing inventions in machine designs and standard screw threads. His famous micrometer could measure a millionth-part of an inch. Whitworth's machine tools, steel armor, and armaments made him rich and famous. Oversize forgings were the specialty of James Nasmyth (1808–90), whose huge steam-powered hammer has become an icon of the industrial era. His Bridgewater Foundry, near George Stephenson's Liverpool & Manchester Railway and the Bridgewater Canal, employed as many as 1,500 workers in building steam engines, machines, and locomotives. Nasmyth tirelessly promoted the superiority of machine tools, especially self-acting ones, over skilled workmen. These machine tools, he wrote, "never got drunk; their hands never shook from excess; they were never absent from work; they did not strike for wages; they were unfailing in their accuracy and regularity."[35]

Roberts' self-acting machines, invented at the bidding of strike-bound manufacturers, as well as Nasmyth's continual strife with his workmen (he retired, as he told a friend, to be free from "this continually threatening trade union volcano") underscores that more was at issue than the efficient production of goods. Manchester witnessed stark social conflict. On 18 August 1819, the nerve-strung local militia in Manchester opened fire on a mass meeting of Chartists, working-class political reformers, killing eleven and injuring hundreds. In the charged aftermath of the massacre, a leading London banker pointed to "the desire and policy of men engaged in trade . . . to screw down the price of labour as low as possible." The social conflict caused by industrialism spilled out onto the streets. "The smashing of machinery, the destruction of mills and other property, and the assaulting of 'blacklegs,' occurred with alarming frequency." The assaults on factories during the spinners' strike of 1812, the organized machine wrecking in 1826, and the "plug riots" of 1842—in which discontented workers disabled factories' steam boilers—amounted to what one author called "a campaign in a social war."[36]

Clearly, something unprecedented was happening. "Manchester is the chimney of the world," observed Major-General Sir Charles Napier, temporarily home from India in 1839. "What a place! the entrance to hell realized." Charles Dickens agreed. "What I have seen has disgusted me and astonished me beyond all measure," he wrote to a friend, after his first visit to Manchester in 1838. Dickens' classic industrial novel, *Hard Times*, describes the physical and moral degradation brought by industrialism, a theme also of Elizabeth Gaskell's *Mary Barton*, set in Manchester. So widely known were the stock images—"the forest of chimneys pouring forth volumes of steam and smoke, forming an inky canopy which seemed to embrace and involve the entire place"—that one American commentator published his "firsthand" observations of Manchester without bothering to visit England at all.[37]

Not all visitors were dismayed or disgusted. "A precious substance . . . , no dream but a reality, lies hidden in that noisome wrappage," observed Thomas Carlyle, typically remembered as an arch-conservative, after a visit in 1838. "Hast thou heard, with sound ears, the awakening of a Manchester, on Monday morning, at half-past five by the clock; the rushing-off of its thousand mills, like the boom of an Atlantic tide, ten-thousand times ten-thousand spools and spindles all set humming there,— it is perhaps, if thou knew it well, sublime as a Niagara, or more so." Benjamin Disraeli in *Coningsby* (1844) also drew attention to this "new world, pregnant with new ideas and suggestive of new trains of thought and feeling." He stated, "Rightly understood, Manchester is as great a human exploit as Athens."[38]

All this commotion attracted young Friedrich Engels to Manchester. Engels arrived in December 1842 and took detailed notes during his twenty-one-month stay. Engels unlike most literary visitors knew industrial textiles from personal experience. He had already apprenticed in his father's textile mill in Germany and sharpened his business skills at a textile-exporting firm. Just before leaving Germany, he met the editor of the *Rheinische Zeitung*, one Karl Marx, and promised him firsthand essays on English industry. For Engels, Manchester was ground zero for the industrial revolution (he wrote specifically of "industriellen Umwälzung"). His duties in Manchester at the Ermen and Engels mill, partly owned by his father, left plenty of time for observing the city's working-class districts. Engels deliberately disdained "the dinner parties, the champagne, and the port-wine of the middle-classes." Workers' doors were opened for him by a young Irish factory girl, Mary Burns, with whom he would live for the next twenty years.

Returning to Germany, Engels put the final touches to *The Condition of the Working Class in England* (1845).[39] It aimed to be an archetype. Engels saw Manchester as the "masterpiece" of the industrial revolution and the "mainspring of all the workers' movements." At Manchester, he wrote, "the essence of modern industry" was plainly in view: water and steam power replacing hand power, power looms and self-acting mules replacing the hand loom and spinning wheel, and the division of labor "pushed to its furthest limits." Accordingly, the "inevitable consequences of industrialisation," especially as they affected the working classes, were "most strikingly evident." Not only did Manchester reveal the degradation of workers brought by the introduction of steam power, machinery, and the division of labor, but one could see there, as Engels put it, "the strenuous efforts of the proletariat to raise themselves" up (50).

Engels in his writings said little about the working conditions inside the textile mills besides noting, "I cannot recall having seen a single healthy and well-built girl in the throstle room of the mill in Manchester in which I worked" (185). His point was to shock readers with the city's horrible living conditions: Tumbledown dwellings in one place were built to within six feet of "a narrow, coal-black stinking river full of filth and rubbish" fouled by runoff from tanneries, dye-works, bone mills, gasworks, sewers, and privies. The poorest workers rented dirt-floor cellars that were below the river level and always damp. Pig breeders rented the courts of the apartment buildings, where garbage thrown from the windows fattened the pigs, and stinking privies fouled the air. Engels keeps count of the number of persons using a single privy (380 is tops). "Only industry," Engels concludes, "has made it possible for workers who have barely emerged from a state of serfdom to be again treated as chattels and not as human beings" (60–64).

The very worst of Manchester was Little Ireland, whose 4,000 inhabitants were hemmed in by the River Medlock, railroad tracks, cotton mills, a gasworks, and an iron foundry. "Heaps of refuse, offal and sickening filth are everywhere interspersed with pools of stagnant liquid. The atmosphere is polluted by the stench and is darkened by the thick smoke of a dozen factory chimneys. A horde of ragged women and children swarm about the streets and they are just as dirty as the pigs which wallow happily on the heaps of garbage and in the pools of filth" (71). Engels was himself shocked at the damp one-room cellars "in whose pestilential atmosphere from twelve to sixteen persons were crowded." Numerous instances in which a man shared a bed with both his wife and his adult sister-in-law, however, were too much for his (dare one say) middle-class sensibilities, as was the mixing of the sexes in common lodging houses, where "in every room five or seven beds are made up on the floor and human beings of both sexes are packed into them indiscriminately" resulting in "much conduct of an unnatural and revolting character" (77–78).

Chased out of Germany after he and Marx published the *Communist Manifesto* in 1848, Engels returned to Manchester, where he worked again for the Ermen and Engels Company, eventually rising to full partner. For two decades (1850–69) Engels lived a split existence: capitalist by day and working-class radical by night. Marx during these years lived in London and, with financial support from Engels, spent his days at the British Museum writing *Das Kapital*. That profits from a capitalist textile firm so directly supported the greatest critic of capitalism lends a fine irony to the famous line in the *Communist Manifesto*, "What the bourgeoisie . . . produces, above all, is its own grave-diggers."[40] Marx, with no firsthand industrial experience of his own, took Engels' description of Manchester as the paradigm of capitalist industry. Neither of them noticed a quite different mode of industry forming in Sheffield.

"ONE HUGE WORKSHOP FOR STEEL": SHEFFIELD

"Our journey between Manchester and Sheffield was not through a rich tract of country, but along a valley walled by bleak, ridgy hills, extending straight as a rampart, and across black moorlands, with here and there a plantation of trees," wrote novelist Nathaniel Hawthorne, who lived in nearby Liverpool during the mid-1850s. "The train stopped a minute or two, to allow the tickets to be taken, just before entering the Sheffield station, and thence I had a glimpse of the famous town of razors and pen knives, enveloped in a cloud of its own diffusing. My impressions of it are extremely vague and misty—or, rather, smoky—: for Sheffield seems to me smokier than Manchester, Liverpool, or Birmingham; smokier than all England besides, unless Newcastle be the exception. It might have been Pluto's own metropolis, shrouded in sulphurous vapour."[41]

Sheffield was internationally known as a center for high-quality steel and high-priced steel products. Sheffielders transformed Swedish bar iron into crucible steel, the only steel available in quantity before Henry Bessemer's steel inventions of the 1850s (see chapter 6). While many of Sheffield's craftsmen specialized in steelmaking, a much larger number worked in the numerous trades that cut, forged, and ground the bars of steel into pocketknives, cutlery, saws, and files. Sheffielders exported to the United States up to a third of their total production, finding a ready market there for steel traps, agricultural implements, hunting knives, and unfinished bars. Not concentrated Manchester-style factories but decentralized networks of skilled workers typified Sheffield's mode of industry.

Like London's port and Manchester's moist air and coal, Sheffield's geography shaped its industry. Coal from the vast south Pennines field, scant miles away, fueled its furnaces. Sandstone for grinding wheels as well as the fire-resistant "refractories" for clay pots and bricks also came from nearby. Four sizable rivers, including the Sheaf, which named the town, drained into the River Don, forming 30 miles of river frontage where waterpower could drive industry. Some of the region's 115 water mill sites were built for milling grain, making paper, or grinding snuff, but most were used for grinding, forging, and rolling the region's characteristic steel products.

Already Britain's eighth largest town in 1775, Sheffield nearly tripled in size during the years from 1801 to 1851, when the population reached 135,000. Its early growth was built on a flourishing cottage industry. By the early seventeenth century its workshops hummed with the making of scissors, shears, files, nails, forks, and razors. In 1624 Sheffield's skilled artisans formed the Cutlers' Company to organize, promote, and protect their trade. Artisans looking for better-quality markets specialized in making small metal boxes for tobacco, snuff, and trinkets. In the 1740s, two famous inventions secured the region's technical advantage for nearly a century: crucible steel, examined later, and "Sheffield plate." Sheffield plate married the luster of silver with the economy of copper by fusing silver foil onto copper dishes and utensils. Makers of saws, anvils, lancets, and household cutlery, easily the city's most famous early trade, were prominent well before 1800 (fig. 3.6).

A chronicler of the city in 1824 identified sixty-two high-profile trades. In an alphabetical list of them, the B's alone include bayonets, bellows, boilers for steam, bone scales, brace bits, brass bolsters, brass in general, Britannia metal, butchers' steels, buttons, and button molds. The city's ten cutlery trades, led by the makers of table knives and pocketknives, employed more than 10,000. Some of the cutlery trade unions were tiny, such as the 50 souls devoted to plate-, spoon-, and fork-filing. In 1841, even though women were excluded from the apprenticeships necessary for

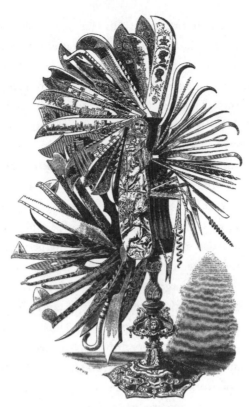

FIG. 3.6. SHEFFIELD CUTLERY AT CRYSTAL PALACE, 1851

Rodgers & Sons (Sheffield) sportsman's knife, 12 inches long, "containing eighty blades and other instruments . . . with gold inlaying, etching and engraving, representing various subjects, including views of the Exhibition Building, Windsor Castle, Osborne House, the Britannia Bridge, etc."

Christopher Hobhouse, *1851 and the Crystal Palace* (London: John Murray, 1937), 94.

skilled work, there were 159 women cutlers, 158 scissors makers, 123 file makers, 42 fork makers, as well as numerous hafters and bone cutters. The high-level skills encouraged by this specialization drew ever more workers (and manufacturers and buyers) to the district. By 1841 Sheffield had 54 percent of Britain's file makers, 60 percent of its cutlers, and 80 percent of its saw makers.[42]

Sheffield's world-famous crucible steel, the second notable invention of the 1740s, owes its creation to a clockmaker named Benjamin Huntsman. In an effort to make suitable clock springs, he increased the heat of a coal-fired, brass-melting furnace so it was able to melt pieces of blister steel, iron that had been baked in charcoal for a week. Huntsman's success came in developing clay pots, or crucibles, that remained strong in the extreme heat. In the early 1770s Huntsman built a steelmaking works using his inventions. By 1800 the Sheffield district comprised nine firms making blister steel, eleven firms melting it to cast ingots of crucible steel, and a number of separate water-powered mills hammering the cast steel ingots into rods and bars. With diverse local, national, and overseas markets, Sheffield's crucible steel industry grew steadily. By the 1840s the Sheffield region's steelmaking industry employed about 5,000 workers, who made 90 percent of Britain's steel output and around half of the world's.

It is necessary to understand that the factory system so important in Manchester was absent in Sheffield. "The manufacturers, for the most part," wrote one visitor in 1831, "are carried on in an unostentatious way, in small scattered workshops, and nowhere make the noise and bustle of a great ironworks." One famous razor-making works evidently confounded an American visitor, who, perhaps expecting a huge Manchester-style factory, observed that instead its workmen were scattered across town. Knife blades were rough forged in one building, taken across the street for finishing, ground and polished at a third works "some distance off," and sent to a fourth building to be joined to handles, wrapped, packed, and finally dispatched to the warehouse-salesroom. Even the largest cutlery firms, such as Joseph Rodgers and George Wostenholm, each employing more than 500 workers by the 1850s, depended on the region's characteristic outwork system. Cutlers were not really full-time employees, for they worked for more than one firm when it suited them. A cutler might work at a single kind of knife all his life. "Ask him to do anything else and there'd be trouble. He'd get moody, throw off his apron and get drunk for the rest of the day." A visitor in 1844 observed that, whether large or small, "each class of manufacturers is so dependent on the others, and there is such a chain of links connecting them all, that we have found it convenient to speak of Sheffield as one huge workshop for steel goods."[43]

Joining the ranks of manufacturers in Sheffield did not require building, buying, or even renting a factory. It was possible to be "a full-blown manufacturer with nothing more than a stamp with your name on it and a tiny office or room in your house." All the skilled labor for making a knife or edge tool—forging, hardening, grinding, hafting, even making the identifying stamp—could be arranged by visiting skilled artisans' shops up and down the street. An enterprising artisan might, by obtaining raw materials on credit and hiring his fellow artisans, become a "little mester," a step up from wage laborer. Some firms did nothing but coordinate such "hire-work" and market the finished goods, at home or overseas. These firms had the advantages of low capital, quick turnover, and the flexibility to "pick and choose to fit things in with whatever you were doing."[44]

For decades the Sheffield steelmaking industry, too, was typified by small- and medium-sized enterprises linked into a regional network. Certain firms specialized in converting bar iron into blister steel, others in melting (or "refining") blister steel into cast ingots of crucible steel. Specialized tilting (hammering) and rolling mills formed steel ingots into useful bars and rods, even for the larger steel firms.[45] Beginning in the 1840s Sheffield steelmakers moved east of town to a new district along the Don River valley, with cheap land and access to the new railroad. There in the 1850s Jessop's Brightside works, with 10 converting furnaces and 120 crucible melting holes, briefly held honors as the country's largest steelmaker. Other steelmakers opening works in this district included Cammell's Cyclops works in 1846, Firth's Norfolk works in 1852, John Brown's Atlas works in 1855, and Vickers' River Don works in 1863.

These large firms, already producing a majority of the region's steel, grew even larger with the advent of Henry Bessemer's process. Bessemer built a steel plant at Sheffield in 1858 with the purpose, as he put it, "not to work my process as a monopoly but simply to force the trade to adopt it by underselling them in their own market . . . while still retaining a very high rate of profit on all that was produced."[46] Bessemer in the early 1860s licensed his process to John Brown, Charles Cammell, and Samuel Fox, who each constructed large-scale steel works. In 1864 John Brown's invested capital was nearly five times larger than the leading crucible-only firm (Vickers') while its workforce was more than three times larger. Brown's Atlas works covered 21 acres. In the latter part of the nineteenth century, these large steel mills and oversize forging shops represented a second generation of Sheffield's steel industry. A string of new high-tech alloy steels kept Sheffield at the forefront of high-quality steelmaking well into the twentieth century.

The shift from waterpower to steam upended the region's industrial geography, especially its power-dependent trades. Foremost among these was the grinding trade, segmented into distinct branches for scissors, forks, needles, razors, penknives, table blades, saws, files, and scythes. Writes Sidney Pollard, in his definitive history of Sheffield's industrial workers, "The grinders had been the last to leave the countryside, . . . where they worked along the rivers whose water-power they used, 'a law unto themselves,' with their own habits, customs, and traditions."[47] Grinders hacked and turned rough-cut sandstone from the quarry into wheels. Then, sitting astride the wheel, as on a horse saddle, the grinder pressed the article to be smoothed or sharpened against the rotating stone. The work was physically demanding and, as we will see, hazardous beyond belief.

TABLE 3.1. STEAM AND WATER POWER IN SHEFFIELD STEEL GRINDING, 1770–1865

| | WATER POWER | | STEAM POWER | |
	WHEELS	TROUGHS	WHEELS	TROUGHS
1770	133	896		
1794	83	1,415	3	320
1841	40		50	
1857	16*		80	
1865	32		132	

Note: Wheel = building containing multiple workrooms; *trough* = work station of one grinder and one grinding stone.

* Pollard noted that this figure was probably an underestimate, omitting the smaller wheels.

Source: Sidney Pollard, *A History of Labour in Sheffield* (Liverpool: Liverpool University Press, 1959), 53.

The application of steam power was the only "industrial" technology that affected the Sheffield trades before the steel works' expansion of the 1850s. First applied to a grinding "wheel" in 1786, steam power grew steadily, as table 3.1 shows. (In Sheffield a "wheel" referred to a building containing a number of workrooms, or "hulls," each of which had as many as six individual "troughs" where a grinder worked his stone.) The application of steam power, as Pollard describes it, "changed the way of life of grinders, who became town-dwelling, full-time industrial workers instead of members of a part-time rural industry. The application of steam to cutting, glazing and drilling similarly drew cutlers, ivory cutters, hafters and others from their homes or the small lean-to's, in which work had previously been carried on, into the town workshops provided with power." By 1854, when Sheffield had 109 steam engines, the grinders consumed 58 percent of the city's steam power, with cutlery and toolmakers accounting for most of the rest.[48]

Even though grinding was centralized in town, the possibility of becoming a "little mester" persisted. The "little mesters" employed fellow grinders in hope of their growing into full-fledged manufacturers. In-workers used a wheel owned by a manufacturer and worked for set piece rates. Out-workers might work for more than one manufacturer at a time, renting an individual trough at a public wheel whose owner supplied space and power to his tenants for a weekly rent. An ambitious out-worker might become a "little mester" by purchasing raw materials on his own account, arranging with a middleman to take the completed articles, and completing

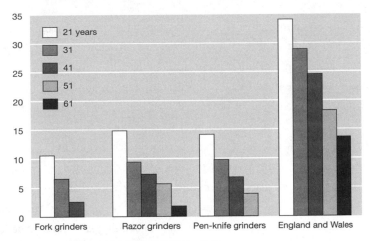

FIG. 3.7. MORTALITY OF SHEFFIELD GRINDERS, 1841

Life expectancy (in years) of Sheffield grinders at different ages. Last group, "England and Wales," is the average population. For example, an average person in England or Wales at age 21 could expect to live 34 more years; at age 41 they could expect to live 25 more years. But a Sheffield fork grinder at age 21 could expect to live 11 more years; if he reached age 41, just 2.5 more years.

Data from Sidney Pollard, *A History of Labour in Sheffield* (Liverpool: Liverpool University Press, 1959), 328.

the work by renting several troughs and employing fellow grinders. Figures from one Sheffield wheel in 1824 indicate the spectrum ranging from wage laborer to "little mester": 35 grinders rented a single trough, 30 grinders rented between 1.5 and 3 troughs, while 1 grinder rented 4 troughs and another 6 troughs.

Steam grinding killed an alarming number of grinders. "Till steam-power was introduced in the trade . . . , the grinders' disease was scarcely known," observed one employment report from the mid-1860s. Previously, grinders had worked part-time at water-driven wheels in the countryside, with substantial breaks for farming or gardening. With the expansion of their work to a full-time, year-round occupation, many grinders fell sick and died from what we would today identify as silicosis and tuberculosis. The most dangerous grinding was done "dry" (without water on the wheel). Dry grinding removed more steel and allowed the article to be watched constantly, and it long remained the common practice for grinding needles, forks, spindles, scissors, and razors. "During the process of dry grinding (and in dry grinding the greatest danger to the workmen arises), the dust which is created from the stone and the metal pervades the room in considerable quantities," noted one medical doctor.[49]

The life expectancy for several of these Sheffield trades was terrifyingly short.

During 1820–1840, a majority of the general population of the surrounding Midland counties (62%) could expect to live beyond age 50. However, a chilling 58 percent of fork and needle grinders were dead by age 30, while 75 percent of razor and file grinders were dead by age 40. Files although ground "wet" were cut with sharp chisels over a cushion of powdered lead, commonly resulting in "colic and paralysis" among the 2,000 men and boys who practiced this trade. Only the table blade, scythe, and saw grinders had any reasonable chance of living into their 50s. (Figure 3.7 shows the same dreadful picture, using calculations of life expectancy.) Fork grinding and razor grinding remained so dangerous that these grinders were excluded from other grinders' workspace and from the "sick clubs" that helped pay for medical care. One fork grinder related, "I shall be thirty-six years old next month; and you know measter, that's getting a very old man in our trade." Grinders of saws, scythes, and table blades, for whom wet grinding was common, suffered less from lung diseases. By 1840 penknives and pocket blades were also wet ground. For all grinders, there was a significant danger that their wheels would blow apart while rotating. One grinder testified that in eighteen years no fewer than ten stones had burst under him. These dire conditions led to descriptive articles in the *Times of London*, the *Illustrated London News*, the *British Medical Journal*, and elsewhere, as well as testimony to royal commissions and the British Parliament.[50]

Steam not only directly killed many grinders through dangerous working conditions, but also indirectly brought the deaths of many residents crammed into the poorest central districts. In the 1840s sanitary reformers praised Sheffield for combating cholera epidemics, draining burial grounds, and cleaning streets. With better living conditions, the death rate for children under five years dropped from a bleak 32.7 percent in 1841 to 10.9 percent in 1851. But these modest sanitary improvements were simply overwhelmed by the city's central-area building boom in the 1850s. The conversion of many grinding, tilting, and rolling mills to steam power and their relocation from outlying areas packed the city's central area with factories. The construction of large steam-powered steel plants in the city's eastern district contributed to crowding, too. Many open areas in the city center were now covered with buildings, and some houses with low roofs and inadequate ventilation were converted to factories. "Sheffield, in all matters relating to sanitary appliances, is behind them all," wrote a qualified observer in 1861.

> The three rivers sluggishly flowing through the town are made the conduits
> of all imaginable filth, and at one spot . . . positively run blood. These rivers,

that should water Sheffield so pleasantly, are polluted with dirt, dust, dung and carrion; the embankments are ragged and ruined; here and there over-hung with privies; and often the site of ash and offal heaps—most desolate and sickening objects. No hope of health for people compelled to breathe so large an amount of putrefying refuse.[51]

Sheffield's dire sanitary conditions resembled those of London or Manchester for much the same reason: the city's densely packed population lived in filthy conditions and lacked clean water. Sheffield's water started from a clean and clear reservoir outside town, but after a mile of iron piping it reached town unfiltered, clouded, and rusty. For the lucky 75 percent of houses in the borough (19,000 of 25,000) connected to the system in 1843, water could be drawn from standpipes in the yard for two hours on three days each week. For other times families had to store water in tubs, barrels, or cisterns. The 25 percent of families without city water took their chances with wells, and the poorest sank wells in their building's walled-in court-yards, the location also of privies serving up to a dozen households. Even allowing for some exaggeration on the part of reformers, sanitary conditions in Sheffield were bad and got worse. In the mid-1860s there were credible reports that certain privies "did duty" for up to sixty persons. One correspondent counted twenty-eight rotting dog carcasses piled under a city bridge. Between the 1860s and 1890s the city's adult mortality rates (still above national averages) improved somewhat. Infant mortality did not. During these decades one in *six* Sheffield infants died before their first birthday.[52]

The geographies of industry surveyed in this chapter—multidimensional urban networks in London, factory systems in Manchester, and sector-specific regional networks in Sheffield—clinch the argument that there were many "paths" to the industrial revolution. No single mode of industry transformed Britain in lockstep fashion. London, driven by population growth, its port activities, and the imperatives of a capital city, industrialized along many dimensions at once, often in spatially compact districts and always tied together by ancillary industries such as those that grew up around porter brewing. Brewing, shipbuilding, engineering, and construction were among the chief innovative centers, but London would not have industrialized (and in turn would not have had such sustained population growth and market impact) absent its innumerable ancillary industries. The city showcased the country's industrial attainments at the Exhibition of 1851 in the Crystal Palace (fig. 3.8).

FIG. 3.8. BUILDING THE CRYSTAL PALACE, 1851

The Crystal Palace, built for the 1851 Exposition in London, expressed Britain's commercial and industrial might. At the top are the glazing wagons workers used to place the 300,000 panes of glass, 12 by 49 inches square, that formed the roof.

Christopher Hobhouse, *1851 and the Crystal Palace* (London: John Murray, 1937), 48.

In sharp counterpoint to London's multidimensional growth is Manchester's single-industry cotton factory system. There, truly, cotton was king. As cotton textiles spread into the surrounding Lancashire district, Manchester expanded the ancillary industries of bleaching, printing, and dyeing. The region's demand for textile machines attracted the London machine builders, who moved to Manchester, bringing a crucial ancillary industry, and subsequently made machine tools, bridges, and locomotives. Sheffield, too, centered on a single sector, high-quality steel products. In a sense, the entire city constituted a spatially compact network of ancillary industries—"one huge workshop for steel goods"—providing all the specialized skills needed to make bars of steel into pocketknives, cutlery, saws, files, and other valuable items. Sheffield's network model, and especially the persistence of the small-shop "little mester" system, provided much greater scope for occupational mobility than Manchester's capital-intensive factories. Nor were Sheffielders time-bound preindustrial artisans, as attested by their innovations in silver plate and crucible steel, their adoption of Bessemer steel, and the city's international fame in top-quality specialty steels. Sheffield was a model for economist Alfred Marshall's dynamic "industrial districts," embodying such traits as hereditary knowledge, subsidiary trades, highly specialized techniques, local markets for special skills, and industrial leadership.[53]

Yet the shocking filth and poverty evident in each of these three industrializing sites was fuel to the fires for critics of industrial society. It is striking indeed that distinct paths to industrial revolution led to similarly dire environmental outcomes. There was the centralizing power of steam. Workers in steam-driven occupations, whether in London, Manchester, Sheffield, or other regions, were less likely to live in the country, to eat fresh food, to drink clean water, and (especially if female) to be skilled and have reasonable wages.

Steam driven factories drew into urban areas, where coal was most easily delivered, not only the factory workers themselves but also the ancillary industries' workers and the host of needed shopkeepers and service-sector workers. In this way steam was responsible for packing numerous residents into the industrial cities. This pattern of spatial concentration continued as the higher thermal efficiencies of larger steam engines gave profit-minded industrialists a relentless incentive to build correspondingly larger factories. Indeed, the unmistakable social problems created by steam-driven urban factories inspired Rudolf Diesel to invent an internal combustion engine that, he hoped, would return power, literally and figuratively, to decentralized, small-scale workshops that might be located in the countryside.[54]

The industrializing cities were certainly striking to the senses and sensibilities, but of course they were not the whole picture. Indeed, the industrial revolution did not so much create a single homogenized "industrial society" during these decades as generate ominous *differences* within British society. Visits to industrial urban slums shocked many commentators because they came with their sensibilities from town or rural life still very much in place. Traditional, rural ways of life persisted throughout the industrial period, often quite near cities. London's wealthy retreated to their rural estates within a day's horse ride, while the city's poor were stuck there in "dark satanic mills." It is striking that even Friedrich Engels framed his critique of Manchester's squalid living conditions through a moralistic, middle-class set of values. Manchester provoked a storm of criticism because it, and the other industrial centers, appeared to be dividing Britain into "two nations" split by irreparable differences. By comparison, working and living conditions in rural-based traditional industries, such as the woolen industry, passed largely without critical comment.

Looking farther afield, we can see that any complete view of the industrial revolution needs to incorporate not only the contributions of countries other than England but also their hugely varied experiences with industrialization. Here I can give only the briefest sketch. To begin, the classic industrial technologies were typically the result of varied national contributions. The textile industries took form with continual interchange and technology transfer among England, France, the United States, and other countries. Artisans in India made grades of cotton calicos, printed chintzes, and fine muslins using bright colorfast dyes that were the envy of the world. While Americans drew heavily on British textile machines, they built integrated textile factories arguably a decade earlier than did Manchester (with the pioneering Lowell-system mills in Waltham [1814–15] and Lowell [1821], Massachusetts). Similarly, one might see the distinctive "American system" of manufacturing as replicating and mechanizing the French achievement of manufacturing muskets, as early as the 1780s, that featured interchangeable parts. Another distinctive American achievement, the steel industry, depended on the transfer of production processes from Britain, France, Germany, and other countries.[55]

Different countries took several distinct paths to industrial revolution. Countries like Sweden and the United States were rich in wood and water power, and consequently they were largely spared the pressing problems of draining coal and mineral mines that had driven the development of steam engines in Britain. Svante Lindqvist provides a rough index of Swedish technological priorities in the late eighteenth century: out of 212 technical models held by the Royal Chamber of Models, there were numerous agriculture machines (43), wood-conserving fireplaces (29), mining machines (30), handicraft and textile machines (33), and various hydraulic works (35), but just *one* steam engine.[56] (Curiously, early attempts in both Sweden and America to transfer Newcomen steam engines from England ended in outright failure.) What is more, compared to Britain, coal-fired iron smelting came later to Sweden and the United States for the forceful reason that they had ready supplies of wood-based charcoal for fuel. Charcoal was used in making Sweden's famed bar iron, exported to Sheffield and around the world. Even more so than Britain, these countries depended on water power, which in turn, because water power sites were spread out along rivers, lessened the rush to steam-and-coal-centered cities.

While variations in availability of raw materials obviously influenced a country's industrial path, cultural and political preferences mattered, too. France did not follow the model of a British-style industrial revolution but rather pursued artisan-based industries, which used a skilled and flexible labor force and were adapted to regional markets. Despite its population's being roughly twice that of Britain's, France had just one-third the number of large cities (three versus nine cities above 100,000). In the already urbanized Netherlands, there emerged a distinct mix of both craft-oriented and mass-production industries. In Germany, politically united only in the 1870s, industrialization was generally later to develop but featured rapid growth of industrial cities, such as those in the Ruhr, built on heavy industries of iron, steel, and coal (and soon enough, as we will see in chapter 5, on electricity and chemicals).

Since the 1970s, historians have emphasized "variety" in industry and the sources of economic growth. In these accounts, cotton was forcefully dethroned as "king" in the British industrial revolution, in favor of substantial variety in wool, leather, building and even in the factory system as well as the compelling view of "alternative paths" taken by different countries to industrialization. By contrast, the recent history of capitalism restores cotton as king. In his prize-winning *Empire of Cotton* (2014), Sven Beckert returns cotton to the center of world history. For Beckert, "cotton was the most important human manufacturing activity" for 900 years of human history, neatly pushing aside spices, gold, silver, wool, sugar, iron, steel, and even steam power. He notes, "the global cotton industry connected technologies from China and India to labor from Africa, capital from Europe to expropriated lands in the Americas, and consumers from Brazil to manufacturers in Great Britain." And cotton's was a sordid violence-soaked history of coercive domination. In summary, one reviewer offered this haiku: "Cotton was central / to global capitalism / which hurt most people."[57]

In chapters to come we'll see the differentiation of industrial technologies in the eras of systems, modernism, warfare, and globalization. Each of these eras was built on industrial technologies and yet led to their transformation. The next chapter follows up on Britain's industrialists, technologists, and government officials, whose eyes were "wandering over the map of the world for new markets" in a new, imperialist era for technology.

CHAPTER 4

1840 – 1914

Instruments of Empire

British technology, propelled forward by the industrial revolution, reached something of a plateau by the mid-nineteenth century. The display of Britain's mechanical marvels at London's Crystal Palace exposition in 1851 stirred the imagination, but now British industry faced a host of rising competitors in Europe and North America. (And as chapter 5 describes it was Germany and the United States that would spearhead a "second" industrial revolution in the decades to come.) At midcentury in Britain, and soon across much of the industrialized world, a new technological era took shape as colonial powers attended to the unparalleled problems of far-flung overseas empires. To a striking extent, inventors, engineers, traders, financiers, and government officials turned their attentions from blast furnaces and textile factories at home to steamships, telegraphs, and railway lines in the colonies. These technologies, supported and guided by empire, made possible the dramatic expansion of Western political power and economic influence around the globe.

Britain was of course no newcomer to empire. In 1763, after the Seven Years' War, Britain had gained most of France's holdings in India and North America. In the next decade, however, American colonists rejected new taxes to pay for the heavy costs of this war and initiated their rebellion, which, in the view of one Englishman, would have failed if only Britain had possessed effective communications technology.[1] British rule in India was the next dilemma. The East India Company, a creature of Britain's mercantilist past, ran afoul of Britain's rising industrialists. Owing greatly to their relentless antimercantile lobbying, Parliament in 1813 stripped the East India Company of its monopoly of the lucrative Indian trade and, by rechartering the company in 1833, ended its control over private British traders within India and terminated its sole remaining commercial monopoly, that of trade with China. New players crowded in. The Peninsular & Oriental Steam Navigation Company, the legendary "P&O," gained a valuable mail contract between England and Egypt in 1840, with an onward connection to India. The rise of "free trade" in the 1840s, also promoted by British industrialists keen to secure raw materials for their factories and cheap food for their factory workers, led to a wild scramble in the Far East. At least

FIG. 4.1. INDIAN MUTINY OF 1857–1858

"Attack of the Mutineers on the Redan Battery at Lucknow, July 30th 1857." British accounts inevitably stressed the "atrocities" committed by the rebel Indian soldiers, but in his *Memories of the Mutiny* (London, 1894, pp. 273–80), Col. F. C. Maude detailed his own part in desecrating the bodies of executed Indian prisoners at Bussarat Gunj.

Illustration from Charles Ball, *The History of the Indian Mutiny* (London: London Printing and Publishing, 1858–59), vol. 2, facing p. 9.

sixty British trading firms in China clamored for military assistance to uphold "free trade" there. The British government intervened, and this led to the first opium war (1840–42).

New technologies were critical to both the penetration phase of empire, in which the British deployed steam-powered gunboats and malaria-suppressing quinine to establish settlements inland beyond the coastal trading zones, and in the subsequent consolidation phase that stressed the maintenance and control of imperial outposts through a complex of public works.[2] Effective military technologies such as steam-powered gunboats, breechloading rifles, and later the fearsome rapid-firing machine guns helped the British extend their control over the Indian subcontinent and quell repeated uprisings. Even before the Indian Mutiny of 1857–58, which was a hard-fought battle against insurgent Indian troops who ranged across much of the northern and central regions of India and whose defeat cost the staggering sum of £40 million, there were major military campaigns nearly every decade (fig. 4.1). These included three Mahrattan wars (1775 to 1817), two Mysore wars in the 1780s and 1790s, the Gurkha war of 1814–15 in Nepal, two Anglo-Burmese wars in the 1820s and 1852, and the first of two opium wars in China. In the 1840s alone the British conducted a three-year military occupation to subdue and secure Afghanistan, which failed in 1842, a swift campaign crushing the Sinds in what is now Pakistan the following year, and the bloody Sikh wars over the Punjab in 1845 and 1848.

The tremendous cost of these military campaigns as well as the ongoing expenses for transporting, lodging, provisioning, and pensioning imperial officials simply ate up the profits of empire. We noted in chapter 3 that the East India Company put on the imperial payroll 1,200 workers in London alone. On balance, these collateral expenses of British empire completely absorbed the sizable profits of imperial trade. Imperialism in India did not generate wealth. Rather it shifted wealth from taxpayers in India and Britain to prominent traders, investors, military officers, and imperial officials, who became its principal beneficiaries.[3] This point is important to emphasize because critics have long taken for granted that the imperatives of capitalism required imperialism (for acquiring cheap raw materials and disposing of surplus factory-made goods) and that the machinery of imperialism made money. Equally important, the wealth-*consuming* nature of imperial technologies sets off the imperial era from the earlier wealth-generating ones of commerce and industry. Imperial officials, and the visionary technology promoters they funded, spared no expense in developing instruments of empire that promised to achieve rapid and comfortable transportation, quick and secure communication, and above all sufficient and effective military power.

STEAM AND OPIUM

Steam entered India innocently enough in 1817 when an 8-horsepower steam engine was brought to Calcutta in a short-lived attempt to dredge the Hooghly River. This plan was revived in 1822 when the East India Company bought the engine and again used it to clear a channel up the Hooghly to speed the passage of sailing vessels to Calcutta proper, some 50 miles inland from the ocean. At the time, Calcutta was the chief British port in India and the principal seat of its political power. The second city, Bombay, on India's western coast, was actually closer as the crow flies to London, but favorable winds made it quicker to sail to Calcutta via the African cape. The initial experiments with steam engines around Calcutta were, from a commercial point of view, rather disappointing. The steamship *Diana* worked on the Hooghly for a year as a passage boat while the more substantial *Enterprise*, a 140-foot wooden vessel with two 60-horsepower engines, a joint project of Maudslay & Field, the great

London engineers, and Gordon & Company, was the first steam vessel to reach India under its own power, having steamed around the African cape in early 1825. The trip took a discouraging 113 days. The huge amount of fuel required by early steamers made them commercially viable only where abundant supplies of fuel were readily at hand, such as along the Mississippi River, or where their use secured some special advantage that covered their astronomical operating costs.[4]

An early indication of the significance of steamers in imperial India came in the first Anglo-Burmese war (1824–26). At first the war went badly for the British, who hoped to claim valuable tea-growing lands recently annexed by Burma. Britain's hope for a quick victory literally bogged down—in the narrow twisted channels of the lower Irrawaddy River. Britain's majestic sailing vessels were no match for the Burmese *prau*, a speedy wooden craft of around 90 feet in length propelled by up to seventy oarsmen and armed with heavy guns fixed to its bow. The British lost three-quarters of their force, mostly to disease, in the swamps of the Irrawaddy. The tide turned abruptly, however, when the British ordered up three steam vessels whose shallow draft and rapid maneuverability altered the balance of power. The *Enterprise* rapidly brought in reinforcements from Calcutta, while the *Pluto* and *Diana* directly towed British sailing ships into militarily advantageous positions. Another steam vessel that saw action later in the war was the *Irrawaddy*, arguably the first steam gunboat, with its complement of ten 9-pound guns and one swivel-mounted 12-pound gun. The defining image of this war was of the *Diana*, known to the Burmese as "fire devil," tirelessly running down the *praus* and their exhausted oarsmen. The king of Burma surrendered when the British force, assisted by the *Diana*, reached 400 miles upstream and threatened his capital.

Following this imposing performance in Burma, steamboat ventures proliferated in the next decade. The East India Company deployed steamers to tow sailing vessels between Calcutta and the ocean and dispatched the pioneering steamer up the Hooghly and onward to the Ganges River. Accurately mapping the Ganges had been a necessary first step in transforming the vague territorial boundaries assumed by the company into a well-defined colonial state. To this end one could say that the first imperial technology deployed on the Ganges was James Rennell's detailed *Map of Hindoostan* (1782). Rennell also published the equally valuable *Bengal Atlas* (1779).[5] Regular steam service on the Ganges between Calcutta and upstream Allahabad began in 1834; the journey took between twenty and twenty-four days depending on the season. The river journey bore a high price. A cabin on a steamer between Calcutta and Allahabad cost 400 rupees, or £30, about half the cost of the entire journey from London to India and completely beyond the means of ordinary Indians. Freight rates from £6 to £20 per ton effectively limited cargoes to precious goods like silk and opium, in addition to the personal belongings of traveling officials and the necessary imperial supplies such as guns, medicines, stationery, official documents, and tax receipts. In the strange accounting of imperialism, however, even these whopping fares may have been a bargain for the East India Company, since in the latter 1820s it was paying a half-million rupees annually for hiring boats just to ferry European troops up and down the Ganges. Quicker river journeys also trimmed the generous traveling allowances paid to military officers. The economics of river transport were not only administrative ones. General Sir Charles Napier, who led the military campaign in the 1830s to open up for steam navigation the Indus River, India's second most important inland artery, pointed out direct commercial consequences. He wrote, "India should suck English manufacturers up her great rivers, and pour down these rivers her own varied products."[6]

Steam also promised to tighten up the imperial tie with London. Before the 1830s a letter traveled by way of a sailing vessel from London around the African cape and could take five to six months to arrive in India. And because no ship captain would venture into the Indian Ocean's seasonal monsoons, a reply might not arrive back in London for a full two years. Given these lengthy delays, India was not really in London's control. British residents in Madras urged, "Nothing will tend so materially to develop the resources of India . . . and to secure to the Crown . . . the integrity of its empire over India, as the rapid and continued intercourse between the two countries by means of steam." Merchants and colonial administrators in Bombay were eager to secure similar benefits. In the new age of steam Bombay's proximity to the Red Sea was a distinct advantage. From there the Mediterranean could be reached by a desert crossing between Suez and Cairo and then down the Nile River to the port at Alexandria. Efforts of the Bombay steam lobby resulted in the 1829 launch of the *Hugh Lindsay*, powered by twin 80-horsepower engines, which carried its first load of passengers and mail from Bombay to Suez in twenty-one days. Even adding the onward link to London, the *Hugh Lindsay* halved the transit time for mail. So valuable was the Red Sea route that the British became by far the largest users of the French-funded, Egyptian-built Suez Canal, opened in 1869. In 1875 Britain purchased the Egyptian ruler's entire share in the canal company for £4 million and in 1882 invaded Egypt to maintain control over this vital imperial lifeline. The 102-mile-long canal cast a long shadow in the region. British troops remained in Egypt over seven decades until after the Suez Crisis in 1954.

Already by 1840 steamboats in several ways had shown substantial capacity for knitting together the wayward strands of the British Empire. In the opium war of 1840–42 they proved their value in projecting raw imperial power. The opium wars, in the early 1840s and again in the late 1850s, were triggered by China's desperate attempts to restrain "free trade" in opium. Opium was grown on the East India Company's lands in Bengal, auctioned in Calcutta, and then shipped by private traders to China. The large-scale trade in opium closed a yawning trade gap with the Celestial Empire, for whose tea Britain's citizens had developed an insatiable thirst. British exports of factory-made cotton to India completed the trade triangle. Opium, like all narcotics, is highly addictive. (One of the most chilling scenes in the Sherlock Holmes series is a visit to one of London's numerous opium dens, in "The Man with the Twisted Lip," where "through the gloom one could dimly catch a glimpse of bodies lying in strange fantastic poses . . . there glimmered little red circles of light, now bright, now faint, as the burning poison waxed or waned in the bowls of the metal pipes.") The opium war began when China took determined steps to ban importation of the destructive substance, and the British government, acting on the demand of Britain's sixty trading firms with business in China, insisted on maintaining "free trade" in opium and dispatched a fleet of warships to China.

Steamers played a decisive role in the opium war. The British fleet was able to do little more than harass coastal towns until the steamer *Nemesis* arrived in China in November 1840, after a grueling eight-month voyage around the African cape. The *Nemesis*, at 184 feet in length, was not merely the largest iron vessel of the time. It was, more to the point, built as a gunboat with twin 60-horsepower engines, shallow 5 foot draft, two swiveling 32-pound guns fore and aft, along with fifteen smaller cannon. The *Nemesis* was central in the 1841 campaign that seized the major city of Canton. *Nemesis* sank or captured numerous Chinese war "junks" half its size, took possession of a 1,000-ton American-built trading ship recently purchased by the Chinese, towed out of the way deadly oil-and-gunpowder "fire rafts," and attacked fortifications along the river passage up to Canton. The *Nemesis*, wrote its captain, "does the whole of the advanced work for the Expedition and what with towing transports, frigates, large junks, and carrying cargoes of provisions, troops and sailors, and repeatedly coming into contact with sunken junks—rocks, sand banks, and fishing stakes in these unknown waters, which we are obliged to navigate by night as well as by day, she must be the strongest of the strong to stand it."[7]

In 1842 the *Nemesis*, now leading a fleet comprising ten steamers, including its sister ship *Phlegethon*, eight sailing warships, and fifty smaller vessels, carried the campaign up the Yangtze River. At one battle, *Nemesis* positioned an eighteen-gun warship, whose guns dispersed the Chinese fleet, including three human-powered paddle wheelers. The steamers promptly overtook them. The steamers also hauled the sailing vessels far up the river, over sandbars and mud, to take control of Chinkiang (today's Zhenjiang) at the junction of the Yangtze River and the Grand Canal. The Grand Canal was China's own imperial lifeline, linking the capital, Beijing, in the north to the rice-growing districts in the south. The Chinese had little choice but to accept British terms. In 1869, in the aftermath of a second opium war, the Chinese Foreign Office pleaded with the British government to curtail the deadly trade:

> The Chinese merchant supplies your country with his goodly tea and silk, conferring thereby a benefit upon her; but the English merchant empoisons China with pestilent opium. Such conduct is unrighteous. Who can justify it? What wonder if officials and people say that England is willfully working out China's ruin, and has no real friendly feeling for her? The wealth and generosity of England are spoken by all; she is anxious to prevent and anticipate all injury to her commercial interests. How is it, then, she can hesitate to remove an acknowledged evil? Indeed, it cannot be that England still holds to this evil business, earning the hatred of the officials and people of China, and making herself a reproach among the nations, because she would lose a little revenue were she to forfeit the cultivation of the poppy.[8]

Unfortunately for the Chinese people more than "a little revenue" was at play. Opium was a financial mainstay of the British Empire, accounting for one-seventh of the total revenues of British India. Repeated attempts by humanitarian reformers to eliminate the opium trade ran square into the sorry fact that British India was hooked on opium. While opium addiction was a severe problem in some districts of India, the imperial system depended on the flow of opium money. Annual net opium revenues—the export taxes on Malwa opium grown in western India added to the operating profits from growing opium in Bengal in the east, manufacturing it in government factories at Patna and Ghazipur, and exporting the product to China—were just over £3.5 million in 1907–8, with another £981,000 being added in the excise tax on opium consumption in India.[9] In 1907 Britain officially agreed over a ten-year period to wind down the odious enterprise. This was difficult. In April 1917 the Shanghai Opium Combine, a group of private British traders, was left holding around 420,000 pounds of opium and with no official legal market; a corrupt vice president of China purchased the lot for $20 million, purportedly for "medical purposes." Five years later, humanitarian reformers in Peking organized an international anti-opium association to enforce existing international laws and crack down on "vast quantities of morphia . . . manufactured in, and transported across, the United States" to evade existing laws and agreements. Finally, after much additional pain and suffering, in the 1950s the People's Republic of China suppressed opium with brutal force.[10]

TELEGRAPHS AND PUBLIC WORKS

In the industrializing countries of Western Europe and North America, telegraph systems grew up alongside railroads. Telegraph lines literally followed railway lines, since telegraph companies frequently erected their poles in railroad right-of-ways. Telegraphs in these countries not only directed railroad traffic, a critical safety task because all railways had two-way traffic but most had only a single track; telegraphs also became the information pipelines between commercial centers, carrying all manner of market-moving news. In India, by contrast, the driving force behind the telegraph network was not commerce or industry but empire. As one Englishman phrased it, "the unity of feeling and of action which constitutes imperialism would scarcely have been possible without the telegraph."[11]

Telegraph lines were so important for imperialism in India that they were built in advance of railway lines (fig. 4.2). The driving figure in this endeavor was the Marquis of Dalhousie. As governor-general over India from 1848 to 1856, Dalhousie presided over an energetic campaign to bring Western ideas and Western technology to India. Dalhousie's territorial annexations in these years increased by almost half the size of British India and added substantially to the administrative complexity of governing it. The new possessions included the Punjab in the far northwest, the province of Pegu in Burma, and five native states including Satara, Sambalpur, Nagpur, Jhansi, and Oudh. The addition of Nagpur was especially welcomed by Lancashire cotton industrialists eager to secure alternative sources of raw cotton (to lessen their dependence on the American South as the Civil War loomed); colonial troops deployed to Nagpur helped fortify the continent-spanning road between Bombay and Calcutta. To help consolidate these far-flung holdings Dalhousie launched or completed a number of technological ventures, including the Grand Trunk Road and the Ganges Canal, in addition to the railroad and wide-ranging Public Works Department discussed later. His first priority was the telegraph.

Dalhousie shaped India's telegraph network to fulfill the administrative and

FIG. 4.2. ERECTING THE INDIAN TELEGRAPH

The telegraph network across India as well as between India and Europe depended on native Indian labor to erect, operate, and maintain the lines.

Frederic John Goldsmid, *Telegraph and Travel* (London: Macmillan, 1874), frontispiece.

military imperatives of empire. The first experimental line was built in two phases and ran from Calcutta to the mouth of the Hooghly River at Kedgeree. Events immediately proved its significance. News of the outbreak of the second Anglo-Burmese war arrived by ship at Kedgeree on 14 February 1852 and was telegraphed at once to Dalhousie at Calcutta. "If additional proof of its political value were required," Dalhousie wrote in the midst of war two months later, "it would be found in recent events when the existence of an electric telegraph would have gained for us days when even hours were precious instead of being dependent for the conveyance of a material portion of our orders upon the poor pace of a dak foot-runner."[12] In December Dalhousie delineated his nationwide plan. His top priority was "a telegraph line connecting Calcutta, Benaras, Allahabad, Agra, Amballa, Lahore and Peshawar" to link up all locations "in which the occurrence of political events was at all likely." After the line to Peshawar, at the Afghan border in the far west, came a line to Bombay, through which a line to London might be completed. Of lesser importance was a line to Madras, considered politically reliable. Moreover, to connect Calcutta with Madras Dalhousie planned not a direct telegraph line south down the eastern coastline but a much longer, indirect connection via Bombay. From Bombay to Madras this line down the western coast passed through the military outposts at Poona, Bellary, Bangalore, and Arcot. The India Office in London quickly approved funds for the entire 3,150-mile network outlined in Dalhousie's plan.

Construction on the telegraph network began in November 1853, after a team of sixty installers was trained under the supervision of William O'Shaughnessy. O'Shaughnessy, a self-taught electrical expert and formerly deputy assay master of the Calcutta Mint, pushed the lines forward with breakneck speed. At first his installers placed a "flying line" of 5/16-inch iron rod on uninsulated bamboo poles for immediate military use, later transferring it to insulated poles of stone, ironwood, or teak. Within five months the first trunk line, running the 800 miles from Calcutta to Agra, was opened; and by the end of 1854 the entire national backbone was complete, with links to Peshawar, Bombay, and Madras. Two years later, all the major military stations in India were interconnected by telegraph. Dalhousie, who aimed to mobilize 150,000 troops in one hour with the telegraph, had become acutely worried about increasing signs of military and civil discontent around him by the time he left India in 1856. ("Those who have travelled on an Indian line, or loitered at a Hindoo railway station, have seen the most persuasive missionary at work that ever preached in the East," wrote his biographer. "Thirty miles an hour is fatal to the slow deities of paganism."[13]) The wrenching cultural changes brought by his rapid-fire Westernization touched off a major nationwide rebellion.

The outbreak of the Indian Mutiny, on 10 May 1857, began a nerve-wracking "week of telegraphs." Earlier that spring native Indian troops near Calcutta had in several instances been openly disobedient, refusing to load cartridges in the new Enfield rifle. Loading the rifle required soldiers to bite open cartridges that they believed to be coated with beef or pork fat, substances deeply offensive to Hindus and Muslims. (The Woolwich factory in England had indeed used beef fat, while the Dum-Dum factory in India had apparently not.) On 10 May native troops in Meerut, close to Delhi in the north central part of the country, seized control of that station, after eighty-five native members of a cavalry troop stationed there had been court-marshaled, publicly stripped of their uniforms, and imprisoned for refusing the suspect Enfield cartridges. The rebels marched on Delhi, took control of the city, and proclaimed a new Mogul emperor of India. By destroying the surrounding telegraph lines, the rebels cut off communication with the Punjab to the north but not before a telegraph message had gone out on the morning of 12 May warning the British officers in the Punjab that certain native troops planned a rebellion there the following evening. Officers at Punjab quickly disarmed the native regiments before they got word of the uprising, which had been sent by foot runner. "Under Providence, the electric telegraph saved us," affirmed one British official. Calcutta heard by telegraph of the fall of Delhi on 14 May and immediately dispatched messages requesting reinforcements for Delhi and Agra and inquiring about numerous potential trouble spots. "All is quiet here but affairs are critical," Calcutta heard on 16 May from Lucknow station. "Get every European you can from China, Ceylon, and elsewhere; also all the Goorkas from the hills; time is everything."[14]

Swift use of the telegraph saved not merely the British in Punjab but arguably the rest of British India as well. The telegraph made possible massive troop movements aimed at the most serious sites of rebellion against British rule. British and loyal native troops returning from Persia were directed to trouble spots in Calcutta, Delhi, and the Punjab; loyal native troops from Madras moved to reinforce Calcutta; while British troops in Ceylon, Burma, and Singapore were called in as well. Officials in Calcutta coordinated steamers and sailing vessels and in short order resolved numerous logistical difficulties, all by telegraph. In June and July the Mutiny spread across the northern and central provinces of India. But by then the deployment of numerous regiments loyal to the British prevented the Indian rebels from gaining ground. The promptness of the British responses astonished them. In the field campaigns that followed, the most famous use of telegraphs was in the Cawnpore–Lucknow "flying line" that aided the British troops in their assaults on the beleaguered Lucknow in November 1857 and March 1858. Isolated acts of rebellion continued until the capture of an important rebel leader in April 1859, and memories of atrocities poisoned relations between rulers and ruled for decades. One rebel on his way to execution pointed out the telegraph wire overhead as "the accursed string that strangles me."[15]

News of the Mutiny took forty days to arrive in London, traveling by steamers, camels, and European telegraphs. Consequently, imperial officials there were helpless bystanders as the conflict unfolded, was fought, and eventually ended. Insistent calls for action in the wake of this exasperating situation led to the inevitable government subsidies, but the first attempts to lay undersea telegraph cables or to use landlines across the Middle East proved expensive, slow, and unreliable. Messages relayed by non-English-speaking telegraph clerks might arrive a month after being sent and be totally unreadable. Not until 1870 was there a reliable telegraph connection between London and India (see fig. 4.3). The first line to open, a double landline running across Europe to Tehran, where it connected with a cable to Karachi via the Persian Gulf, was built by the German firm of Siemens and Halske, a leader in the second industrial revolution (see chapter 5). A second telegraph, also opened

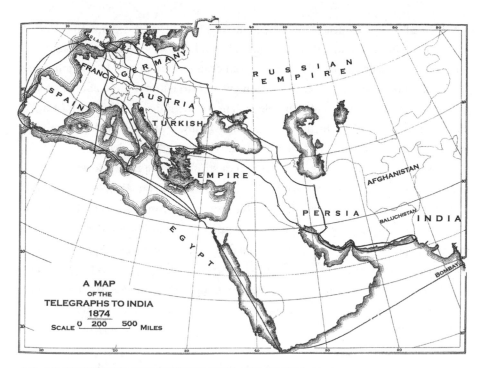

FIG. 4.3. TELEGRAPH LINES BETWEEN INDIA AND EUROPE, 1874

A high-technology imperial lifeline connecting Britain with India was established in 1870. Two overland telegraph lines ran from London, through Europe and the Middle Eastern countries, while a third, all-underwater line went through the Mediterranean Sea, the Suez Canal (opened in 1869), the Red Sea, and on to the Indian Ocean. The telegraph cut the time needed to send a message between London and India from months to hours.

Frederic John Goldsmid, *Telegraph and Travel* (London: Macmillan, 1874), facing p. 325.

in 1870, went wholly by undersea cables from England to Bombay via the Atlantic, Mediterranean, and Red Sea. By 1873 telegraph messages between England and India took three hours. The British went on to lay undersea telegraph cables literally around the world, culminating with its famous "all red" route—named for the color of imperial possessions in the official maps—completed in 1902.[16]

By the time of the 1857 Mutiny, British rule in India had become dependent on telegraphs, steamships, roads, and irrigation works; soon to come was an expanded campaign of railway building prompted by the Mutiny itself. Yet, as hinted in the training of those sixty telegraph assistants, the British also became dependent on a technical cadre of native assistants to construct, operate, and maintain the instruments of empire. Budgetary constraints made it impossible to pay the high cost of importing British technicians for these numerous lower-level positions. The colonial government in India had little choice but to begin large-scale educational programs to train native technicians. These pressures became acute in the decade following the Mutiny because the colonial government embarked on a large and expensive program of roads, canals, and railroads designed to reinforce its rule. The East India Company was dissolved in the wake of the Mutiny, and the British government assumed direct rule through top-level officials in London and Calcutta.

During these same years, Lancashire industrialists were frantic to secure alternative supplies of cotton imperiled by the American Civil War. Their well-organized lobbying in this instance prevailed on the home government in London, which directed the colonial government in India to open up India's interior cotton-growing regions to nearby ports. A wide-ranging public works campaign might have led to balanced economic development, but the effect of British policy was to discourage the development of Indian industry. The prevailing view was neatly summarized by Thomas Bazley, president of the Manchester Chamber of Commerce and member of Parliament for Manchester: "The great interest of India was to be agricultural rather than manufacturing and mechanical."[17]

One can discern a decidedly nonmechanical slant in the structure of the Public Works Department (PWD) itself, the technical education it presided over, and not least the public works projects that it helped construct. The PWD, founded in 1854 to coordinate Dalhousie's numerous transportation and infrastructure projects, dominated state-sponsored technology in India. (Quite separately, wealthy Indian traders from Bombay revived the cotton textile industry that had flourished in the eighteenth century around that city, leading the Lancashire cotton lobby to redouble its effort in the 1870s to secure "free trade" in cotton.) Most immediately the PWD set the agenda for technology in India through large construction efforts that included roads, canals, and irrigation projects, often—explicitly—with a view toward increasing exports of cotton or wheat.[18]

The character of the PWD's projects was no accident. The department reported to the British colonial officials in Calcutta, who could sometimes engage in creative interpretation of directives from the London-based secretary of state for India. But the policy was set in London, and there the officials responsible for India policy were receptive to domestic pressure groups such as the Lancashire textile industrialists. The secretary of state's office brimmed with letters, petitions, and all manner of insistent appeals from the Manchester Chamber of Commerce. In 1863, fearful of "the insurrection of cotton people," Charles Wood, secretary of state for India from 1859 to 1866, directed his colonial colleague in India to build "cotton roads" at great haste. "I cannot write too strongly on this point," he said. "The sensible cotton people say they acquit us of any serious neglect . . . but that we must make roads."

The first of two large ventures taking shape in this political climate was a road-and-harbor project to link an inland region southeast of Bombay (Dharwar) to a new harbor site at Karwar, about 100 miles distant. Dharwar was of particular interest to the Manchester Cotton Company, a recently formed joint-stock company that aimed to ship new sources of raw Indian cotton to Lancashire, because the region grew the desirable long-staple cotton previously obtained from the American South. In October 1862 the British governor of Bombay, apprehensive that the complex project was being pushed so rapidly that proper engineering and cost estimates had not been made, nevertheless endorsed it: "The money value to India is very great, but its value to England cannot be told in money, and every thousand bales which we can get down to the sea coast before the season closes in June 1863 may not only save a score of weavers from starvation or crime but may play an important part in ensuring peace and prosperity to the manufacturing districts of more than one country in Europe."[19]

Even larger than the Dharwar–Karwar project, which cost a total of £225,000, was a grandiose plan to turn the 400-mile Godavari River into a major transportation artery linking the central Deccan plain with the eastern coast. The plan would send to the coast the cotton from Nagpur and Berar in central India, and this time the Cotton Supply Association was the principal Lancashire supporter. Work on this rock-studded, cholera-ridden river proved a vast money sink, however. By 1865, when the Lancashire lobby quietly gave up its campaign for Indian cotton and returned to peacetime American supplies, the Godavari scheme had cost £286,000 with little result. In July 1868 the first 200-mile stretch was opened, for limited traffic, and by the time the ill-conceived project was canceled in 1872 it had cost the grand sum of £750,000. Such terrific expenditures guaranteed that imperialism soaked up investments rather than generated profits. As it turned out, the Lancashire lobby threw its support behind the Great Indian Peninsula Railway that connected cotton-rich Nagpur with the western port of Bombay.

The PWD's leading role also stamped an imperial seal on technical education in India. The four principal engineering schools in India, founded between 1847 and 1866 at Roorkee, Calcutta, Madras, and Poona, had a heavy emphasis on civil engineering. The PWD was not only the source of many faculty at the engineering schools and of the examiners who approved their graduates but also far and away the leading employer of engineers in India. Indian courses were "unduly encumbered with subjects that are of little educational value for engineers, but which are possibly calculated to add to the immediate utility of the student in routine matters when he first goes on apprenticeship to the PWD," observed a witness before the Public Works Reorganization Committee in 1917. Another witness stated, "mechanical engineering has been greatly neglected."

The development of a well-rounded system of technical education in India was further hampered by the elite Royal Indian Engineering College, located, conveniently enough, twenty miles west of London at a country estate in Surrey called Cooper's Hill. It was founded in 1870 explicitly to prepare young, well-educated British gentlemen for supervisory engineering posts in India. Successful applicants had to pass a rigorous examination in mathematics and physical science; Latin, Greek, French, and German; the works of Shakespeare, Johnson, Scott, and Byron; and English history from 1688 to 1756. The college was permitted to enroll two "natives of India" each year "if there is room." Its founding president evidently was a worried soul, writing in a grim novella, "And what was there left to us to live for? Stripped of our colonies . . . India lost forever." At the PWD in 1886, natives of India accounted for just 86 of 1,015 engineers, although they filled many of the lower ("upper subordinate" and "lower subordinate") grades. Indian prospects for technical education improved somewhat with the closing of the flagrantly discriminatory Cooper's Hill in 1903 and the founding of the native-directed Bengal Technical Institute (1906) and Indian Institute of Science (1911). By the 1930s Indian students could gain degrees in electrical, mechanical, and metallurgical engineering in India.[20]

RAILWAY IMPERIALISM

Railroads in countries throughout Western Europe and North America were powerful agents of economic, political, and social change. Their immense capital requirements led to fundamental changes in the business structures of all those countries and in the financial markets that increasingly spanned them. Building and operating the railroads consumed large amounts of coal, iron, and steel, leading to rapid growth in heavy industries. Their ability to move goods cheaply led to the creation of national markets, while their ability to move troops rapidly strengthened the nation-states that possessed them (fig. 4.4). In the 1860s superior railway systems shaped military victories by railroad-rich Prussia over Austria and the northern US states over the southern Confederacy.

But in the imperial arenas, the dynamics of empire subtly but perceptibly altered railways and the changes they brought. Where imperial officials were essential in

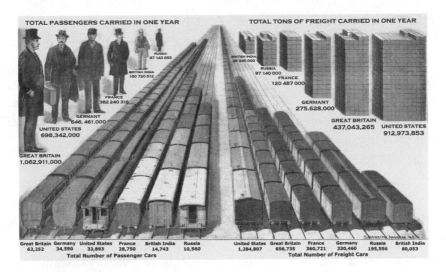

FIG. 4.4. WORLD LEADERS IN RAILWAYS, 1899

By the turn of the century, the Indian railway was the fifth largest in the world in passenger travel and sixth largest in freight.

"Railways of the World Compared," *Scientific American* (23 December 1899): 401. Courtesy of Illinois Institute of Technology Special Collections.

arranging financing, their military and administrative priorities shaped the timing, pace, and routes of colonial railroads. Colonial railroads also reflected the economic priorities of bankers in London and other financial centers who floated huge loans for their construction; nearly always these bankers preferred open "free trade" markets to closed high-tariff ones, strong central colonial governments to divided regional ones, and easily collected import-and-export taxes. For all these reasons, railway imperialism typically led toward political centralization, economic concentration, and extractive development. This is not to say that railway imperialists always got what they wanted. For both the colonies and the metropoles, railroads were above all a way of conducting "politics by other means" often involving contests between local and global imperatives and powers. A survey of railway imperialism in India, North America, and South Africa rounds out this chapter.

In India the political and military utilities of a wide-ranging railroad network were painfully obvious to the railroad promoters, since their thunder had been stolen by the prior construction of the telegraph network. Yet even as commands for troop movement could be sent down an iron wire at nearly the speed of light, the troops themselves required a less ethereal mode of transport. Railways constituted political power in such a sprawling and politically unsteady place as colonial India. "It is not," wrote one railway economist in 1845, "with any hope of inspiring the company of British merchants trading to India with an expensive sympathy for the social and moral advancement of their millions of native subjects that we urge the formation of a well-considered means of railway communication,—but as a necessary means of giving strength, efficiency, and compactness to their political rule in those territories."[21]

Imperial priorities informed the two British engineers who planned the pioneering lines, Rowland M. Stephenson and John Chapman. Stephenson came from a family long associated with Indian commercial and political affairs, with no evident relation to the railroad pioneer George Stephenson. After becoming a civil engineer in the 1830s, Stephenson promoted various Indian steam ventures to investors in London. Having seen the success of the Peninsular & Oriental's steamship venture (he briefly served as secretary for one of its rivals and later for the P&O itself), Stephenson journeyed to Calcutta. Writing in the *Englishman* of Calcutta in 1844, Stephenson proposed a 5,000-mile network consisting of six major lines. "The first consideration is as a military measure for the better security with less outlay of the entire territory," he wrote of his plan. "The second is a commercial point of view, in which the chief object is to provide the means of conveyance to the nearest shipping ports of the rich and varied productions of the country, and to transmit back manufactured goods of Great Britain, salt, etc., in exchange." Developing native Indian industry, which his plan would simultaneously deprive of homegrown raw materials and overwhelm with British manufactured goods, was assuredly not among his goals. In 1845 Stephenson joined the newly formed East Indian Railway Company as managing director. Stephenson's plan was given official sanction by Dalhousie's 1853 "Minute" on railroads, which set railroad policy for decades. John Chapman was the chief technical figure of the Great Indian Peninsula Railway, or GIP, formed also in 1845. Its projected line, originating at the port of Bombay, climbed steeply up the Western Ghats to the central Deccan plateau, a prime cotton-growing area as noted above. Three years later, Chapman observed that the Lancashire merchants thought of the GIP as "nothing more than an extension of their own line from Manchester to Liverpool."[22]

The first generation of Indian railways took form under a peculiar and still-controversial imperial arrangement. Under a precedent-setting 1849 contract with the East India Company (EIC, the statelike entity responsible for governing India until 1858), the pioneering East Indian Railway and the Great Indian Peninsula Railway turned over to the EIC their entire paid-in capital. As the railroads planned routes that met the EIC's criteria, including specific routings, single- or twin-tracking, and various engineering standards, the EIC disbursed the "allowed" expenditures to the respective railroad. The EIC leased generous swaths of land to the railroads without cost for ninety-nine years. At any time through the ninety-eighth year, the railroads could turn over their companies to the state and demand full compensation; in the ninety-ninth year, just possibly too late, the state could claim the right to take over the roads without compensation.

The controversial point was the guaranteed return to investors. The EIC, which held the railroads' capital in Calcutta, promised interest payments to investors of 5 percent. Operating profits up to 5 percent went to the EIC to offset its guarantee payments, while any profits above 5 percent were split equally between the EIC and the railroad or, if the operating revenues had completely covered the EIC's 5 percent payments, with no backlog from previous years, the profits above 5 percent went entirely to the railroad company. The guaranteed interest payments rested ultimately on the EIC's ability to collect money from Indian taxpayers. The rub was that through 1870 the roads consistently made average annual profits of only around 3 percent while the EIC-backed investors collecting 5 percent were overwhelmingly British. (In 1868 less than 1 percent of shareholders were Indians, 397 out of 50,000—understandably enough, since shares traded only in London.) The scheme thus transferred money from Indian taxpayers to British investors. The finance minister of India, William Massie (1863–69), saw the problem clearly: "All the money came from the English capitalist, and so long as he was guaranteed 5 per cent on the revenues of India, it was immaterial to him whether the funds that he lent were thrown into the Hooghly or converted into bricks and mortar."[23]

In fact, however, since nearly all of the locomotives, most of the rails, and even some of the coal were imported from Britain, fully two-fifths of the money raised in Britain was spent in Britain. The Indian railroads' high expenses were the result not so much of flagrant corruption as of the subtle but again perceptible way in which the dynamics of empire shaped their form. Imperial considerations most obviously structured the financing and routing of the Indian railways. A vision of empire also inspired the vast, overbuilt railway terminals, such as Bombay's Victoria Station. One can furthermore see empire inscribed in the railroads' technical details, including their track gauges, bridge construction, and locomotive designs.

In 1846 the British Parliament passed the Act for Regulating the Gauge of Railways, setting what would become the standard gauge in Europe and North

FIG. 4.5. GOKTEIK VIADUCT IN UPPER BURMA

The railway track is 825 feet above the level of the Chungzoune River, which flows through a tunnel beneath the bridge.

Frederick Talbot, *The Railway Conquest of the World* (London, Heinemann, 1911), following p. 256. Courtesy of Illinois Institute of Technology Special Collections.

America: 4 feet 8½ inches (or 1.435 meters). Nevertheless, the EIC's Court of Directors set the gauge of India's railroads at 5 feet 5 inches (or 1.676 meters) and furthermore decided in 1847 that all tunnels, bridges, and excavations must be made wide enough for double tracking. The mandate for a wide gauge and double tracking inflated construction costs: all bridge superstructures had to be extra wide, while all curves had to be extra spacious. The Great Indian Peninsula Railway's double-track construction up the Western Ghats, a steep mountain range rising to the cotton-rich central Deccan plain, required an ungainly arrangement of reversing stations, in addition to numerous tunnels, viaducts, and bridges to gain the needed elevation. Some of the vast sums spent on bridges can be fairly traced to India's wide, deep, and at times fast-flowing rivers and their fearsome monsoon flooding (figs. 4.5 and 4.6). Adding

FIG. 4.6. BRIDGING THE GANGES RIVER AT ALLAHABAD

In building the Curzon Bridge to cross the mighty Ganges River at Allahabad in northern India, native laborers moved 50 million cubic feet of earth to narrow the river from its high-water clay riverbanks (3 miles) to the width of its low-water channel (about 1¼ miles). The railway bridge's span at 3,000 feet was opened in 1905.

Frederick Talbot, *The Railway Conquest of the World* (London: Heinemann, 1911), following p. 254. Courtesy of Illinois Institute of Technology Special Collections. See *Scientific American* (26 September 1908): 204–206 at https://ia801607. us.archive.org/1/items/scientific-american-1908-09-26/scientific-american-v99-n13-1908-09-26.pdf.

to the costs, however, was the British engineers' preference for expensive designs: no rough-and-ready timber trestles like those American railroad engineers were building. Instead, Indian bridges mostly conformed to British designs, built of wrought-iron trusses over masonry piers. Numbered bridge parts sent from Britain were riveted together into spans by Indian craftsmen and the spans placed by elephant-powered hoists. In 1862 the second largest bridge in the world was opened across the Sone River near Delhi, and it cost the astounding sum of £430,000.

British-constructed Indian locomotives were also built for the ages and were correspondingly expensive. The most common class of locomotive in India, the Scindia, was built on a rigid frame with six sets of driving wheels (known as a 0-6-0 configuration) and featured copper fireboxes, copper or brass boiler tubes, forged valves, and inside cylinders with cranked axles. These were the Rolls-Royces of locomotives. By contrast, North American locomotives of this era had steel fireboxes and boiler tubes, cast-iron valves, and external cylinders, as well as leading "bogie" wheels that improved steering on sharp turns (4-4-0 or 2-8-0 configuration). Although India's railroad shops constructed approximately 700 locomotives before independence in 1947, the vast majority (80%) were built in Britain, supplemented by imports from America and Germany. During these years Indian railroads bought fully one-fifth of the British locomotive industry's total output.[24]

Railway construction under the first guarantee system picked up pace after the Mutiny of 1857–58, when there were just 200 miles of rail line. In 1870 the Indian colonial government, reeling under the budget-breaking costs of the 5,000 miles of privately constructed but publicly financed railroads, embarked on a phase of state-built railroads featuring a narrow (meter-wide) gauge. The Indian colonial government built 2,200 miles of these roads at half the cost per mile of the guaranteed railroads. But in 1879 the secretary of state for India mandated that the Indian government only build railroads to the strategically sensitive northern (Afghan) frontier. Private companies, under a new guarantee scheme negotiated with the government in 1874, took up the sharp boom of railroad building across the rest of the country. At the turn of the century India had nearly 25,000 miles of railroad track. India's railway mileage surpassed Britain's in 1895, France's in 1900, and Germany's in 1920, by which point only the United States, Soviet Union, and Canada had more mileage. Unfortunately for the Indian treasury, the roads built under the second guarantee scheme were also money pits (only the East Indian Railway's trunk line from Calcutta to Delhi consistently made money). Guarantee payments to the railroads between 1850 and 1900 totaled a whopping £50 million.[25] By the 1920s, Indian railroads, by then run-down for lack of investment, became a prime target of Indian nationalists agitating for the end of British rule.

Compared with India, railway imperialism in North America was a more complicated venture, not least because two imperial powers, Britain and the United States, had various claims on the continent. Railway building in the eastern half of the United States reflected commercial and industrial impulses. Merchants and traders in Baltimore and Philadelphia, for example, backed two huge railroad-building schemes to capture a share of the agricultural bounty that was flowing east via the Erie Canal to New York City. By 1860 a network of railroads from the Atlantic Ocean to the Mississippi River had created an industrial heartland that extended to Chicago. The United States, with 30,000 miles, had more than three times the railroad track of second-place Britain and nearly five times the mileage of third-place Germany.

Construction of the transcontinental railroads from the Mississippi to the Pacific Ocean during the next four decades (1860s–1900) boosted the country's railroad trackage to 260,000 miles. The defining governmental action—the Pacific

Railroad Act of 1862—granted huge blocks of land to the transcontinental railroads for building their lines and for developing traffic for their lines by selling land to settlers. All but one of the transcontinental lines—including the Union Pacific, Central Pacific, Northern Pacific, Kansas Pacific, Southern Pacific, and Atchison, Topeka & Santa Fe railroads—were beneficiaries of these land grants. The Illinois Central had pioneered the federal land grant, receiving 2.6 million acres in 1850 for its rail line south to New Orleans.

The strange economics of empire also came into play in North America. The US government faced a heavy financial burden in mobilizing and provisioning sufficient army units to safeguard settlers and railroad construction crews from the Native Americans who were determined not to give up their buffalo hunting grounds without a fight. In 1867 the *Omaha Weekly Herald* claimed that it cost the "large sum" of $500,000 for each Indian killed in the recurrent prairie battles. The same year General William T. Sherman stated: "The more we can kill this year, the less will have to be killed in the next war, for the more I see of these Indians the more convinced I am that they all have to be killed or maintained as a species of paupers." Railway promoters pointed out that the increased military mobility brought by the railroad, cutting the needed number of military units, would dramatically reduce the high cost of projecting power across these many sparsely settled miles. In this respect, the railroads in British India and the American West have more than a casual similarity. "The construction of the road virtually solved the Indian problem," stated one American railroad official two years prior to the infamous massacre of 300 Lakota Sioux at Wounded Knee, South Dakota, in 1890.[26]

Citizens of British North America reacted with some alarm to these territorial developments. At the time, Canada was not a unified nation but a set of independent provinces. Yet not even the fear of being annexed to the United States, as was nearly half of Mexico in the 1840s, united them. Merchants and traders along the St. Lawrence–Great Lakes canal system, a huge project of the 1840s, looked naturally to the shipping ports of the south; some even favored joining the United States. A railroad boom in the 1850s resulted in 2,000 miles of disconnected lines whose operating deficits emptied colonial Canadian treasuries and whose north-south orientation drained off Canadian products to the south. Would-be railway imperialists still lacked the necessary east-west lines that might bring economic and political cohesion to the provinces. Worse, during the Civil War the United States started a trade war with the Canadian provinces and made threats, deemed serious by many Canadians, to invade their western lands. Equally worrisome, US economic domination might formally detach British Columbia from the British Empire and, on the other side of the continent, informally control the maritime provinces.

A generous program of imperial railway subsidies was the glue that fixed the slippery provinces into place. In the 1860s a series of labyrinthine negotiations between Canadian colonial officials, British imperial officials, and London financiers arrived at this formula: London financiers would support large railway loans if the provinces were politically united; the independent provinces would agree to confederation if the Colonial Office sweetened the deal with government guarantees for railway construction (valuable not least for the patronage jobs). So, the Colonial Office in London duly arranged government guarantees for the railway loans. Thus was the Dominion of Canada created in 1867 as a federation of Canada West and Canada East with the maritime provinces of New Brunswick and Nova Scotia. Railroads figured explicitly. The maritime provinces made their assent to the confederation agreement conditional on the building of an intercolonial railroad in addition to the Halifax–Quebec railroad, already planned. Furthermore, expansion-minded citizens of Canada West received promises of amiably settling the Hudson Bay Company's preemptive claim on western lands. By 1874 the *British* government had made guaranteed loans and grants totaling £8 million for the intercolonial lines and transcontinental Canadian Pacific Railway. Of the Canadian Pacific the first premier of the Canadian confederation commented, "Until that road is built to British Columbia and the Pacific, this Dominion is a mere geographical expression, and not one great Dominion; until bound by the iron link, as we have bound Nova Scotia and New Brunswick with the Intercolonial Railway, we are not a Dominion in Fact."[27]

Railway imperialism in Mexico affords a glimpse of what might have happened to Canada absent the countervailing power of imperial Britain. The Mexican railroad system took shape under the long rule of the autocratic Porfirio Díaz (1876–1911). As early as the 1870s the Southern Pacific and Santa Fe railroads, then building extensive lines in the southwestern United States, began planning routes south into Mexico. Mexico at the time consisted of fourteen provinces whose disorganized finances left no hope of gaining external financing from London or other international money centers. With few options in sight, and a hope that railroads might bring "order and progress" to Mexico, Díaz gave concessions to the US railroads to build five lines totaling 2,500 miles of track. Something like free-trade imperialism followed. In 1881 the US secretary of state informed the Mexican government that it would need to get rid of the "local complicated . . . tariff regulations which obtain between the different Mexican States themselves" and sign a reciprocal free-trade agreement with the United States. Díaz prevailed upon the Mexican Congress to ratify the free-trade pact, as the US Senate had done, but the agreement did not go into effect owing to a quirk of US domestic politics. US railroad and mining promoters flooded south all the same. Mexico, as one railroad promoter effused in 1884, was "one magnificent but undeveloped mine—our India in commercial importance."[28]

Trade between Mexico and the United States increased sevenfold from 1880 to 1910. Total US investment in Mexico soared to over $1 billion, an amount greater than all its other foreign investment combined and indeed greater than Mexico's own internal investment. Fully 62 percent of US investment was in railroads; 24 percent was in mines.[29] By 1911 US firms owned or controlled most of the 15,000 miles of railroad lines; three-quarters of mining and smelting concerns processing silver, zinc, lead, and copper; and more than half of oil lands, ranches, and plantations. Four great trunk lines shipped Mexican products north to the border. But the extractive mining boom brought wrenching social and political changes. The blatant US domination of railroads inflamed the sensibilities of Mexican nationalists. The Díaz government nationalized two-thirds of the country's railroads in 1910, but the aging dictator was overthrown the next year. The legacy of railway imperialism in Mexico was not the "order and progress" that Díaz had aimed for but a confusing period of civil strife (1911–17) and a transportation system designed for an extractive economy.

In South Africa railroads at first had some of the centralizing and integrating effects that we have noted in India, Canada, the United States, and Mexico, but railway imperialism there ran headlong into a countervailing force, "railway republicanism." The result was not railroad-made confederation along Canadian lines, as many in Britain hoped, but a political fracturing of the region that ignited the second Anglo-Boer War (1899–1902). Southern Africa even before the railway age was divided into four distinct political units: two acknowledged colonies of Britain (the Cape Colony, at the southern-most tip of the continent, and Natal, up the eastern coast) and two Boer republics (the inland Orange Free State, and the landlocked Transvaal republic) over which Britain from the mid-1880s claimed suzerainty. Britain understood this subcolonial status as a step to full integration into the British Empire, whereas the fiercely independent Boers, descendants of seventeenth-century Dutch settlers who had trekked inland earlier in the nineteenth century to escape British rule, saw it as one step from complete independence.

Imperial concerns mounted with the waves of British citizens brought to the region by the discovery of diamonds (1867) at Kimberley in eastern Orange Free State (fig. 4.7) and gold (1886) in the central Transvaal. One of the British newcomers

FIG. 4.7. KIMBERLEY DIAMOND MINE, SOUTH AFRICA, CA. 1880

Southern Africa's premier diamond mine, with its 31-square-foot claims, was covered by hundreds of wire ropes connecting each claim to the unexcavated "reef" visible in the background. "Each claim was to all intents and purposes a separate mine."

Scientific American (27 January 1900): 56. Courtesy of Illinois Institute of Technology Special Collections.

was Cecil Rhodes who made his fortune in the Kimberley fields in the 1870s, formed the De Beers Mining Company in the 1880s (a successor of which set worldwide raw diamond prices throughout the twentieth century), and secured a wide-ranging royal charter for his British South Africa Company in 1889. The story goes that Rhodes, his hands on a map of Africa, had gestured: "This is my dream, all English."[30] Rhodes, along with other centralizing imperialists, hoped to form southern Africa's diverse linguistic and ethnic groups into a single, unified colony dependent on Britain. Rhodes used the promise of building railroads with his chartered company to secure the political backing of the Cape Colony's Afrikaner majority. As the Cape's prime minister from 1890 to 1896 he led the railway-building campaign north across the Orange Free State and into the Transvaal, not least by arranging financing in London. In getting his rails into the Transvaal Rhodes hoped to preempt that republic's plans to revive a defunct railway to the Portuguese port city of Lorenço Marques (now Mobutu, Mozambique). The early railway campaigns brought a degree of cooperation, through cross-traffic railroad agreements and a customs union, between the Cape Colony and Orange Free State, on the one hand, and the Natal Colony and Transvaal on the other.

Ironically, railroads thwarted the imperial dream in South Africa. Rhodes found his match in the "railway republican" Paul Kruger, president of the Transvaal, and political leader of the region's Afrikaner Boers. Nature had dealt Kruger's Transvaal the supreme trump card: the massive Witwatersrand gold reef near Johannesburg, discovered in 1891 and about which would pivot much of the region's turbulent twentieth-century history. Mining the Rand's deep gold veins required heavy machinery and led to large-scale industrial development around Johannesburg. Accordingly, three of the region's four railroads had their terminals there.

Although Rhodes had hoped that his Cape Colony railroad's extension to the Transvaal would give him leverage over the inland republic's leader, quite the opposite happened. Kruger contested the imperialist's plan and gained the upper hand by appealing directly to London financiers himself. With a £3 million loan from the Rothschilds in 1892, Kruger completed the Transvaal's independent rail link to the port at Lorenço Marques.[31] From then on, Kruger could bestow as he saw fit the Rand's lucrative traffic among three railroads (his own, Natal's, and the Cape's). The tremendous fixed investment of these railroads, along with the light traffic elsewhere, gave whoever controlled the Rand traffic tremendous clout. Having failed to achieve anything like political union by railway imperialism, Rhodes tried it the old-fashioned way—militarily. In the Drifts Crisis of 1895, sparked by a dispute over railway freight charges around Johannesburg, the Cape Colony called for British military intervention against the Transvaal. With this crisis barely settled, Rhodes launched the Jameson Raid, an ill-conceived military invasion aimed at overthrowing Kruger

By this time Britain may well have wished to wash its hands of the region's bitter disputes, but there remained the matter of protecting its investors' £28 million in colonial railway debts. In the face of Kruger's open hostility toward its citizens, Britain launched the Second Boer War (1899–1902), during which its army crushed the independence-minded Boer republics. The same Transvaal railway that carried gold out to the ocean also carried the British army in to the Rand and carried Kruger away to exile. The region's railroad problem was a high priority for the British high commissioner charged with rebuilding the region after the war: "On the manner and spirit in which the peoples and Parliaments of South Africa handle this railway question depends the eventual answer to the larger question, whether South Africa is to contain one great and united people . . . or a congeries of separate and constantly quarreling little states," he wrote in 1905.[32] In his 1907 blueprint for federating the South African colonies, fully two-thirds of his 90,000-word text dealt with railroad issues. Railway union, he saw, was imperative, since independent provincial rail-

FIG. 4.8. BUILDING THE CAPE TOWN TO CAIRO RAILWAY

Native laborers at work on Cecil Rhodes' grand imperial dream, a railroad that would connect the southern tip of Africa with a Mediterranean port. Here, workers set a continental record, laying 5¾ miles of track in ten hours.

Frederick Talbot, *The Railway Conquest of the World* (London: Heinemann, 1911), following p. 144. Courtesy of Illinois Institute of Technology Special Collections.

roads would continually be instruments for sectional strife. The South Africa Act of 1909 created the region-wide South African Railways and Harbours Administration and helped unify the shaky Republic of South Africa, formed the next year. In 1914 the center of a unified railway administration was fittingly enough relocated to Johannesburg, the seat of railway power.

Although his political career ended with the failed Jameson Raid, Rhodes in his remaining years turned his considerable promotional abilities toward that most grandiose of all imperial schemes, the Cape Town to Cairo Railway, which aimed to stitch together a patchwork of mining and settlement ventures northward through Africa (fig. 4.8). Rhodes' British South Africa Company had no realistic chance of laying rails all the way to Egypt, not least because the scheme ran square into the determination of the Belgian Congo to control the rich Katanga copper belt in central Africa and of German Southeast Africa to dominate the territory from there east to the Indian Ocean. Rhodes' scheme did hasten the European settlement of the land-locked Rhodesias. The railroads just completed through Southern Rhodesia (now Zimbabwe) prevented mass starvation when *rinderpest* decimated the region's draft animals in the late 1890s, while in 1911 Northern Rhodesia (now Zambia) was created through and circumscribed by its mining and railroad activities (fig. 4.9). Perhaps the farthest-reaching achievement of the Cape-to-Cairo scheme was in durably linking the midcontinent copper belt with South Africa well into the postcolonial decades.[33]

The legacies of imperialism remain fraught with controversy in our own post-colonial time. Indeed the arguments have sharpened with the debate on globaliza-tion, since many of its critics denounce globalization as little more than an updated imperialism (discussed in chapter 8). Few today have sympathy for the "civilizing mission" of the imperialist era, with the assumption that selling opium or string-ing telegraphs or building railroads would bring the unalloyed benefits of Western civilization to the native populations of Asia, Africa, or South America. From our present-day sensibilities we see only too clearly the gunboats, rifle cartridges, and machine guns that these ventures entailed. The title of Jared Diamond's prize-win-ning *Guns, Germs, and Steel* (1997) gives a three-word explanation for the rise of Western global power through technology.

Imperialism in the nineteenth century left a long shadow over the globe. We have seen how key transport nodes like the Suez Canal (as well as the Panama Canal completed in 1914) fell captive to imperial powers. The fraught history of Hong Kong can be traced to the first opium war in the 1840s after which Britain annexed this compact but valuable port and then, years later, signed a ninety-nine-year lease with China; in the interim, China itself went through political upheavals so that today Hong Kong is at the center of battles over the "authoritarian internet" (see chapter 10). In the early twentieth century, Great Britain and the United States carved up the Middle East's immense but untapped oil reserves, tying that region of the world to geopolitical instabilities and systemic risk (see chapter 9).

Both before and since the anticolonial independence movements of the late 1940s through 1960s, nationalists in Asia, Latin America, and Africa condemned the schemes that brought imperialist-dominated "development" to their countries. "It is difficult to measure the harm that Manchester has done to us. It is due to Manchester [and cotton machinery] that Indian handicraft has all but disappeared,"

FIG. 4.9. SPANNING THE ZAMBESI RIVER AT VICTORIA FALLS

The 500-foot span over the Zambesi River just below the stunning Victoria Falls. Trains passed across the gorge 420 feet above low water. The net beneath the bridge was "to catch falling tools and workmen."

Frederick Talbot, *The Railway Conquest of the World* (London, 1911), following p. 144. Courtesy of Illinois Institute of Technology Special Collections.

wrote Mohandas Gandhi.[34] During the long decades of the Cold War (1947–89), the superpowers imposed a marked preference for capital-intensive, centralized, extractive industry that tied their clients' economies and politics to the superpowers. Even today one can discern a shadow of the imperialist era in railroad maps of North America (look carefully at Canada, the western United States, and Mexico), in the prestige structure of technical education, and in the policy preferences of the mainline development agencies in the United States and Europe.

Even for the dominant countries, imperialism was a venture with mixed economic results. In the aggregate Britain as a country did not profit by imperialism. While many of its traders and entrepreneurs, as well as the technological visionaries who tapped the imperial impulse, made their individual fortunes, the profits of the imperial economy were simply not large enough to pay for the heavy expenses of sending imperial military forces overseas, maintaining the imperial bureaucracy, and funding the high-priced imperial technologies. We now have insight into why imperialism did not make money. Profit was simply not the point of imperial technologies: the expensive steam vessels, the goldbricked locomotives, the double-tracked wide-gauge railways, the far-flung telegraph and cable networks.

In this respect, we can see that imperialism was not merely a continuation of the eras of commerce or industry; rather, to a significant extent, imperialism competed with and in some circumstances displaced commerce and industry as the primary focus of technologists. By creating a captive overseas market for British steamships, machine tools, locomotives, steel, and cotton textiles, imperialism insulated British industrialists in these sectors from upstart rivals and, in the long run, may even have hastened their decline in worldwide competitiveness.[35] Is it only a coincidence that Britain, a leader in the eras of industry and imperialism, was distinctly a follower behind Germany and the United States in the subsequent science-and-systems era? At the least, we must allow that the imperialist era had a distinct vision for social and economic development, dominated by the Western countries, and distinctive goals for technologies.

CHAPTER 5

1870 – 1930

Science and Systems

*I*n the half-century after 1870 a "second" industrial revolution, built from science-based technologies, altered not merely how goods were produced and consumed but also how industrial society evolved. The new industries included synthetic chemicals, electric light and power, refrigeration, mechanized food production, automobiles, and many others. By transforming curiosities of the laboratory into widespread consumer products, through product innovation and energetic advertising, science-based industry helped create a mass consumer society. A related development was the rise of corporate industry and its new relationships with research universities and government agencies.

The economic imperatives of science and systems, not least the rise of corporate industry, decisively changed the character of industrial society. In surveying the strains and instabilities of the first industrial revolution, Karl Marx and Friedrich Engels prophesied that capitalism's contradictions would bring about its destruction, that (in their memorable phrase) the bourgeoisie would produce its own grave-diggers. The recurrent economic panics—no fewer than nine major depressions between 1819 and 1929 in the United States, almost always linked to the British and European economies—lent force to their diagnosis that industrial capitalism was dangerous to society and doomed to failure. But the emergence of a technologically stabilized "organized capitalism" escaped their view. The new organized capitalism resulted in distinct economic and cultural dynamics for society with technical change at its core.

Technologists in this era, especially those working in the science-based industries, increasingly focused on improving, stabilizing, and entrenching existing systems rather than inventing entirely new ones. Their system-stabilizing efforts—buttressed by patent laws, corporate ownership, industrial research laboratories, and engineering education—transformed industrial society. In these same decades *technology* took on its present-day meaning as a *set* of devices, a complex of industry, and an abstract society-changing force in itself. Technology and culture were, once again, entangled and transformed.[1]

British industry largely failed in the science-based fields, even though British scientists discovered much of the pertinent science. As chapter 3 recounted, London beer brewers pioneered factory scale, steam power, mechanization, and temperature control. "A brewery," wrote one industry expert in 1838, "may appropriately be termed a brewing chemical laboratory."[2] In electricity, it was the English chemist Humphrey Davy who first experimented with incandescent and arc lighting (1808) and the English physicist Michael Faraday who discovered electromagnetic induction and first built a dynamo, or direct-current generator (1831). In synthetic chemicals, the English chemist William Perkin first synthesized an artificial dye (1856). Yet technical and commercial leadership in each of these industries slipped through British hands and landed in German and American ones.

THE BUSINESS OF SCIENCE

The first science-based industry started, improbably enough, with a sticky black tar derived from coal. By the time Charles Dickens set *Hard Times* (1854) in the fictional "Coketown," Britain was mining some 50 million tons of coal each year. Coke, a product of baking coal in the absence of oxygen, had been used for decades as a high-grade industrial fuel for smelting iron and roasting barley malt, while the resulting coal gas found ready use in the lighting of factories (from the 1790s) and city streets, including starting with London (1807) and Baltimore (1817). The baking of coal into valuable gas and coke in London alone employed 1,700 persons by mid-century. Sticky black coal tar was the unwanted by-product. One could say that the synthetic chemical industry was born as an ancillary industry.

Fittingly enough, it was in coal-fouled London that coal tar attracted the attention of an expatriate German chemist. August von Hofmann set his laboratory students at the Royal College of Chemistry the task of finding something useful in all that surplus coal tar. One student successfully made benzene and toluene but, tragically, burned himself to death when his distillation still caught fire. William Perkin (1838–1907) was luckier. Hoping to work up coal tar into quinine (an antimalarial agent of great relevance to empire) Perkin instead made one dark blob after another. One day in 1856, his chemical sink turned bright purple. Upon further testing he found that the purple crystals could turn cotton cloth purple, too. His purple dye became popular in Paris as *mauve*, and was the first in a famous family of aniline dyes.

With this success, Perkin at age eighteen quit the Royal College to set up a dye factory with his father and brother. Textile industrialists soon clamored for more dyes in eye-catching colors and with fabric-bonding fastness. Dye factories sprang up in Manchester, in Lyons, France, and in Ludwigshaven, Germany. Like Perkin, their chemists found new dyes by an empirical, or cut-and-try, approach. Promising compounds, like the coal tar–derived aniline, were treated with likely reagents, among which were arsenic compounds that caused major health problems.[3] French chemists achieved a striking dye now known as the color "magenta" (and the "m" in CMYK color printing) and a cluster of aniline blues, while Hofmann in London worked up a range of purples. The chemical structure of these early dyes was unknown at the time. Before long, German chemists—based in universities and with close ties to industry—discovered their chemical structures and set the stage for a science-based industry.

Perhaps distracted by the cotton shortages of the 1860s (see chapter 4), Britain's textile industrialists fumbled away a major ancillary industry. Between 1861 and 1867 German industrialists founded four of the five most important German chemical companies: Bayer, Hoechst, Badische Analin- & Soda-Fabrik (BASF), and Aktien-Gesellschaft für Anilinfabrik (AGFA). The fifth company, Casella, already in existence as a vegetable-dye dealer, entered the aniline dye field in 1867.[4] These firms, at first ancillary to the textile industry, soon created and then dominated the world-wide synthetic chemicals industry. Even today consumers of aspirin, recording tape, and photographic film may recognize their names. In the 1860s German chemists who had gone abroad flocked back home. Hofmann in 1865 took up a prestigious chair of organic chemistry at the University of Berlin. Among his returning students were such up-and-coming leaders of the German chemical industry as Carl Martius, Otto Witt, and Heinrich Caro, who became, respectively, a founder of AGFA, chief chemist of Casella, and chief chemist and director of BASF. Losing so many talented German chemists was a severe blow to England's struggling dye industry.

Germany's science establishment, although vigorous, was not originally designed to support science-based industry. German state universities trained young men for the traditional professions, the government bureaucracy, and teaching in classical secondary schools. The universities emphasized "pure" research, leaving such practical topics as pharmacy, dentistry, agriculture, and technology to lower-prestige colleges. Technical colleges, intended originally to train engineers and technicians, gained the privilege of awarding doctorate degrees only in 1899. By 1905 there were ten such technical universities in Germany, located in Berlin, Munich, Karlsruhe, and other industrial centers. A third component to the German science system was a network of state-sponsored research institutes. These included institutes for physics and technology (established in 1887), agriculture and forestry (1905), biology (1905), chemistry (1911), and coal (1912).

Another major hire by a German university in the 1860s was August Kekulé. Kekulé, like Hofmann, had gone abroad to Belgium after his training with the famed chemist Justus Liebig at the University of Giessen. Kekulé created modern structural organic chemistry by theorizing that each carbon atom forms four single bonds and that benzene is a six-sided ring. He proposed these theories in 1857 and 1865, respectively. Kekulé's structural insights contrast markedly with Perkin's empirical method. Instead of "tinkering with aniline," chemists with Kekulé's concepts were now "looking for the 'mother-substance,'" or skeleton, of the dyestuff's molecule.[5]

Kekulé's concepts helped chemists synthesize such naturally occurring dyes as alizarin and indigo, but the real payoff came with the creation of an entirely new class of chemicals. Preparing such commercial blockbusters as aniline yellow and Bismarck brown became the objective of a fiercely competitive science race. Chemists in several countries worked furiously on a compound believed to be intermediate between those two colors. In 1877 Hofmann publicly announced the general structure of this dye and proposed a vast new class of "azo" dyes by pointing to the "coupling reaction" that produced their distinctive chemical structure (two benzene rings linked up by a pair of nitrogen atoms). Each of the twin benzene rings could be decorated with numerous chemical additions, and all such dyes had bright colors. BASF's Caro saw it as "an endless combination game." Exploiting the estimated 100 million possibilities would require a new type of scientific work, however, which he termed "scientific mass-labor" (*wissenschaftliche Massenarbeit*). Inventive activity in the industrial research laboratories set up to conduct this new type of chemical work was no longer, as it had been for Perkin, a matter of individual luck and inspired tinkering. As Caro saw it, now chemical invention was "construction bound to rules." The results were impressive.[6] By 1897 there were 4,000 German chemists *outside* the universities. Together with their colleagues in universities, they created the scientific field of modern organic chemistry. The number of known structural formulas for carbon-containing compounds more than tripled (to 74,000) during the sixteen years before 1899 and would double again by 1910.[7]

The patent laws in Germany, by erecting a legal shield around the azo-dye field and preserving the scientific mass-labor, reinforced a system-stabilizing mode of innovation. Hofmann's public disclosure of the azo process in January 1877, six months before a major new patent law came into effect, was a practical argument in favor of public disclosure and patent protection. From the viewpoint of the chemical companies the 1877 law had a severe shortcoming, however, in that the law explicitly did not cover new chemical substances. Chemical "inventions" could be patented only if they concerned "a particular process" for manufacturing them. So it seemed doubtful that each of the individual dyes resulting from the innumerable azo-coupling reactions could be patented. In this climate, lawyers typically drafted azo-dye patents to cover a broad class of related substances, perhaps 200 or more, to bolster the patent's legal standing. A revised patent law in 1887 required patentees to hand over physical samples of *each* of the numerous substances claimed in the patent. The 1887 law obviously hampered individual inventors and smaller firms, which rarely had the scientific mass-labor to prepare hundreds of samples and accordingly favored companies that did have such staffing.

Two years later, in the precedent-setting "Congo red" legal decision of 1889, Germany's highest court forcefully endorsed the emerging paradigm of system stability. Two chemical giants, AGFA and Bayer, were suing a smaller rival for infringing on their patent. The smaller firm countered with a claim denying the validity of the Congo red patent itself. To support its claim, it called as an expert witness BASF's Caro, who testified that his laboratory assistant needed to see only the patent's *title* to prepare the dye. The court was swayed, however, by an impassioned speech by Bayer's Carl Duisberg, a rising star in the industry. If the court accepted Caro's line of reasoning, argued Duisberg, it would "deliver the death-blow to the whole chemistry of azo dyes and thereby [bring] most of the patents of [Caro's] own firm to the brink of the abyss."[8] Duisberg's argument carried the day. Under a freshly minted legal doctrine of "new technical effect," the court sided with the two chemical giants and upheld the dubious Congo red patent. The court admitted the patent lacked inventiveness but found that it had significant commercial value, and that was enough. The new legal doctrine saved the scientific mass-labor, entrenching the system-stabilizing mode of innovation. "Mass production methods which dominate modern economic life have also penetrated experimental science," the chemist Emil Fischer stated in his Nobel Prize lecture in 1902. "Consequently the progress of science today is not so much determined by brilliant achievements of individual workers, but rather by the planned collaboration of many observers."[9] Duisberg put the same point more succinctly: "Nowhere any trace of a flash of genius."[10]

Patent statistics provide a ready measure of the German chemical industry's worldwide dominance. Germany, already holding 50 percent of the world market for dyes in the 1870s, achieved a full 88 percent by 1913. In 1907 Germany's only rivals in dyestuffs, England and Switzerland, together, managed to register just 35 dye patents while Germany registered 134—at the *British* patent office. At the German patent office the count was even more lopsided, 14 to 285.[11] Owing to its strong

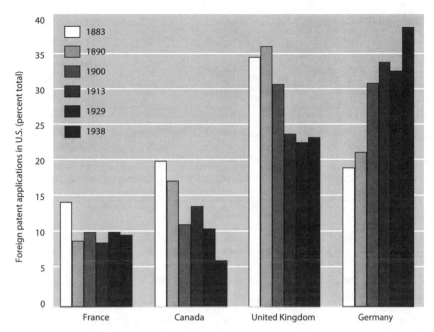

FIG. 5.1. FOREIGN APPLICATIONS FOR US PATENTS, 1883–1938

After these four countries, no other received more than 4 percent of the patents granted during these years.

Data from Keith Pavitt and Luc Soete, "International Differences in Economic Growth and the International Location of Innovation," in Herbert Giersch, ed., *Emerging Technologies* (Tübingen: Mohr, 1982), 109.

positions in chemicals and electricity, Germany's share of total foreign patents in the United States surpassed France's in 1883 and pulled ahead of Canada's in 1890 and England's by 1900. In 1938 Germany's US patents equaled those of the other three countries combined (fig. 5.1).

The synthetic-chemical empire led to the German chemical industry's direct involvement with the First World War and its appalling entanglement in the second. In World War I, popularly known as the chemist's war, chemists were directly involved in poison gas manufacture. Chemistry also shaped the trench warfare of the Western Front and the economic collapse of Germany that ended the war. The outbreak of war in the summer of 1914 cut off Germany from its imported sources of nitrates, required for agricultural fertilizers and military explosives alike. Germany's on-hand stock of nitrates was enough only to supply the army with explosives until June 1915. BASF kept the army in the field beyond this date with its high-pressure reactor vessels that chemically "fixed" atmospheric nitrogen (as ammonia), which could then be made into explosives or fertilizer. Of necessity the German army planned its offensives to match synthetic gunpowder production. For both sides, the mass production of explosives was a material precondition to keeping the machine guns firing, on which trench warfare grimly depended.

Chemistry also featured prominently in the attempt to break the stalemate of trench warfare. On 22 April 1915 German soldiers attacking the village of Ypres, in Flanders, opened 6,000 cylinders of chlorine (a familiar product BASF had supplied to textile bleachers for decades) and watched a yellow-green toxic cloud scatter the French defenders. Soon chemical factories—on both sides—were turning out quantities of chlorine and other poisonous gases, as well as the blister-inducing mustard gas. As one poet-soldier recorded the experience, "Then smartly, poison hit us in the face. . . . / Dim, through the misty panes and heavy light, / As under a dark sea, I saw him drowning. / In all my dreams, before my helpless sight / He lunges at me, guttering, choking, drowning."[12]

Horrific as the gas clouds, gas shells, and gas mortars were—and poison gas caused at least 500,000 casualties—chemical warfare was not a breakthrough "winning weapon," because soldiers' defensive measures (such as the poet's "misty panes") were surprisingly effective, if cumbersome and uncomfortable. Only civilians, animals, and the Italian army faced poison gas unprotected. In just one instance, late in the conflict, did gas affect the course of war. After its failed offensive in spring 1918, the German army used poison gas to cover an orderly retreat.[13] At the time, Germany still had plenty of poison gas, but it had run out of clothes, rubber, fuel, and food, a consequence of sending all fixed-nitrogen production into munitions while starving the agricultural sector.

The entanglement of the German chemical industry with the Nazis' Third Reich also has much to do with the system-stabilizing innovation and the corporate and political forms required for its perpetuation. By 1916, Germany's leading chemical companies, anticipating the end of the war, were already seeking a method of halting the system-destabilizing price competition that had plagued the industry before the war. That year the six leading concerns transformed their two trade associations (Dreibund and Dreiverband) into a "community of interest," together with two other independent firms. The eight companies pooled their profits and coordinated their activities in acquiring new factories, raising capital, and exchanging knowledge (fig. 5.2). Later, in the tumult of the 1920s, these eight companies in effect merged into one single corporation, I.G. Farben (1925–45). By 1930 the combine's 100 or so factories and mines employed about 120,000 people and accounted for 100 percent of German dyes, nearly all its explosives, 90 percent of its mineral acids, around 75

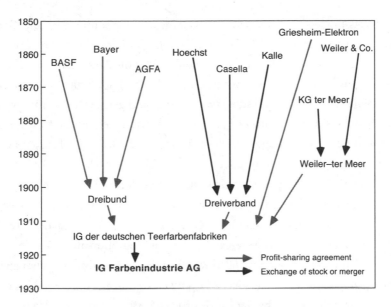

FIG. 5.2. CONSOLIDATION OF THE GERMAN CHEMICAL INDUSTRY, 1860–1925

Data from L. F. Haber, *The Chemical Industry during the Nineteenth Century* (Oxford: Clarendon Press, 1958), 128–36, 170–80; L. F. Haber, *The Chemical Industry 1900–1930* (Oxford: Clarendon Press, 1971), 121–28.

percent of its fixed nitrogen, 40 percent of its pharmaceuticals, and 30 percent of its synthetic rayon. Farben also had significant holdings in coal, banking, oil, and metals. With these weighty investments, Farben's executives felt they had little choice but to conform with Hitler's mad agenda after he seized power in 1933. Not Nazis themselves—one-fourth of the top-level supervisory board were Jews, until the Aryanization laws of 1938—they nevertheless became complicit in the murderous regime. During World War II Farben made synthetic explosives, synthetic fuels, and synthetic rubber for the National Socialist war effort. Many have never forgiven its provision to the Nazis of Zyklon B (the death-camp gas), its nerve-gas experiments on camp inmates, and its use of up to 35,000 slave laborers to build a synthetic rubber complex at Auschwitz.[14]

FLASHES OF GENIUS

The extreme routinization evident in the German dye laboratories with their "scientific mass-labor" was not so prominent in the field of electricity. In both Germany and the United States, a dynamic sector of electrical manufacturers and electric utilities featured corporate consolidation and research-driven patent strategies. In the

United States the leading concerns were George Westinghouse's firm and General Electric, the successor to the Edison companies. In Germany, the leading firms were Siemens & Halske and German Edison, an offshoot of the American company.

The singular career of Thomas Edison (1847–1931) aptly illustrates the subtle but profound difference separating system-originating inventions from system-stabilizing ones. Edison positioned his own inventive career at the fringe of large corporate concerns, including Western Union and General Electric. Edison's intensely competitive style of invention led to a lifetime total of 1,093 US patents, among them an array of system-originating ones in electric lighting, phonographs, and motion pictures. His style of invention, however, ill-suited him to delivering the system-stabilizing inventions sought by General Electric. The company hired new talent and adopted an industrial-research model of innovation after 1900.

While Edison's boyhood was filled with farming, railroads, and newspapers, it was the telegraph industry that launched his inventive career. A generation removed from the pioneering work by Samuel Morse and Moses Farmer, telegraphers in Edison's time already confronted a nationwide system that had literally grown up alongside the railroads. Telegraph operators formed an early "virtual" community, tapping out tips, gossip, and professional news during slack times on the wire. Edison's first significant patents were a result of the telegraph industry's transition from an earlier city-to-city system, which Western Union had dominated, to a with-in-city phase that demanded rapid on-site delivery of market-moving information. While Western Union's intercity lines might deliver such news from other cities to its central New York City office, a legion of bankers, lawyers, stockbrokers, corporation executives, and newspaper editors wanted that news relayed instantly to their own offices. Responding to this opportunity, Edison moved to New York and made his mark as an inventor of printing telegraphs, automatic telegraphs, and stock tickers. As a telegraph inventor Edison's principal financial backers were Western Union, the telegraph combine, and Gold & Stock, a leading financial information provider.

Western Union continued to be a force in Edison's life even after he resigned from that firm in 1869 to pursue full-time invention. Working as an independent professional inventor saddled Edison with the responsibility of paying his own bills. For years, he struggled to balance inventive freedom with adequate funding. Manufacturing his telegraph equipment in Newark, New Jersey, Edison at one time oversaw as many as five factory establishments and 150 skilled workmen. His funding came from working on inventions deemed important by, among others, Western Union. In the 1870s Western Union engaged Edison's inventive talents to help it master the "patent intricacy" of duplex telegraphs and, as Edison dryly remarked, "as an insurance against other parties using them."[15] Duplex telegraphs, which sent two messages on one line, and quadruplex telegraphs, which sent four, were especially attractive to a large firm like Western Union seeking to wring additional efficiency from already-built lines. Even though he resented the "small-brained capitalists" who commissioned this work, he still needed them to finance his inventive ventures.

In the spring of 1876 Edison completed his "invention factory" at Menlo Park, located in the New Jersey countryside midway between New York and Philadelphia. There, as he famously promised, he would achieve "a minor invention every ten days and a big thing every six months or so."[16] In the past seven years, Edison had achieved an average of forty patents each year, or one every nine days. He was granted 107 patents during 1882, his most productive year at Menlo Park, in the midst of the electric light campaign described later. An inventory of Menlo Park two years after its opening itemizes a well-equipped electrical laboratory: galvanometers, static generators, Leyden jars, induction coils, a Rühmkorff coil capable of producing an 8-inch spark, condensers, Wheatstone bridges, and several types of specialized galvanometers. There was soon a well-stocked chemical laboratory, too.[17] At Menlo Park Edison directed a team of about twenty model makers, precision craftsmen, and skilled workmen, as well as a physicist and chemist. In 1877 Edison invented a tinfoil-cylinder "talking machine" (an early phonograph) and a carbon-diaphragm microphone for the telephone, invented a year earlier by Alexander Graham Bell. Publicizing the talking machine kept Edison in the public eye.

As Edison reviewed the technical literature he discovered that the outstanding problem in electrical lighting was that arc lights, popular for illuminating storefronts, were far too bright for indoor use, while incandescent bulbs were far too short-lived. Since the 1840s, inventors in several countries had patented twenty or more incandescent bulbs. For achieving lower-intensity arc lighting, one alternative was placing three or four arc lights on a single circuit. Edison conceived of something similar for incandescent lights. His dramatic boast in September 1878 ("have struck a bonanza in electric light—infinite subdivision of light") was not an accurate technical claim. It was a commercial claim, fed to newspaper reporters, to generate investors' enthusiasm.

October 1878 was a busy month even for Edison. On the fifth he applied for a patent on a platinum-filament bulb, on the fifteenth he formed the Edison Electric Light Company to develop the invention and license his patents, and on the twentieth he announced his plans for a complete lighting system in the *New York Sun*. In fact, absent practical incandescent bulbs, suitable generators, and a reliable system of distribution, Edison was months away from a complete system. Edison wrote that fall to an associate: "I have the right principle and am on the right track, but time, hard work and some good luck are necessary too. It has been just so in all of my inventions. The first step is an intuition, and comes with a burst, then difficulties arise—this thing gives out and [it is] then that 'Bugs'—as such little faults and difficulties are called—show themselves and months of intense watching, study and labor are requisite before commercial success or failure is certainly reached."[18]

Science arrived at Edison's laboratory in the person of Francis Upton, whom Edison hired in December 1878. Upton had an elite education from Phillips Andover Academy, Bowdoin College, and Princeton University and had recently returned from physics lectures in Berlin. Edison wanted his electric lighting system to be cost competitive with gas lighting and knew that the direct-current system he planned was viable only in a densely populated urban center. Using Ohm's and Joule's laws of electricity allowed Upton and Edison to achieve these techno-economic goals. While Edison was already familiar with Ohm's law, Joule's law first appears in his notebooks just after Upton's arrival. Ohm's law states that electrical resistance equals voltage divided by current ($R = V/I$). Joule's law states that losses due to heat equal voltage times current, or (combining with Ohm's law) current-squared times resistance (heat $= V \times I$ or $I^2 \times R$).

The critical problem these equations spotlighted was reducing the energy lost in transmitting direct-current electricity. If Edison and his researchers reduced the electrical resistance of the copper wire by increasing its diameter, the costs for copper would skyrocket and doom the project financially. If they reduced the current, the bulbs would dim. But if they *raised* the voltage in the circuit, the current needed to deliver the same amount of electrical energy to the bulb would drop, thereby cutting the wasteful heat losses. Upton's calculations, then, indicated that Edison's system needed a light-bulb filament with high electrical resistance. Later, Joule's law also allowed Upton to make the counterintuitive suggestion of lowering the internal electrical resistance of the Edison system's dynamo. Other electricians had their dynamo's internal resistance equal to the external circuit's resistance, and these dynamos converted into usable electricity only about 60 percent of the mechanical energy (supplied typically by a steam engine). In comparison, by 1880 Edison's low-resistance dynamos achieved 82 percent efficiency.[19]

On 4 November 1879, after two months of intensive experimenting on filaments, Edison filed his classic electric-bulb patent. Its claims to novelty included the "light-giving body of carbon wire or sheets," with high resistance and low surface area, as well as the filament's being enclosed in a "nearly perfect vacuum to prevent oxidation and injury to the conductor by the atmosphere."[20] These individual features were not particularly novel. The big news was that Edison's bulb burned brightly for an extended period, it cost little to make, and was part of an economical operating system of electric lighting. No other bulbs could compare. When Edison tested his system in January 1881 he used a 16-candlepower bulb at 104 volts, with resistance of 114 ohms and current of 0.9 amps. The US standard of 110 volts thus has its roots in Edison's precedent-setting early systems. (Similarly, the European standard of 220 volts can be traced to the decision in 1899 of the leading utility in Berlin [BEW] to cut its distribution costs by doubling the voltage it supplied to consumers, from 110 to 220 volts. BEW paid for the costs of converting consumers' appliances and motors to the higher voltage.)[21]

While Edison himself, the "Wizard of Menlo Park," cultivated newspaper reporters, it was Grosvenor P. Lowrey that steered Edison's lighting venture through the more complicated arena of New York City finance and politics. A leading corporate lawyer, Lowrey numbered Wells Fargo, the Baltimore & Ohio Railroad, and Western Union among his clients. He first met Edison in the mid-1860s in connection with telegraph-patent litigation for Western Union and became his attorney in 1877. After experiencing the dazzling reception given to a new system of arc lighting in Paris in 1878, Lowrey pressed Edison to focus on electric lighting. While many figures were clamoring for Edison to take up electric lighting, Lowrey arranged financing for his inventive effort from the Vanderbilt family, several Western Union officers, and Drexel, Morgan and Company. The Edison Electric Light Company, initially capitalized at $300,000, was Lowrey's creation.

So was a brilliant lobbying event that Lowrey staged on 20 December 1880 to help secure a city franchise. The city's go-ahead was necessary for the Edison company to dig up the city streets and install its lines underground. (The profusion of aboveground telegraph wires was already a major nuisance in the city's financial district, where Edison intended to build his system.) Edison did not clearly see the wisdom of winning over certain aldermen aligned with rival gas-lighting interests. Lowrey saw this plainly and devised an ingenious plan. He hired a special train that brought to Menlo Park eight New York aldermen, the superintendent of gas and lamps, as well as the parks and excise commissioners. They received a full two-hour tour of the Edison facilities followed by a lavish dinner, dramatically lit up by Edison lights. "For half an hour only the clatter of dishes and the popping of champagne corks could be heard, and then the wine began to work and the Aldermen, true to

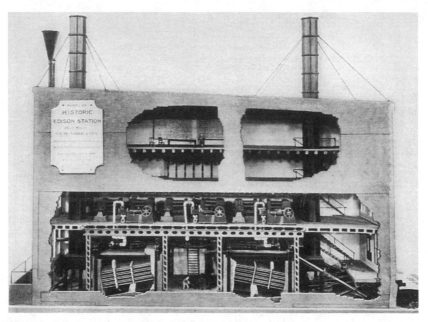

FIG. 5.3. EDISON'S PEARL STREET STATION

This model of Thomas Edison's first central station (1882–94) was exhibited at the St. Louis Exposition in 1904.

T. C. Martin and S. L. Coles, *The Story of Electricity* (New York: Marcy, 1919), 1:85.

their political instincts, began to howl 'Speech, speech,'" wrote one bemused newspaper reporter.[22] Having put his guests in a receptive mood, Lowrey outlined his case on behalf of Edison. In short order, the franchise was approved.

Edison's first central electric station, at 257 Pearl Street, represented a synthesis of scientific insight, technical research, political talent, and perhaps a bit of Edisonian luck (fig. 5.3). In building the generating station and distribution system and installing the wiring, Edison sustained day-by-day management of construction. He maintained a furious pace of invention. He applied for 60 patents in 1880, 89 in 1881, and a career-record 107 in 1882. On 4 September 1882, after a technician connected the steam-driven generator to the lighting circuit, Edison himself stood at 23 Wall Street—the offices of his financial backer Drexel Morgan—and switched on the first 400 lights. Within a month wiring was in place for 1,600 lamps; within a year that number topped 11,000. Edison still faced a number of daunting technical and financial problems. The Edison utility did not charge for electricity during 1882, and it lost money throughout 1883. By 1886 Edison needed to create a broader customer base, but he simply could not raise the money to do so. A Drexel Morgan syndicate with sizable investments finally "[squeezed] out some money somewhere."[23] Fires from defective wiring were another serious problem. A fire on 2 January 1890 closed the Pearl Street station for eleven days. At J. P. Morgan's personal residence at 219 Madison Avenue an electrical fire damaged his library.[24]

Six years after opening the pioneering Pearl Street station, Edison had established large companies or granted franchises in Philadelphia, New Orleans, Detroit, St. Paul, Chicago, and Brooklyn. In Europe central stations were operating in Milan and Berlin. By 1890 even the once-troubled New York utility reported impressive figures: compared with 1888, its customers had grown from 710 to 1,698; the number of 16-candlepower lamps from 16,000 to 64,000; and the company's net earnings from $116,000 to $229,000.[25] From its origins in 1886 through his death in 1931, Edison was at his West Orange, New Jersey, laboratory working on phonographs, motion pictures, batteries, and other inventions mostly unrelated to electric light. While he retained his directorships in the Edison electric companies, he had withdrawn from the active management of them.

BATTLE OF THE SYSTEMS

The early dominance of Edison's lighting system may obscure the fact that it was a peculiar system devised for a specific environment. Edison had designed his direct-current system for the dense customer load in large urban areas. His DC system was not suited for the smaller cities and rural areas where most Americans still lived. (Only after 1920 would more than half of all Americans live in urban areas larger than 2,500 persons.) Meanwhile, the uses for electricity multiplied during the 1880s to include electric motors for street railways and factories, and there was further expansion in arc lighting. Inventors increasingly turned to alternating current electricity. By using transformers to step *up* the voltage for transmission (cutting I^2R heat losses) AC systems could send electricity over hundreds of miles, then step *down* the voltage for safe local distribution to consumers.

The Edison companies made step-by-step improvements in DC systems, but for the most part left the AC field to others. Beginning in 1886 Edison, at his new West Orange laboratory, sketched numerous AC generators, transformers, and distribution networks, and filed for at least four AC patents before concluding that it was economically unpromising. Edison was wary of the energy losses of transformers, the high capital costs of building large AC stations, and the difficulties of finding insulators that could safely handle 1,000 volts. "The use of the alternating current is unworthy of practical men," he wrote to one financial backer.[26]

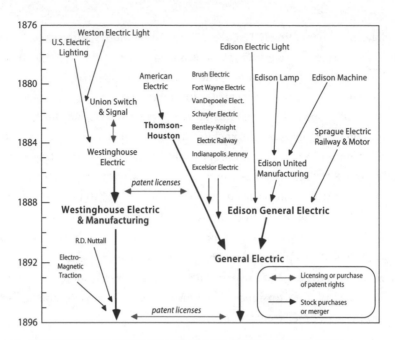

FIG. 5.4. CONSOLIDATION OF THE US ELECTRICAL INDUSTRY, 1876–1896

Adapted from Arthur A. Bright, *The Electric Lamp Industry* (New York: Macmillan, 1949), with data from W. B. Carlson, *Innovation as a Social Process* (Cambridge: Cambridge University Press, 1991).

Arc lighting for streets, AC incandescent systems for smaller towns, AC motors for factories, and the pell-mell world of street railways were among the lucrative fields that Edison's diagnosis discounted. It was precisely these technologies, in addition to DC incandescent lighting for urban areas, that Edison's principal rivals—the Thomson-Houston, Brush, and Westinghouse firms—succeeded at inventing, developing, and selling commercially. These several uses were brought together in the 1890s through the development of a "universal system." It was also during the 1890s that the companies founded by Edison and his principal rivals came increasingly to be steered by financiers. Whereas Edison had lived on the thrill of system-jarring competition, for these financiers such cutthroat competition brought instability to their systems and was undesirable—and avoidable. In the mid-1880s there were twenty firms competing in the field of electric lighting; by 1892, less than a decade later, after a period of rapid consolidation and mergers, there would be just two: Westinghouse and General Electric (fig. 5.4).

In 1889 the several Edison firms and the Sprague Electric Railway & Motor Company were merged by Henry Villard, an investor in the early Edison companies, to form Edison General Electric. Its capital of $12 million came from German banks, Villard himself, and J. P. Morgan. Taking the title of president, Villard left the day-to-day management with Samuel Insull, formerly Edison's personal secretary and later, for many years, head of Chicago's Commonwealth Edison utility (1892–1934). Insull had the difficult job of coordinating three large production facilities—a huge machine works at Schenectady, New York, a lamp factory at Harrison, New Jersey, and a factory in New York City. Edison General Electric contracted with Edison to develop improved DC equipment at his West Orange laboratory. Insull even hoped that Edison would develop a full-fledged AC system, but Edison instead focused his inventive energies on improving his phonograph and refining iron ore.

In 1878, when Edison was starting his electric lighting project at Menlo Park, Elihu Thomson was entering the field of arc lighting. Just six years younger than Edison, Thomson (1853–1937) would distinguish himself as the nation's third most prolific inventor, with a career total of 696 patents. In the mid-1880s Thomson turned his inventive efforts to incandescent lighting and AC systems. His other notable inventions include electric welding, street railway components, improved transformers, energy meters, and induction motors. These inventions were among the necessary technical components of the universal system of the 1890s.

To finance his inventions Thomson had struggled through a series of ill-funded partnerships. The capital of American Electric Company ($0.5 million), formed to market Thomson's award-winning arc lights, was a fraction of the Edison concern ($3.8 million). In 1882 Charles A. Coffin, a self-made Boston entrepreneur, launched Thomson's career. While he had earlier averaged at best twenty patents a year, Thomson was soon applying for forty patents a year, equal to Edison's boast of an invention every ten days. Coffin and his investment syndicate understood that Thomson was most valuable to the company while inventing, and they encouraged Thomson to set up a special "Model Room" at the company's principal factory, at Lynn, Massachusetts.

Coffin and his investors also recognized that real success in the electrical field would require equal strengths in innovation, marketing, and manufacturing. A string of successfully marketed inventions propelled the firm into the big league. Coffin also bought up no fewer than seven rivals, including the Brush firm, at a cost of $4 million. To raise such large amounts of money, Coffin turned to Lee, Higginson & Company, a Boston brokerage house that pioneered in the marketing of industrial securities. By 1892, the Thomson-Houston company was capitalized at $15 million, had 3,500 employees, and made annual profits of $1.5 million. By then, some 800 companies were using Thomson-Houston's arc and incandescent lighting equipment, while in the rapidly growing street-railway field an additional 204 companies were using its equipment.

The only serious rival to Edison General Electric and Thomson-Houston was the concern founded by George Westinghouse. Westinghouse (1846–1914) was already rich and famous from licensing his patents on railroad air brakes. In the mid-1880s he purchased key patents and hired the talent needed to work on AC lighting. Westinghouse employed William Stanley as an in-house inventor; in 1886 Stanley perfected an AC transformer. That same year, Westinghouse installed its first AC lighting system, in Buffalo, New York. By 1889 he had secured US licenses from independent inventor Nikola Tesla for an AC induction motor and related AC polyphase equipment (strengthening his company's emerging "universal system") while concluding a patent-sharing agreement with Thomson-Houston. Under its terms, Westinghouse could sell Thomson's arc lighting, while Thomson-Houston could manufacture and sell its own AC systems without fear of the Westinghouse AC patents.[27] By 1890, thanks to the popularity of its AC system, Westinghouse had sales of $4 million and total assets of $11 million.

The relentless competition of these three huge firms for customers, inventors, and patents played out in the race for contracts, through legal battles, and in a feeding frenzy to buy up smaller rivals. Rivals disappeared at a quick pace between 1888 and 1891, when Thomson-Houston acquired seven, Westinghouse three, and Edison General Electric two. By this strategy the big three put pesky rivals out of business, used their patents, and gained the services of their inventors and skilled workforce. Competition could also take a more bizarre form. Edison General Electric, flagging technically in AC systems, used its political connections to launch a lurid campaign to denounce AC as the "death current." The Edison firm backed the efforts of Harold Brown, once a salesman for Edison's electric pens and now an imaginative opponent of AC. With the quiet help of Edison's laboratory staff, Brown staged public spectacles in which he used AC current to electrocute dogs, calves, and horses. Typically, Brown would first apply a mild DC current, suggesting its relative safety, then dispatch the animal with a hefty jolt of the "deadly" AC. In 1890 he arranged for a 1,000-volt Westinghouse generator to be used, despite that company's opposition, in the first public electrocution at New York State's Sing Sing prison. The Edison publicity machine then warned against the Westinghouse-supplied "executioner's current." (The "electric chair" at Sing Sing continued in use until 1963.) Edison General Electric also lobbied state legislatures in New York, Ohio, and Virginia to limit the maximum voltage of electrical systems to 300, a legal move to scuttle the principal advantages of AC systems.[28]

FIG. 5.5. THE ELECTRIC CITY

Inspecting electric arc street lighting using a special electric vehicle.

T. C. Martin and S. L. Coles, *The Story of Electricity* (NewYork: Marcy, 1922), 2:32.

Edison himself, smarting from the rough battle over patents, derailed the first serious attempt at consolidating his concern with the Thomson-Houston firm. When he heard of the merger plans in 1889, he fumed, "My usefulness as an inventor would be gone. My services wouldn't be worth a penny. I can only invent under powerful incentive. No competition means no invention." But the financiers that ran both firms increasingly favored merger and consolidation. J. P. Morgan, soon to reorganize a quarter of the nation's railroads, expressed alarm at the electrical manufacturers' boundless hunger for capital. "What we all want," wrote Charles Fairfield (a partner at Lee, Higginson) in 1891, "is the union of the large Electrical Companies." An attempt that year to merge Westinghouse with Thomson-Houston failed. George Westinghouse, while a brilliant technician and able factory manager, lacked the skills of a financier. "He irritates his rivals beyond endurance," Fairfield said.[29]

General Electric, a creation of the nation's leading financiers, took form in April 1892. After several months of negotiations, the Thomson-Houston and Edison General Electric companies exchanged their existing stock for shares in the new company, capitalized at $50 million. This figure made it the second largest merger at the time. General Electric's board of directors included two men each from the Thomson and Edison companies, with a Morgan associate as chairman. (Thomas Edison was on the board only nominally; Thomson outright declined a directorship in order to continue his technical work.) Coffin became GE's founding president. Insull, offered the job of second vice president in the new company, instead departed for Chicago, where he developed and installed one of the early polyphase AC, or universal systems, and pioneered managerial and financial techniques in building a regional utility empire (fig. 5.5).

TENDERS OF TECHNOLOGICAL SYSTEMS

Edison fought it, Thomson denied it, and Insull embraced it: a new pattern of technological change focused on stabilizing large-scale systems rather than inventing wholly new ones. In the most capital-intensive industries, including railroads, steel, chemicals, and electrical manufacturing, financiers like J. P. Morgan and Lee, Higginson in effect ended the ceaseless competition and fierce pace of freewheeling technological innovation. In doing so, they were the chief agents in giving birth to a stable organized capitalism. In the second industrial revolution, somewhat paradoxically, technological change became evolutionary. The arch-apostle of stability, U.S. Steel, extinguished at least two important innovations in steel making, the oversize H-beam mill of Henry Grey and its own experiments in the continuous rolling of thin sheets. (And, in a delightful irony, the German chemical and steel industries both looked to U.S. Steel as a harbinger of what was modern and progressive in organizing industry, modeled themselves after its sprawling bureaucracy, and got much the same unwieldy result.) Exhausting battles with corporations contributed to the tragic suicides of chemical inventor Wallace Carothers and radio inventor Edwin Armstrong. Tesla, like Edison, left electric light and power to focus his inventive efforts and flamboyant publicity on long-distance wireless transmission. But the well-funded Guglielmo Marconi scooped up the key US patents on radio; Tesla died, penniless, in 1943. Ironically, just when independent inventors were pushed to the margins, Thomas Edison was lionized at home and around the world for his past and legendary inventive feats. The United States celebrates National Inventors' Day on Edison's birthday of 11 February.[30]

Besides financiers, the most important agents of stabilizing existing technological systems were scientists and engineers. Industrial scientists and science-based engineers stabilized the large systems by striving to fit into them and, most importantly, by solving technical problems deemed crucial to their orderly expansion. Neither of these professions existed in anything like their modern form as recently as 1870. Before then, engineers had been mostly either military or "civil" engineers who built fortifications, bridges, canals, and railways. During the second industrial revolution, engineering emerged as a profession. National societies were founded in the United States by mining engineers (1871), mechanical engineers (1880), electrical engineers (1884), and chemical engineers (1908). Industrial scientists too were created in these decades. In the German chemical industry, as we saw, the "scientific mass-labor" needed to synthesize and patent new dyes in the 1880s led to the large-scale deployment of scientists in industry. Such a model of industrial research appeared somewhat later in the United States, with full-scale industrial research laboratories organized by General Electric in 1901, DuPont in 1902, and AT&T in 1911.

In 1900 there were hundreds of laboratories in US industry but no "research and development" laboratories as we know them. Edison's laboratories at Menlo Park and West Orange had been expressions of Edison's personal drive for innovation. They foundered whenever Edison was physically absent, and they went dormant in the early 1890s when Edison left them to pursue an ill-fated iron-ore venture. From the 1870s, the railroad, steel, and heavy chemical industries had located laboratories on factory sites and sometimes in factory buildings, and their staffs were responsible for analyzing raw materials and testing products. By contrast, industrial research laboratories grew only in the largest companies, and mostly in the chemical and electrical sectors of science-based industry; additionally, these research laboratories were isolated—often physically—from the companies' production facilities.

The desire for new products and new processes, frequently touted as the chief inspiration for industrial research, is only part of the story. For example, DuPont invested heavily in industrial research beginning in the 1920s and loudly publicized such results as neoprene and nylon ("Better Things for Better Living through Chemistry" went its advertising slogan). But acquiring competitors, rather than conducting research, remained DuPont's most important source of new technology for decades. A National Bureau of Economic Research study found that of twenty-five major product and process innovations at DuPont between 1920 and 1950, only ten resulted from its in-house research while fifteen resulted from acquiring outside inventions or companies.[31]

In 1901 General Electric saw its laboratory as a way to internalize innovation. Having suffered the expiration of its Edison-era patents on carbon-filament light bulbs, and facing determined competitors with new and promising bulb designs, GE founded its laboratory to head off future technological surprises. Initially, GE hired sixteen chemists and physicists; their task, directed by Willis Whitney, was to build a better light bulb. In 1904 the laboratory produced a major improvement in the carbon filament, dubbed "General Electric Metalized." Yet even with the GEM bulbs, the company had to purchase patent rights from outside inventors; in 1907 it paid two German companies $350,000 for patent rights and in 1909 an additional $490,000 for the use of tungsten filaments. This expensive situation persisted until 1912 when, after years of concerted work, one of the laboratory's chemists patented an improved method for making long-lived tungsten filaments.

The success of fitting science into industry at GE must be largely credited to Willis Whitney. Whitney came to GE with a German doctorate in physical chemistry, a lucrative (patented) process of his own for recovering solvents, and an unsatisfactory experience teaching at MIT. He arranged for his laboratory's PhD scientists to have plentiful equipment, ample intellectual leeway, and a stimulating research environment. Whitney's strong encouragement, especially for research work relevant to the company's commercial concerns, was legendary. Yet, like the German dye laboratories, the results were not the scientist's own. "Whatever invention results from his work becomes the property of the company," Whitney stated in 1909. "I believe that no other way is practicable."[32] Whitney took substantial pride in hiring Irving Langmuir, easily GE's most-renowned scientist. Langmuir, after teaching at Stevens Institute of Technology, began research at GE in 1910 to perfect and patent a gas-filled light bulb. The so-called Mazda bulb brought GE stunning profits of $30 million a year. In 1928, thanks to its Mazda bulb sales, GE held 96 percent of the US market for incandescent bulbs. Langmuir's follow-up research on the surface chemistry of such bulbs brought him a Nobel Prize in 1932.

Industrial research became a source of competitive advantage for the largest firms, such as General Electric and General Motors. Companies that could mount well-funded research efforts gained technical, patent, and legal advantages over those that could not. Independent inventors, formerly the nation's leading source of new technology, either were squeezed out of promising market areas targeted by the large science-based firms or went to work for them solving problems of the companies' choosing. In the electrical industry, 82 percent of research personnel were employed by just one-quarter of the companies. By the 1930s, when General

Electric and AT&T between them employed 40 percent of the entire membership of the American Physical Society, the industrial research model became a dominant mode for organizing innovation and employing scientists. By 1940 DuPont alone employed 2,500 chemists. At the time fully 80 percent of the nation's R&D funding ($250 million) came from industry.

The search for stability involved not only financiers, corporations, engineering societies, and research laboratories. The very content of engineering was at play, as we can see in the electrical engineering program at the Massachusetts Institute of Technology around 1900. Achieving stability had a precise technical meaning for electrical engineers in this period, since instabilities or transients in the growing regional electrical transmission systems were major obstacles to their orderly and safe development. The industrial orientation of electrical engineering at MIT from around 1900 into the 1930s contrasts markedly with its more scientific and military orientation during and after the Second World War (see chapter 7).

Despite its support of Alexander Graham Bell's experiments on telephony in the mid-1870s, MIT did not found a four-year course in electrical engineering until 1882, and for years the program was mostly an applied offshoot of the physics faculty. In 1902, when a separate electrical engineering department was organized, Elihu Thomson joined the faculty as a special nonresident professor of applied electricity. From then on MIT would enjoy close relations with General Electric's massive Lynn factory, formerly the location of Thomson's Model Room, and would send students there for practical training and receive from GE electrical machinery to supply its laboratories. Thomas Edison and George Westinghouse also gave electrical machinery to the new program. Enrollments in the electrical engineering department grew quickly (it was briefly, in the mid-1890s, the institute's most popular program), and by the turn of the century electrical engineering had joined civil, mechanical, and mining engineering as MIT's dominant offerings.

A sharp departure from the department's origins in physics occurred under the leadership of Dugald Jackson, its chair from 1907 to 1935. Jackson recognized that engineering was not merely applied science. As he put it in a joint address to the American Institute of Electrical Engineers and the Society for the Promotion of Engineering Education in 1903, the engineer was "competent to conceive, organize and direct extended industrial enterprises of a broadly varied character. Such a man . . . must have an extended, and even profound, knowledge of natural laws with their useful applications. Moreover, he must know men and the affairs of men—which is sociology; and he must be acquainted with business methods and the affairs of the business world."[33] Such a formulation clearly framed engineering as a male-only

FIG. 5.6. ELECTRIC TURBINES AND THE "WEALTH OF NATIONS"

Dugald Jackson revamped electrical engineering at MIT to engage the technical and managerial problems of electric power systems, to tap (as General Electric put it) "the wealth of nations generated in turbine wheels." Pictured is Ohio Power Company's station at Philo, Ohio.

General Electric Review (November 1929): cover. Courtesy of Illinois Institute of Technology Special Collections.

profession. Jackson, upon arriving at MIT in February 1907, revamped the physics-laden curriculum in close consultation with a special departmental advisory committee. The imprint of business was unmistakable. Its members, in addition to Elihu Thomson (from General Electric), were Charles Edgar (president of Boston Edison), Hammond Hayes (chief engineer, AT&T), Louis Ferguson (vice president, Chicago Edison), and Charles Scott (consulting engineer, Westinghouse). Harry Clifford competently handled courses in AC electricity and machinery, which Jackson had previously taught at Wisconsin. Jackson himself developed a new fourth-year elective on managerial topics, "Organization and Administration of Public Service Companies" (i.e., electric utilities) (fig. 5.6).

The research, consulting, and teaching activities during the Jackson years aligned MIT's electrical engineering offerings with electrical manufacturers and electrical utilities. Jackson consulted on several leading electrification projects—including the

TABLE 5.1. MIT ELECTRICAL ENGINEERING CURRICULUM FOR 1916

SUBJECT NUMBER	MATERIAL COVERED
	Principles of Electrical Engineering Series
601	Fundamental concepts of electric and magnetic circuits
602	DC machinery
603	Electrostatics, variable and alternating currents
604	AC machinery
605	AC machinery, transmission
	Fourth-Year Electives
632	Transmission equipment
635	Industrial applications of electric power
637	Central stations
642	Electric railways
645–646	Principles of dynamo design
655	Illumination
658	Telephone engineering
	Elective Graduate Subjects
625	AC machinery
627	Power and telephone transmission
634	Organization and administration of public service companies
639	Power stations and distribution systems
643	Electric railways

Source: Wildes and Lindgren, *A Century of Electrical Engineering and Computer Science at MIT, 1882–1982* (Cambridge: MIT Press, 1985), pp. 58–59.

Note: Laboratories in Technical electrical measurement and Dynamo-electric machinery were required in the third and fourth years.

Conowingo hydroelectric project on the Susquehanna River in Maryland, the Fifteen Miles Falls project on the Connecticut River, the Great Northern Railway's electrification of its Cascade Mountain Division, and the Lackawanna Railroad's electrification of its New Jersey suburban service. Even when Jackson taught other subjects, his students reported, "they were learning a great deal about public utility companies and their management."[34] In 1913 Jackson tapped Harold Pender, the author of two influential textbooks, to head the department's research division, funded by General Electric, Public Service Railway, Stone & Webster, and AT&T. In line with the department's research and consulting activities, the curriculum revision in 1916 "was predominantly aimed at the problems of electrical power systems" (table 5.1). Illumination and telephony received some emphasis; the newer fields of radio and electronics did not.[35]

MIT's Technology Plan of 1920 had the effect of implanting Jackson's model of industrial consulting on the institute as a whole. On balance, the plan was not a well-thought-out initiative. In 1916 MIT had vacated its perennially cramped facilities in Boston's Back Bay and, with large gifts from Coleman du Pont and George Eastman, built a new complex of buildings on its present site along the Charles River. The move to Cambridge was advanced by an agreement made in 1914 to share the large bequest given to Harvard University by the shoe manufacturer Gordon McKay. In 1919, however, after a state court struck down the terms of the MIT-Harvard arrangement and the McKay funds disappeared, MIT was plunged into a financial crisis. Making matters worse, effective in 1921 the state legislature cut off its appropriation to MIT of $100,000 a year.

In response to this crisis George Eastman offered $4 million in Eastman Kodak stock—if MIT raised an equal sum by January 1920. Raising such a huge matching gift from individual donors seemed impossible, so president Richard Maclaurin turned to industry. In effect Maclaurin's Technology Plan cashed in on the reputation of MIT's faculty members, laboratories, and libraries and offered them up for solving industrial problems. Each company receiving such access made a financial contribution to MIT. "Had MIT been allowed . . . to 'coast' on the McKay endowment, it would not have been forced to turn full-face to American industry and to become deeply engaged with its real problems of development."[36] While the Technology Plan staved off financial crisis, the narrowly conceived problems deemed important by industry were hardly a satisfactory basis for education and advancement of knowledge. In the early 1920s enrollments plunged in virtually all programs except for electrical engineering. In the 1930s MIT's president Karl Compton struggled to find other sources of funds, ones that did not so narrowly constrain the faculty's problem choices and publication options.

In Jackson's consulting, the electrical-engineering curriculum of 1916, and the Technology Plan, MIT's educational offerings and faculty research were aligned with stabilizing large-scale technological systems. In Harold Hazen's work on the "network analyzer" during the 1920s and 1930s, we can trace the same phenomenon in the very content of engineering research. One line of leading-edge engineering research during these years concerned the behavior and insulation of high voltages. Hazen's research responded to a more subtle problem, that of the instability of highly interconnected electrical systems, then emerging as local electric systems were connected together into regional systems.

The technical problems of large interconnected systems were given sharp focus by the so-called Superpower and Giant Power regional electrification proposals of the 1920s. The problem was not so much that these systems would be geographically spread out, in a network stretching from Boston to Washington with lines running to western Pennsylvania or even beyond to the Midwest, and would feature transmission lines carrying up to 300,000 volts, roughly ten times the prevailing industry standard. The real problem was the system's interconnected nature, carrying multiple loads from multiple generating sources to diverse customers. Transients, or sharp fluctuations in voltage, were of particular concern. No utility would invest in such a mega-system just to have a transient bring it down (something that actually happened in 1965, causing a massive blackout in New York City). In an effort to simulate these systems, engineers at General Electric and Westinghouse as early as 1919 built small-scale models using networks of generators and condensers.

Hazen's work on the "network analyzer" began with his 1924 bachelor's thesis under Vannevar Bush. Bush, a pioneer in analog computing, was working for Jackson's consulting firm studying the Pennsylvania-based Superpower scheme. Hazen's thesis, conducted jointly with Hugh Spencer, involved building a dining table–sized simulator capable of representing three generating stations, 200 miles of line, and six different loads. After graduation Hazen and Spencer joined GE's Schenectady plant, and there worked for Robert Doherty, "who was much concerned with stability studies in power systems." When Hazen returned to MIT as Bush's research assistant in 1925, Bush asked him to simulate an entire urban power system. Simulating polyphase AC power systems involved not only generating multiple-phase currents (Hazen worked at 60 Hz) and shifting their phases to simulate motor loads but also devising highly accurate measuring equipment. For the latter, Hazen worked closely with GE's Lynn factory. By 1929 the measuring problems were solved and GE's Doherty approved the building of a full-scale network analyzer.

Built jointly by GE and MIT and physically located in the third-floor research laboratory in MIT's Building 10, the network analyzer was capable of simulating systems of great complexity. The analyzer was now an entire room full of equipment (fig. 5.7). It featured 8 phase-shifting transformers, 100 variable line resistors, 100 variable reactors, 32 static capacitors, and 40 load units. Electric utilities loved it. In its first ten years, its principal users were American Gas and Electric Service Corporation, General Electric, Canadian General Finance, Jackson & Moreland,

FIG. 5.7. MIT NETWORK ANALYZER.

Harold Hazen at the center of a signature artifact of the science and systems era. Electric utilities simulated complex electric power systems with the network analyzer, from the 1920s through its dismantling in 1953.

Courtesy of Massachusetts Institute of Technology Museum.

Illinois Power and Light, Union Gas and Electric, the Tennessee Valley Authority, and half a dozen other large utilities. During 1939 and 1940 a coalition of nine utilities, then examining a nationwide electrical grid, paid for a massive upgrading and expansion. Used intensively by the electric utilities, the network analyzer became less and less relevant to MIT's educational and research activities. "While our board is still useful to power companies, it no longer serves as an adequate vehicle for creative first-rate graduate or staff research in the power area," stated a 1953 department memorandum recommending its termination. "One can say that it has had a glorious past." The analyzer was sold to Jackson & Moreland and shipped to Puerto Rico.[37] During Hazen's tenure as department head (1938–52), the rise of electronics and the press of military work dramatically changed electrical engineering, MIT itself, and the character of technology (see chapter 7).

Synthetic dyes, poison gases, DC light bulbs, AC systems, and analog computers such as Hazen's network analyzer constituted distinctive artifacts of the science-and-systems era. Broadening our view to include additional or different science-based industries would, I think, alter the particular stories but not change the overall pattern. (One can discern variations on the themes explored in this chapter in the pharmaceutical, automobile, steel, radio, and photographic industries.) The most important pattern was the underlying sociotechnical innovations of research laboratories, patent litigation, and the capital-intensive corporations of science-based industry. For the first time, industrial and university scientists participated equally with inventors, designers, and engineers in developing new technologies. Indeed, the R&D laboratory has become such a landmark that it is sometimes difficult for us to recall that new technologies can (and often are) a product of persons working elsewhere.

A decisive indicator was the decline of British industrial leadership and the rise of German and American primacy in the science-based industries. A neat contrast can be made of the British cotton-textile industry that typified the first industrial revolution and the German synthetic dye industry and American electrical industry that together typified the second. German chemists like August von Hofmann and Heinrich Caro as well as American financiers like Charles Coffin and J. P. Morgan typified this era as much as did the technical artifacts previously listed.

The presence of the financiers, corporations, chemists, and engineers produced a new mode of technical innovation and not coincidentally a new direction in social and cultural innovation. The system-stabilizing mode of technical innovation—"nowhere any trace of a flash of genius"—was actively sought by financiers, who had taken on massive liabilities in financing the large science-based industries, and by those industrial scientists like GE's Willis Whitney and engineers like MIT's Dugald Jackson who welcomed the opportunities the new mode afforded. Edison, who distinctly preferred working at the edge of system-originating inventions, intuitively understood that his glory days as an inventor were over—even as his fame as a cultural icon expanded. General Electric needed system-stabilizing industrial scientists, who might turn light bulbs into Nobel Prizes and profits, more than they needed a brilliant Edisonian inventor. The system-stabilizing innovations, with the heavyweights of industry and finance behind them, also created new mass-consumer markets for electricity, telephones, automobiles, household appliances, home furnishings, radios, and much else in the modern world.

CHAPTER 6

1900 – 1950

Materials of Modernism

"The triumphant progress of science makes profound changes in humanity inevitable, changes which are hacking an abyss between those docile slaves of past tradition and us free moderns, who are confident in the radiant splendor of our future."[1] It jars the ear today to hear this raw modernist language, a reminder that the determinist worldview of the modern movement—here voiced in a 1910 manifesto by a group of Italian Futurist painters—is something of a distant memory. Yet in our age of skepticism about science and technology, it is important to appreciate how science and technology helped create a wide swath of modern culture. Modernism in art and architecture during the first half of the twentieth century can be best understood as a wide-ranging aesthetic movement, floated on the deeper currents of social and economic modernization driven by the science-and-systems technologies. Modernism's articulate promoters recognized the rhetorical force and practical effect of linking their vision to these wider socioeconomic changes. We shall see how these promoters took up a stance of "technological fundamentalism," which asserted the desirability and *necessity* of changing society and culture in the name of technology.

Modernists claimed the twentieth century. They selectively praised the builders of classical Greece and Rome and, closer at hand, the Crystal Palace in London, the Eiffel Tower in Paris, a number of massive grain silos and factories across North America, as well as railway carriages, steamships, power stations, and automobiles. Yet, modernists argued, cultural development had failed to keep pace with the new materials and machine forms of the twentieth century. Bruno Taut, a determined booster of steel-and-glass buildings and a leading modern architect himself, condemned the traditional architect's "confused juggling with outward forms and styles" (fig. 6.1). Physically surrounded and artistically suffocated by the monuments of imperial Vienna, Adolf Loos in his polemical *Ornament and Crime* (1908) argued, "the evolution of culture marches with the elimination of ornament from useful objects." Ornament, he wrote, was no longer a valid expression of contemporary culture.[2] An architecture that expressed the new aesthetic possibilities in material form,

FIG. 6.1. FLAT ROOFS AND RIBBON WINDOWS

Bruno Taut's modernist apartment block in Berlin's Neukölln district.

Bruno Taut, *Modern Architecture* (London. The Studio, 1929), 111.

as Taut, Loos, and their fellow modernists advanced the case, would result in better schools, factories, housing, offices, and cities—indeed a better, modern society.

This chapter situates the development of aesthetic modernism in art and architecture in the deeper currents of social, technological, and economic modernization. It first tells a material history of modernism, stressing the influence of new factory-made materials—especially steel and glass—on discourse about what was "modern."[3] It then gives an intellectual and social history of modernism, especially in art and architecture, again centering on modern materials but also contextualizing concepts drawn from abstract art. The account spotlights personal and intellectual interactions among three leading European movements: the Futurists in Italy, de Stijl in the Netherlands, and the Bauhaus in Germany. Finally, the chapter evaluates the often ironic consequences of the modernist movement for building styles, household labor, and the rise of consumer society. Modernism cascaded outward into the world owing to intentional innovations in architectural schools, professional networks, and tireless promotional efforts on behalf of its well-placed practitioners.

MATERIALS FOR MODERNISM

The materials that modernists deemed expressive of the new era—steel, glass, and concrete—were not new. Glass was truly ancient, while concrete dated to Roman times. Steel was a mere 500 years old; for centuries skilled metalworkers in India, the Middle East, and Japan had hammered bars of high-quality steel (called *wootz, Damascus,* and *tatara,* respectively) into razor-sharp daggers and fearsome swords. Beginning in the sixteenth century gunmakers in Turkey even coiled and forged Damascus steel bars into gun barrels. Europeans first made homegrown steel in the eighteenth century when the Englishman Benjamin Huntsman perfected his crucible steelmaking process (see chapter 3). Huntsman began with iron bars that he had baked in carbon until their surfaces had absorbed the small, but crucial amount of carbon that gave steel its desirable properties: it was tough, flexible on impact, and able to be hardened when quickly cooled from a high temperature. He then packed these bars—carbon-rich steel on the outside while plain iron inside—into closed clay crucibles, put them into a hot coal-fired oven that melted the metal, and finally cast ingots of steel. Huntsman's crucible steel was used in Sheffield and elsewhere for making cutlery, piano wire, scissors, scientific instruments, and umbrellas, but its high cost limited wider uses. Throughout the early industrial era, textile machines, factory framing, bridges, locomotives, and railroad tracks continued to be made mostly of wrought iron or cast iron.

The first mass-produced steel in the world came from another Englishman, Henry Bessemer. Bessemer, as we related in chapter 3, was a talented London inventor with a string of inventions already to his credit when he turned to iron and steel. A French artillery officer had challenged Bessemer to make a metal that could withstand the explosive force concentrated in a cannon barrel. Cannons could be forged from strips of wrought iron, but the process to make wrought iron required many hours of skilled labor, and their seams sometimes split apart after repeated firings. Cannons were also cast whole from molten brass or cast iron, without seams, but brass was expensive and cast iron, while cheap, had problems of its own. Cast iron was brittle. Since cast-iron cannons might blow open without any warning, gunnery officers hated them. Bessemer set about to remedy this situation. He wanted a metal that was malleable like wrought iron but at low cost. In the 1850s he experimented with several ways of blowing air through liquid iron. If conditions were right, oxygen in the air combined with carbon in the iron, and the resulting combustion made a towering white-hot blast. Again, as with Huntsman's crucibles, the hoped-for result was iron with just enough carbon to make it into steel.

Bessemer's dramatic process slashed the heavy costs to steelmakers for tons of coal and hours of skilled work. For fuel, he simply used the carbon in the iron reacting with air (his 1856 paper claiming "The Manufacture of Malleable Iron and Steel without Fuel" brought laughter from experienced ironmasters), while his patented machinery displaced skilled labor. It turned out that Bessemer's process also could be made large—very large indeed. While his early converting vessel held around 900 pounds of fluid metal, the converters he built in 1858 at his factory in Sheffield's burgeoning steel district held more than ten times as much, 5 tons. In time, 10- and even 25-ton converters were built in England, Germany, and the United States. By comparison makers of crucible steel were limited to batches that could be hoisted out of the melting furnace by hand, around 120 pounds.

The huge volume of Bessemer steel was a boon to large users like the railroads. In the 1870s, nearly 90 percent of all Bessemer steel in the United States was made into rails, and the transcontinental railroads that were built in the next two decades depended heavily on Bessemer steel. But American steelmakers, by focusing so single-mindedly on achieving large *volume* of production with the Bessemer process, failed to achieve satisfactory *quality*. In Chicago, one defective Bessemer beam from the Carnegie mills cracked neatly in two while being delivered by horsecart to the building site. Consequently, structural engineers effectively banned Bessemer steel from skyscrapers and bridges. In the 1890s the railroads themselves experienced dangerous cracks and splits in their Bessemer steel rails.

The successful structural use of steel was a result of European metallurgists' work to improve *quality* rather than maximize output. Finding iron ores with chemical characteristics suitable for the Bessemer process proved a difficult task in Europe. Since the original Bessemer process, which used chemically acid converter linings, could not use the commonly available high-phosphorus iron, European steelmakers developed the Thomas process. It used chemically basic linings in a Bessemer-like converter to rid the steel of the phosphorus that caused it to be brittle. Metallurgists soon found that open-hearth furnaces, too, could be lined with the chemically basic firebricks. This trick allowed steelmakers on both sides of the Atlantic to produce a reliable and cost-effective structural steel. Europeans had the Thomas process. Makers of structural steel in the United States favored open-hearth furnaces. These required from twelve to twenty-four hours to refine a batch of steel, so they were free from the relentless production drive of Bessemer mills, where a blow might take a scant ten minutes. From the 1890s on, architects on both sides of the Atlantic had a practicable structural steel.

Glass is by far the oldest of the "modern" materials. Early glass vases, statues, cups, coins, and jewelry from Egypt and Syria are at least 5,000 years old. Phoenicians and later Romans brought glassmaking to the dominions of their empires, and from the Renaissance onward Venice was renowned as a center for fine glassmaking. By the eighteenth century, Bohemia and Germany had become leading producers of window glass. Glassmaking involved no complicated chemistry and no violent Bessemer blasts but only the careful melting of quartz sand with lead salts to add desired coloring. The manufacture of both steel and glass required extremely high temperatures; it is no coincidence that Bessemer had worked on glass-melting furnaces just prior to his steelmaking experiments. But melting was only the start. Workers making glass needed considerable strength and special skills for pressing or blowing the thick mass of molten material into useful or decorative shapes. Initially, most glass for windows was made by blowing a globe of glass then allowing it to collapse flat on itself. In 1688 French glassmakers began casting and polishing large flat sheets of "plate" glass, up to 84 by 50 inches. By the mid-nineteenth century, the window-glass industry comprised four specialized trades. "Blowers" took iron pipes prepared by "gatherers" and created cylinders of glass (often using brass or iron forms); then "cutters" and "flatteners" split open the newly blown cylinders into flat sheets of window glass.[4]

Glass through most of the nineteenth century was in several ways similar to steel before Bessemer. It was an enormously useful material whose manufacture required much fuel and many hours of skilled labor and whose application was limited by its high cost. Beginning in the 1890s, however, a series of mechanical inventions transformed glassmaking into a highly mechanized, mass production industry. Belgian, English, French, and American glassmakers all took part in this achievement. First, coal-fired pots were replaced by gas-fired continuous-tank melting furnaces; then window-glass making was mechanized along "batch" lines; and finally plate-glass making was made into a wholly continuous process by Henry Ford's automobile engineers. (Broadly similar changes occurred in making glass containers and light bulbs.) By the 1920s the modernists, perhaps even more than they knew, had found in glass a material expressive of their fascination with machine production and continuous flow.

Window glass was mechanized early on. By 1880 Belgian window-glass makers were using so-called tank furnaces fired by artificial gas, and the first American instal-

FIG. 6.2. MASS-PRODUCED WINDOW GLASS

Machine blowing cylinders of glass at Pilkington Brothers' St. Helens works in England. Cylinders were blown up to 40 feet in height before they were detached, cut into lengths, split open, and flattened into huge sheets of window glass.

Raymond McGrath and A. C. Frost, *Glass in Architecture and Decoration* (London: Architectural Press, 1937), 61.

lations of tank furnaces using natural gas soon followed. By 1895 fully 60 percent of American window glass was melted in tank furnaces. Around 1900 a continuous-tank factory in Pennsylvania, with three furnaces, required just seven workers to unload the raw materials, feed them into the hoppers, and stir and prepare the melted glass. Each of the three tank furnaces produced enough glass for ten skilled glass blowers and their numerous helpers, who gathered, blew, flattened, and cut the glass into window sheets. Melting window glass by the batch in pots persisted well into the twentieth century, however. While Pennsylvania accounted for two-fifths of America's total glass production, the discovery of cheap natural gas in Ohio, Indiana, Illinois, and Missouri led to the profusion in these states of small pot-melting furnaces.

The first window-glass blowing machine did not reduce the total number of glassmakers, but did dramatically alter the needed skills. John Lubbers, a window-glass flattener in the employ of the American Window Glass Company—an 1899 merger of firms that accounted for 70 percent of the nation's window-glass capacity and nearly all its tank furnaces—experimented for seven years until 1903 when his batch process was able to compete successfully with handblown window glass. (French, Belgian, and other American inventors had in earlier decades experimented with rival schemes for mechanical blowing.) Lubbers' machine produced huge cylinders of glass, up to twice the diameter and five times as long as the handblown cylinders.

With its clanking and puffing, Lubbers' mechanical monster must have been something to watch (fig. 6.2). It performed a classic batch process. A hollow cast-iron cylinder, the "bait" to "catch" the molten glass, was lowered into a waiting vat of hot glass. After a moment, while glass solidified on the inside of the bait, two motors were started, one to slowly raise the bait with its cylinder of glass attached, and the other blowing air into the growing glass cylinder. When the glass cylinder reached its desired diameter, the blowing motor slowed down while the raising motor kept up its work, pulling out a tall cylinder of glass eventually 35 to 40 feet high. At the top of the cycle, the raising motor was briefly speeded up (to thin out the glass wall), then the huge cylinder of glass was cracked off and swung by crane onto a receiving area. There it was sliced into smaller cylinders, which were split and flattened into window sheets as before. This mechanical contraption required a dozen or more tenders, in addition to the cutters and flatteners. But the new mechanical technology obviously displaced the skilled glass *blowers* and, in the United States at least, wiped out their craft union. By 1905 American window-glass makers had installed 124 of these cylinder machines, and by 1915 a total of 284 cylinder machines (of several different types) accounted for two-thirds of US window-glass production.

A continuous process more streamlined than Lubbers' batch machine helped propel the Libbey-Owens Sheet Glass Company to fame. Development of this process required nearly two decades, beginning with the early experiments of Irving W. Colburn, also working in Pennsylvania, and culminating in its commercial success in 1917, when Libbey built a six-unit factory in Charleston, West Virginia. In the patented Colburn machine, the bait was an iron rod, and it was dipped lengthwise into a shallow pan of hot glass. Once the glass had adhered to the rod, it was pulled by motor up from the vat and over a set of rollers, forming a flat sheet of glass directly. Water-cooled side rollers an inch or two above the molten glass kept the sheet at the proper width. When the sheet extended onto the nearby flattening table, a set of mechanical grip bars took over pulling the ever-lengthening sheet from the vat. From the flattening table the sheet passed directly into an annealing oven, through which it moved on 200 asbestos-covered rollers and was then cut, on a moveable cutting table, into suitable-size sheets. Several of Libbey's competitors imported from Belgium the Fourcault sheet-drawing machine, patented in 1902 but not commercialized until after the First World War. It drew glass straight up into a sheet, rather than horizontally over rollers, and reportedly saved even more labor; only cutters were still needed. Across the 1920s these semi-continuous cylinder and sheet machines together produced an ever-larger share of US window glass. (Handblown window glass dropped from 34 to 2 percent of total production between 1919 and 1926.)

Plate glass was thicker than window glass and could safely be made into much larger sheets. Thick plate glass windows were of crucial importance in the tall office buildings going up in US cities, because ample natural illumination was a primary goal of architects laying out office space in the age of hot, expensive incandescent lights. (American-made plate glass was used mostly for windows; high-quality European plate glass was required for mirrors.) Before 1900 American plate-glass making was wholly dependent on English techniques and machinery. Much labor and skill were required to cast the large plates of glass, weighing up to 1,400 pounds, while semiautomatic machines conducted the subsequent rounds of grinding and polishing to obtain a smooth surface. Industry lore has it that all polishing and grinding machinery was imported from England until an American manufacturer, engaging in a bit of industrial espionage, visited the Pilkington company's plant sometime before 1900 and, upon his return to the States, replicated its secrets.

Plate-glass making between 1900 and 1920 underwent a series of evolutionary changes. Engineers made incremental improvements in processing the varied grades of sand used for grinding and on the rouge used for polishing, while electric motors were increasingly used to drive the huge (35-foot-diameter) round polishing tables. Factory designers sought to speed up the time-intensive grinding and polishing stages. Continuous-flow annealing ovens eased the strains in the newly cast glass. A plate of glass might emerge in as little as three hours from a typical 300-foot-long sequence of five ovens. (By comparison batch-process annealing kilns required forty-eight hours just to cool down so that workmen could climb into the kiln and haul out the sheets by rope.) As a result of these mechanical developments, a large plate of glass that might have taken ten days to complete in 1890 could be finished in as little as thirty-six hours in 1923.

Developments after 1920 transformed plate-glass making into a wholly rather than partially continuous-production industry. In that year, while building his massively integrated River Rouge complex, Henry Ford assigned a team of engineers to work on glassmaking. It seems unclear whether Ford desired simply to raise the volume of production or whether he wanted them to focus on laminated safety glass, a technical challenge that established glassmakers had been unwilling to attempt. In the event, Ford's engineers hit on a continuous production process that became widely adopted by the entire plate-glass industry. In 1927 the Pittsburgh Plate Glass Company, a major manufacturer, owned five plate-glass factories and produced 50 percent of the nation's total, but automobile manufacturers owned eight plate-glass factories (including three of the four continuous-production plants) and their output accounted for 35 percent of the total. By 1929 fully half of American plate glass was manufactured by the continuous process.

In the Ford-style continuous-production scheme, glass flowed from a continuous melting tank (the accepted standard) in a stream onto an inclined plane that formed the glass into a flat sheet. A roller pressed the ever-moving sheet to proper thickness and then the hot glass sheet exited onto a moving table and into a continuous annealing oven. At the far end of the oven, the sheet was cut into desired sizes. Previously, fifty or more workers were needed for the casting stage; now just ten workers tended the entire process, from melting through annealing. Ford engineers also introduced assembly-line principles to the grinding and polishing stages, using a long, narrow continuous conveyer that successively ground and polished each sheet of glass. While the largest plates were still made by the traditional batch regime, the Ford-style continuous-plate process worked best on smaller sizes, and soon these smaller plates were widely used for windows.

The appearance of a pane of glass in 1890 compared with one in 1925 was not so very different. And unit prices for window and plate glass were actually 30–50 percent higher in the mid-1920s than in the 1890s, owing to higher demand. What had changed most was the *amount* of window glass. Window-glass production in the U.S. grew threefold during these years, to 567 million square feet, while plate-glass production grew an astounding fifteenfold, to 148 thousand square feet. Just as with steel, the capability to produce glass in large volumes led to its being "discovered" as a modern material.

MANIFESTOS OF MODERNITY

Modernism in architecture depended on mass-produced modern materials. As early as 1929 the German architect Bruno Taut defined modernism as "flat roofs, huge sheets of glass, 'en tout cas' horizontal ribbon-rows of windows with pillars, which strike the eye as little as may be, by reason of black glass or dull paint, more sheets of concrete than are required for practical purposes, etc."[5] This modern style—as a distinctive architectural style itself—was evident first at the Weissenhof housing exposition at Stuttgart in 1927 and was canonized officially by New York's Museum of Modern Art in 1932. It is recognizable instantly in the glass-box "corporate style" skyscrapers that went up in the 1950s and 1960s, but in the decades since it has become dull and hackneyed from mindless repetition. Central to the development of architectural modernism were the interactions among three groups: the Futurists in Italy, who gave modernism an enthusiastic technology-centered worldview; the members of de Stijl in the Netherlands, who articulated an aesthetic for modern materials; and the synthesis of theory and practice in the Bauhaus in Germany.

The Italian Futurists, a "close-knit fighting unit" led by Filippo Marinetti, found a modern aesthetic in the new world of automobiles, factories, and cities. Marinetti's wild enthusiasm for the machine found expression in a series of culture-defining "manifestos." In the years between 1910 and 1916, the Futurists' poets, painters, sculptors, and architects simply blasted a hole through the traditional views of art and architecture. Two of the group's most creative figures—Umberto Boccioni and Antonio Sant'Elia—were killed in World War I, but across the 1920s Marinetti brought their work to the attention of the embryonic modern movement. The legacies of Futurism include Marinetti's insistence that modern materials were the foundation of modern culture: "I leave you with an explosive gift, this image that best completes our thought: 'Nothing is more beautiful than the steel of a house in construction.'"[6]

Marinetti returned from study in Paris to live and write in the northern Italian city of Milan, at the center of a region undergoing its own second industrial revolution. Textile production around Milan tripled between 1900 and 1912, iron and steel production likewise tripled, to 1,000,000 metric tons, while a world-class automobile industry sprang up in Milan with the establishment of Pirelli and Alfa Romeo and the great FIAT complex in nearby Turin. Automobiles are at the center of Marinetti's founding "Manifesto of Futurism," which launched the modernist vision of technology as a revolutionary cultural force. The 1909 manifesto begins with a set piece at his family's house in old Milan and a late-night discussion among friends. They had argued, he says, to the furthest limits of logic and covered sheets of paper with scrawls. In the middle of the night they felt alone, like proud beacons or forward sentries, curiously at one with "the stokers feeding the hellish fires of great ships . . . the red-hot bellies of locomotives." From the past, they heard the old canal, believed to be a work of Leonardo, "muttering its feeble prayers and the creaking bones of palaces dying above their damp green beards." Then, suddenly, beneath the windows the silence was broken by "the famished roar of automobiles."

Marinetti and his friends piled into three waiting automobiles and raced through the early-morning streets. Swerving around two bicyclists, Marinetti flipped the car and landed in a ditch. Dripping with "good factory muck," he climbed out and proclaimed a manifesto to deliver Italy from "its foul gangrene of professors, archeologists, guides and antiquarians." Marinetti's images, arresting and enduring, forthrightly declared a modern aesthetic in the world of modern technology. "We affirm that the world's splendour has been enriched by a new beauty: the beauty of speed. A racing car whose hood is adorned with great pipes, like serpents of explosive breath—a roaring car that seems to ride on grapeshot—is more beautiful than the *Victory of Samothrace*."

> We will sing of great crowds excited by work, by pleasure, and by revolt; we will sing of the multicolored, polyphonic tides of revolution in the modern capitals; we will sing of the vibrant nightly fervour of arsenals and shipyards blazing with violent electric moons; greedy railway stations that devour smoke-plumed serpents; factories hung on clouds by the crooked lines of their smoke; bridges that leap the rivers like giant gymnasts, flashing in the sun with a glitter of knives; adventurous steamers that sniff the horizon; deep-chested locomotives pawing the tracks like enormous steel horses bridled by tubing; and the sleek flight of planes whose propellers chatter in the wind like banners and seem to cheer like an enthusiastic crowd.

The sculptor Boccioni phrased the cultural shift this way: "The era of the great mechanised individuals has begun, and all the rest is Paleontology."[7]

The first result of Marinetti's call for a techno-cultural revolution was a flurry of free verse with such titles as "L'Elettricità," "A un Aviatore," and "Il Canto della Città di Mannheim." A more significant result came with the paintings of Giacomo Balla and the sculpture of Umberto Boccioni. Their challenge was to deliver on the 1910 manifesto of Futurist painting, which had argued that living art must draw its life from the modern world: "Our forebears drew their artistic inspiration from a religious atmosphere which fed their souls; in the same way we must breathe in the tangible miracles of contemporary life—the iron network of speedy communications which envelops the earth, the transatlantic liners, the dreadnoughts, those marvelous flights which furrow our skies, the profound courage of our submarine navigators . . . the frenetic life of our great cities."[8]

Futurism was not about painting pictures of battleships or airplanes. Balla struggled to express in painting such abstract concepts as dynamism and elasticity, while Boccioni argued that sculptors must "destroy the pretended nobility . . . of bronze and marble" and instead use appropriate combinations of glass, cardboard, cement, iron, electric light, and other modern materials. No classical statues or nude models here. Futurist sculpture, such as Boccioni's *Unique Forms of Continuity in Space* (1913), blended stylized human forms with the machine forms of the modern world. Modern objects, with their "marvelous mathematical and geometrical elements," Boccioni wrote, "will be embedded in the muscular lines of a body. We will see, for example, the wheel of a motor projecting from the armpit of a machinist, or the line of a table cutting through the head of a man who is reading, his book in turn subdividing his stomach with the spread fan of its sharp-edged pages."[9]

With their unbounded enthusiasm for modern technology, the Futurists understandably took a dim view of traditional historical styles. Indeed, in their opinion no proper architecture had existed since the eighteenth century, only the "senseless mixture of the different stylistic elements used to mask the skeletons of modern houses." Their architectural manifesto of 1914 hailed the "new beauty of cement and iron" and called on architects to respond constructively to "the multiplying of machinery, the constantly growing needs imposed by the speed of communications, the concentration of population, hygiene, and a hundred other phenomena of modern life." The Futurist concept of a house would embrace "all the resources of technology and science, generously satisfying all the demands of our habits and our spirit." In short, this would be "an architecture whose sole justification lies in the unique conditions of modern life and its aesthetic correspondence to our sensibilities."

Sant'Elia argued that modern architecture must set tradition aside and make a fresh start: "Modern building materials and scientific concepts are absolutely incompatible with the discipline of historical styles, and are the main reason for the grotesque appearance of 'fashionable' buildings where the architect has tried to use the lightness and superb grace of the iron beam, the fragility of reinforced concrete, to render the heavy curve of the arch and the weight of marble." There was a new ideal of beauty, still emerging yet accessible to the masses. "We feel that we no longer belong to cathedrals, palaces and podiums. We are the men of the great hotels, the railway stations, the wide streets, colossal harbors, covered markets, luminous arcades, straight roads and beneficial demolitions."[10]

Sant'Elia summed up his ideas about modern materials and urban form in his great city-planning project, the famed Città Nuova. A futuristic vision of a hyperindustrialized Milan, Città Nuova was first exhibited in 1914 and continued to inspire modernist architects for decades. Sant'Elia poured out a flood of modernistic images:

> We must invent and rebuild *ex novo* our Modern city like an immense and tumultuous shipyard, active, mobile and everywhere dynamic, and the modern building like a gigantic machine. Lifts must no longer hide away like solitary worms in the stairwells, but the stairs—now useless—must be abolished, and the lifts must swarm up the façades like serpents of glass and iron. The house of cement, iron, and glass, without carved or painted ornament, rich only in the inherent beauty of its lines and modelling, extraordinarily brutish in its mechanical simplicity, as big as need dictates, and not merely as zoning rules permit, must rise from the brink of a tumultuous abyss; the street which, itself, will no longer lie like a doormat at the level of the thresholds, but plunge storeys deep into the earth, gathering up the traffic of the metropolis connected for necessary transfers to metal cat-walks and high-speed conveyor belts.

We must create the new architecture, Sant'Elia proclaimed, "with strokes of genius, equipped only with a scientific and technological culture." Sant'Elia (possibly with some help from Marinetti, ever ready with verbal fireworks) finished up with his most widely quoted conclusion: "Things will endure less than us. Every generation must build its own city."[11]

Marinetti's provocative avant-garde stance, frank celebration of violence, and crypto-revolutionary polemics landed the Futurists squarely in the middle of post-war fascism. Violence was in the air, and Italy's liberal democracy was in tatters. More than a dozen groups, ranging from respectable university students to gun-toting street gangs, used *fascio* in their names. As "the new man," the presumed leader of this motley crew, Marinetti for a time rivaled even Mussolini, known chiefly as the editor of the Socialist Party's newspaper before the war, and at the time a stridently *anti*-socialist editor and journalist. For Marinetti, perhaps the high point (if one can call it that) came in April 1919, when he took a mob through the streets of Milan and wrecked the headquarters of the Socialist Party's newspaper, an event later known in the regime's legends as "the first victory of Fascism."

Marinetti and Mussolini saw eye-to-eye on war, violence, women, and airplanes but not the established social and moral order. In 1920, Mussolini achieved national political prominence through a newfound alliance with Italy's right-wing business and religious elites; two years later, following the infamous "march on Rome," he was sworn in as the Fascist prime minister of Italy.[12] While Mussolini solidified his grip on national power, Marinetti energetically took up the cause of international Futurism. At home, he signaled his political irreverence by developing a revolution in Italian cooking. His *La Cucina Futurista* features such delicacies as Car Crash— the middle course of a Dynamic Dinner—consisting of "a hemisphere of pressed anchovies joined to a hemisphere of date puree, the whole wrapped up in a large, very thin slice of ham marinated in Marsala." The Aeropoetic Futurist Dinner is set in the cockpit of a Ford Trimotor flying at 3,000 meters. Such dishes as Nocturnal Love Feast and Italian Breasts in the Sunshine express a hedonistic attitude to bodies and sex, which was (grotesquely) theorized in Valentine de Saint-Point's "Futurist Manifesto of Lust" (1913). Pasta was banned in Marinetti's Futurist cuisine.[13]

During the 1920s the Futurists slipped off the stage of Italian politics but became a serious international cultural movement. The best-known Futurist work of archi- tecture was an automobile factory complex built during 1914–26 for the Italian firm FIAT outside Turin. With its high and expansive "daylight" windows, its long and unornamented "planar" walls, and especially its dramatic roof—the site of a high-banked oval track where finished FIAT cars could be test driven—it became a classic modernist icon and a mandatory waypoint on modernist pilgrimages. Marinetti worked tirelessly to bring Futurist concepts and images, especially the sev- eral manifestos and Sant'Elia's *Città Nuova*, to receptive audiences. Before the war he proclaimed Futurist manifestos in Paris, London, Rotterdam, and Berlin, while a Futurist exhibition held in Paris subsequently traveled to no fewer than eleven European cities. One receptive audience, as early as 1912, was a group of German Expressionists in Berlin, many of whom were recruited to form the Bauhaus school of design in 1919, as we will see below. A second group that brought Futurist ideas into the larger avant-garde movement was the Dutch movement de Stijl (The Style), which also interacted with the Bauhaus, sharing students and staff.

De Stijl was a loosely interacting group of architects and painters; the name was also the title of their influential art magazine, published from 1917 until 1931.

Sensing in de Stijl a kindred spirit, Marinetti in 1917 sent the Futurist architectural manifesto and a selection of Sant'Elia's drawings to Theo van Doesburg, the group's organizer and central figure. In response *De Stijl* published a warm appreciation ("the perfect management of this building taken as a whole, carried out in modern materials . . . gives this work a freshness, a tautness and definiteness of expression") that secured Sant'Elia's international reputation. *De Stijl*'s far-reaching circulation made one particular drawing, through its multiple reproductions across Europe, the best known of all Sant'Elia's work.[14]

Chief among the de Stijl theorists was Piet Mondrian, a pioneer practitioner of abstract, nonfigurative painting. "The life of today's cultured person turns more and more away from nature; it is an increasingly abstract life," he announced. For de Stijl, the terms *nature* and *abstract* were on opposite sides of the "great divide" between tradition and modernity. Mondrian maintained that artists should recognize that there was an ultimate reality hiding behind everyday appearance and that artists should strive to see through the accidental qualities of surface appearance. He looked to the city as inspiration for the emerging modern style: "The genuinely Modern artist sees the metropolis as Abstract living converted into form; it is nearer to him than nature, and is more likely to stir in him the sense of beauty . . . that is why the metropolis is the place where the coming mathematical artistic temperament is being developed, the place where the new style will emerge." One like-minded modernist put the same point more simply: "After electricity, I lost interest in nature."[15]

Van Doesburg took Mondrian's notions about modern life one step further. In a famous lecture in 1922 called "The Will to Style: The New Form Expression of Life, Art and Technology," van Doesburg told audiences in Jena, Weimar, and Berlin: "Perhaps never before has the struggle between nature and spirit been expressed so clearly as in our time." Machinery, for van Doesburg, was among the progressive forces that promised to lift humans above the primitive state of nature and to foster cultural and spiritual development. The task of the artist was to derive a style—or universal collective manner of expression—that took into account the artistic consequences of modern science and technology:

> Concerning the cultural will to style, the machine comes to the fore. The machine represents the very essence of mental discipline. The attitude towards life and art which is called materialism regarded handiwork as the direct expression of the soul. The new concept of an art of the mind not only postulated the machine as a thing of beauty but also acknowledged

immediately its endless opportunities for expression in art. A style which no longer aims to create individual paintings, ornaments or private houses but, rather, aims to study through team-work entire quarters of a town, skyscrapers and airports—as the economic situation prescribes—cannot be concerned with handicraft. This can be achieved only with the aid of the machine, because handicraft represents a distinctly individual attitude which contemporary developments have surpassed. Handicraft debased *man* to the status of a machine; the correct use of the machine (to build up a culture) is the only path leading towards the opposite, social liberation.

Iron bridges, locomotives, automobiles, telescopes, airport hangars, funicular railways, and skyscrapers were among the sites van Doesburg identified where the new style was emerging.[16]

A more striking association of technology with a desired cultural change is difficult to imagine. This is the crucial shift: whereas the Futurists sang enthusiastic hymns to modern technology and the dynamic city, for de Stijl modern technology and the city were desirable because they were a *means* by which "to build up a culture." This involved careful decisions ("the correct use of the machine"), not the Futurists' trusting embrace of automobiles or airplanes.

The buildings and theoretical writings of H. P. Berlage heavily influenced members of de Stijl. Architectural "style" in the modern age was for Berlage an elusive quality that an architect achieved by practicing "truth to materials" ("decoration and ornament are quite inessential") and creating spaces with proper geometrical proportions. Berlage's own Amsterdam Stock Exchange became a famous modernist building, but of equal importance was his early grasp and interpretation of Frank Lloyd Wright. Wright was virtually the only American architect noticed by the European modernists. Both Berlage's Stock Exchange (1902) and Wright's Larkin office building (1905) used a combination of brick and advanced structural techniques to create large open-air halls surrounded by galleries, in effect creating open spaces in the very middle of the buildings. Europeans did not learn much from Berlage about Wright's interest in the vernacular and nature worship. Instead, Berlage emphasized Wright's views on the technological inspiration of modern culture ("The machine is the normal tool of our civilization, give it work that it can do well; nothing is of greater importance"). In effect, Berlage selectively quoted Wright to support the technological framing of modernism sought by de Stijl: "The old structural forms, which up to the present time have been called architecture, are decayed. Their life went from them long ago and new conditions industrially, steel and concrete, and terra-cotta in particular, are prophesying a more plastic art." Berlage probably showed Le Corbusier, the prolific writer and influential architectural theorist, a Dutch-designed "modern villa at Bremen" that was Corbusier's first view of a modernist building.[17]

J. J. P. Oud was the leading practicing architect associated with de Stijl. Oud knew Wright's work and appreciated "the clarity of a higher reality" achieved by Sant'Elia, but his own practical experiences grounded his theory. His early work—including several houses, villas, shops, a block of workers' apartments (fig. 6.3), and a modernist vacation home—received such acclaim that in 1918, at the age of twenty-eight, he was named city architect for Rotterdam. In 1921 he wrote *On Modern Architecture and Its Architectonic Possibilities.* Oud clearly sensed and helped articulate the architectural possibilities of modern technology, but at the same time he avoided the quagmire of technological utopianism: "I bow the knee to the wonders of technology, but I do not believe that a liner can be compared to the Parthenon [contra Futurists]. I long for a house that will satisfy my every demand for comfort, but a house is not for me a living-machine [contra Corbusier]." Disappointingly for him, "the art of building . . . acts as a drag on the necessary progress of life," Oud wrote: "the products of technological progress do not find immediate application in building, but are first scrutinized by the standards of the ruling aesthetic, and if, as usual, found to be in opposition to them, will have difficulty in maintaining themselves against the venerable weight of the architectural profession." To help architects embrace the new building materials, he articulated an aesthetic for plate glass, iron and steel, reinforced concrete, and machine-produced components.

> When iron came in, great hopes were entertained of a new architecture, but it fell aesthetically-speaking into the background through improper application. Because of its visible solidity—unlike plate-glass which is only solid to the touch—we have supposed its destination to be the creation of masses and planes, instead of reflecting that the characteristic feature of iron construction is that it offers the maximum of structural strength with the minimum of material. . . . Its architectural value therefore lies in the creation of voids, not solids, in contrast to mass-walling, not continuing it.

Glass at the time was usually employed in small panes joined by glazing bars, so that the window "optically continues the solidity of the wall over the openings as well," but Oud argued that glass should instead be used in the largest possible

FIG. 6.3. DUTCH MODERNISM BY J. J. P. OUD

"Graceful development in the new tendency . . . modern architecture has definitely won through in Holland," was Bruno Taut's verdict after seeing these workmen's houses at Hook van Holland.

Bruno Taut, *Modern Architecture* (London: The Studio, 1929), 91, 123.

sheet with the smallest possible glazing bars. Reinforced concrete's tensile strength and smooth surface offered the possibility of "extensive horizontal spans and can-tilevers" and finished surfaces of "a strict clean line" and "pure homogenous plane." In his conclusion, Oud called for an architecture "rationally based on the circum-stances of life today." He also catalogued the proper qualities of modern materi-als. The new architecture's "ordained task will be, in perfect devotion to an almost impersonal method of technical creation, to shape organisms of clear form and proper proportions. In place of the natural attractions of uncultivated materials . . . it would unfold the stimulating qualities of sophisticated materials, the limpidity of glass, the shine and roundness of finishes, lustrous and shining colors, the glitter of steel, and so forth."[18]

The enduring contribution of de Stijl, then, was not merely to assert, as the Futurists had done, that modern materials had artistic consequences, but to identify specific consequences and embed these in an overarching aesthetic theory. Architects now could associate factory-made building materials like steel, glass, and reinforced concrete with specific architectural forms, such as open spaces, extensive spans, and clean horizontal planes. Moreover, with the suggestion that architects devote themselves to an "impersonal method of technical creation," Oud took a fateful step by transforming the Futurists' flexible notion that every generation would have its own architecture into a fixed method of architectural design.

"Picasso, Jacobi, Chaplin, Eiffel, Freud, Stravinsky, Edison etc. all really belong to the Bauhaus," wrote one of the school's students in the 1920s. "Bauhaus is a progressive intellectual direction, an attitude of mind that could well be termed a religion."[19] These heady sentiments found practical expression in an advanced school for art and architecture active in Germany from 1919 to 1933. The Bauhaus was founded and grew during the country's fitful struggles to sustain democracy and the disastrous hyperinflation of 1921–23. Originally located in the capital city of Weimar, the Bauhaus relocated first to Dessau and finally to Berlin. After its break-up in 1933 its leading figures—Walter Gropius, Mies van der Rohe, Lazlo Moholy-Nagy— emigrated to the United States and took up distinguished teaching careers in Boston and Chicago. Gropius wrote of the Bauhaus, "Its responsibility is to educate men and women to understand the world in which they live, and to invent and create forms symbolizing that world."[20]

Such a visionary statement might be regarded as yet another wild-eyed manifesto, but by 1919 Gropius was among the progressive German architects, artists, and designers who had substantial experience with industrial design. The Bauhaus began as the fusion of two existing schools, which brought together students of the fine and applied arts. While the fine arts academy's traditions stretched back into the past, the applied arts school had been organized just fifteen years earlier in a wide-ranging campaign to advance the aesthetic awareness of German industry. Other initiatives of that time included the founding by Hermann Muthesius and Peter Behrens of the Deutscher Werkbund, an association of architects, designers, and industrialists that promoted the gospel of industrial design to German industry, as well as the employment of Behrens by the giant firm AEG (Allgemeine Elektricitäts-Gesellschaft, a successor to German Edison). Behrens was in effect AEG's in-house style maven. Between 1907 and 1914, several major figures in modern architecture—including Mies van der Rohe, Bruno Taut, Le Corbusier, and Gropius himself—worked in Behrens' studio. Indeed, many influential modernist buildings can be traced to this Werkbund–Behrens connection, including Behrens' own factory buildings for AEG (1908–12), Gropius' Faguswerke (1911–13, fig. 6.4) and Werkbund Pavilion (1914), as well as Taut's exhibition pavilions for the steel and glass industries (1913–14), which dramatically displayed these modern materials.

The Bauhaus was unusually well positioned to synthesize and transform

FIG. 6.4. THE FIRST "MODERN" FACTORY

The Faguswerke factory (1911–13), designed by Adolf Meyer and Walter Gropius, made humble shoe lasts for shoemakers, but photographs of the building made history. Modernists praised the glass-enclosed corner stairwell as an open and unbounded vision of space.

Bruno Taut, *Modern Architecture* (London: The Studio, 1929), 57.

advanced concepts circulating in the 1920s about materials, space, and design. Many of its early staff members were drawn from Der Sturm, the Expressionist movement in Berlin that had provided an early audience for the Futurists. These avant-garde painters were especially receptive to the abstract machine-inspired Constructivist art emerging in the early Soviet Union, which the Russian El Lissitzky brought to the attention of Western Europeans through his association with de Stijl in the mid-1920s. Other leading members of de Stijl—including van Doesburg, Mondrian, and Oud—either lectured at the Bauhaus or published in its influential book series. Van Doesburg's lectures at the Bauhaus (including his "Will to Style" quoted earlier) helped turn the Bauhaus away from its early focus on Expressionism, oriental mysticism, and vegetarianism and toward an engagement with real-world problems and advanced artistic concepts.

Elementarism was an abstract concept drafted to the cause. Before the Bauhaus turned it into a distinctive architecture concept, *elementarism* had several diverse meanings. As early as 1915 the Russian abstract painter Kasimir Malevich pointed to the fundamental elements, or simple geometrical forms, that were basic units of his compositions; he later expanded his views in the Bauhaus book *Non-objective World*. In 1917, another exemplar, Gerrit Rietveld, even before he associated with de Stijl, made the first of his famous chairs. They were not intended as comfortable places to sit. Rather, Rietveld separated and analyzed the functions of a chair—sitting, enclosing, supporting—and created a design in which each of the elements of his chairs, made of standard dimensional lumber, was visually separated from the other elements and held in a precise location in space. In 1924, van Doesburg pointed out that "the new architecture is *elementary*," in that it develops from the elements of construction understood in the most comprehensive sense, including function, mass, plane, time, space, light, color, and material. In 1925, self-consciously using the ideas of elementarism, Rietveld built the Schroeder house in Utrecht while van Doesburg coauthored a prize-winning plan for the reconstruction of the central district in Berlin.[21] As interpreted by the Bauhaus theorist Moholy-Nagy, himself an abstract painter, such "elements" were conceived of as the fundamental units of structure and space.

By the mid-1920s students at the Bauhaus were studying with some of the most original artists and architects in Europe. Students began with the six-month introductory course, the Vorkurs and then went on to a three-year formal apprenticeship in a particular craft (e.g., metalwork, pottery, weaving, or woodwork) that resulted in a Journeyman's Diploma. Finally, students could elect a variable period of instruction in architecture or research leading to a Master's Diploma. By 1923 Gropius saw the school as preparing students for modern industry. "The Bauhaus believes the machine to be our modern medium of design and seeks to come to terms with it," he wrote. Training in a craft complemented the desirable "feeling for workmanship" always held by artists and prepared students for designing in mass-production industries.[22] The school's reorientation from mysticism to industry was expressed in its 1923 exposition, Art and Technology—A New Unity.

Von Material zu Architektur, by Lazlo Moholy-Nagy, a key theoretical reflection, was one of the principal Bauhaus texts. In one sense, Moholy-Nagy's title (*From Material to Architecture*) suggests the passage of students from the study of materials, properly theorized, to the study of architecture. This educational plan was expressed in a circular diagram, attributed to Gropius, that showed students passing through the outer layer of the Vorkurs, then specializing in the study of a specific material, and at last taking up the study of architecture, which was at the core. Viewing construction materials as the medium of the modern world figured largely in Moholy-Nagy's career. Moholy-Nagy had come to Berlin in 1921, immersed himself in the avant-garde world of Der Sturm, de Stijl, and Russian abstraction. He was appointed to the Bauhaus in 1923 to oversee the metalworking shop; later in that year he also (with Josef Albers) took on the Vorkurs. One of his innovations was to transform the study of materials from an inspection of their inner nature to an objective physical assessment of their various properties. To test a material's strength or flexibility or workability or transparency, he devised a set of "tactile machines" that were used at the Bauhaus.

IRONIES OF MODERNISM

"We aim to create a clear, organic architecture whose inner logic will be radiant and naked," wrote Walter Gropius in *Idee und Aufbau* (1923). "We want an architecture adapted to our world of machines, radios and fast cars . . . with the increasing strength and solidity of the new materials—steel, concrete, glass—and with the new audacity of engineering, the ponderousness of the old methods of building is giving way to a new lightness and airiness."[23] At the time, Gropius was engaged in creating some modernist icons of his own. He had just completed his striking modernist entry for the Chicago Tribune Tower competition (1922). He would soon turn his architectural energies to designing the new Bauhaus buildings at Dessau (discussed later). The rush to proclaim a distinctive modern Bauhaus style provoked one critic to jest: "Tubular steel chairs: Bauhaus style. Lamp with nickel body and white glass shade: Bauhaus style. Wallpaper covered in cubes: Bauhaus style. Wall without pictures: Bauhaus style. Wall with pictures, no idea what it means: Bauhaus style. Printing with sans serif letters and bold rules: Bauhaus style. doing without capitals: bauhaus style."[24]

The efforts during the 1920s to proclaim a modern style in Germany occurred under unusual difficulties. The country's political turmoil, economic crisis, and street violence made postwar Italy's ferment look calm in comparison. During 1923 the German economy collapsed under the strain of reparations payments imposed on it at the end of World War I. In January 1919 it took eight war-weakened German marks to purchase one US dollar. Four years later it took 7,000 marks, and by December 1923 it would take the stupendous sum of 4.2 trillion marks to purchase one US dollar. In February of that disastrous year, Gropius asked the government for 10 million marks (then equivalent to about $1,000) to help fund the art-and-technology exposition. By the end of that year's hyperinflation, the sum of 10 million marks was worth less than one one-hundred-thousandth of a US penny. The Bauhaus took up the surreal task of designing million-mark banknotes so housewives might buy bread without a wagon to transport the necessary bills. With the stabilization of the German currency in 1924 (the war debts were rescheduled) building projects began once again.

The lack of adequate housing especially plagued Germany's industrial cities, which had grown swiftly during the electricity and chemical booms of the second industrial revolution (see chapter 5). City governments in several parts of Germany began schemes to construct affordable workers' housing. Three cities—Dessau, Berlin, and Frankfurt—gave the German modernists their first opportunities for really large-scale building, and they have figured prominently in histories of modernism ever since. The mayor of industrial Dessau attracted the Bauhaus to his city, offering to fund the salaries of the staff and the construction of new school buildings, in exchange for assistance with his city's housing. For years Gropius had dreamed of rationalizing and industrializing the building process. Not only did he advocate the standardization of component parts, the use of capital-intensive special machinery, and the division of labor; he was also a close student of the labor and organizational methods of Henry Ford and efficiency engineer Frederick W. Taylor. And as we will see, such modernistic "rationalization" came with a similar vengeance to the household and the housewife who worked there.

At Dessau Gropius in effect offered the Bauhaus as an experimental laboratory for the housing industry. His commission from the city was to design and build 316 two-story houses, together with a four-story building for the local cooperative. Like Ford, Gropius specially planned the smooth flow of materials. Composite building blocks and reinforced-concrete beams were fabricated at the building site. The houses stood in rows, and rails laid between them carried in the building materials. Such laborsaving machines as concrete mixers, stone crushers, building-block makers, and reinforced-concrete beam fabricators created a factory-like environment. Like Taylor, Gropius had the planners write out detailed schedules and instructions for the building process. Individual workers performed the same tasks repeatedly on each of the standardized houses.[25]

The housing program in Berlin during the 1920s matched Dessau in innovative construction techniques and use of the modern style, but dwarfed Dessau in scale. To deal with the capital city's housing shortage, Martin Wagner, soon to become Berlin's chief city architect, in 1924 founded a building society, Gemeinnützige Heimstätten-Spar-und-Bau A.G. (GEHAG). Wagner, an engineer and member of the Socialist Party, had previously formed cooperatives of building crafts workers; in turn, trade unions financed GEHAG, soon one of Berlin's two largest building societies. GEHAG's twin aims were to develop economical building techniques and to build low-cost housing. In the five-year period 1925–29, the city's building societies together put up nearly 64,000 dwelling units, and around one-fifth of these were

FIG. 6.5. MAY CONSTRUCTION SYSTEM IN FRANKFURT

In the late 1920s Ernst May, the city architect for Frankfurt, oversaw the building of 15,000 dwelling units there. May's research team devised special construction techniques (e.g., the prefabricated "slabs" of concrete shown here) to hold down costs and speed up the building process. The result was a practical demonstration of mass-produced housing at reasonable cost.

Bruno Taut, *Modern Architecture* (London: Studio, 1929), 114.

designed by modernist architects. (Private enterprise added 37,000 more units.) In preparation for this effort, Wagner visited the United States to examine industrialized building techniques while GEHAG's chief architect, Bruno Taut, studied garden cities in the Netherlands.

Berlin in the mid-1920s was something of a modernist mecca. In 1925 Taut began work on the Britz estate, which would encompass 1,480 GEHAG dwellings, the construction of which employed lifting and earth-moving equipment and rational division of labor. Taut used standardized forms to create a distinctive horseshoe-shaped block. He also designed large estates in the Berlin districts of Wedding, Reinickendorf, and Zehlendorf, the last including 1,600 dwellings in three- and four-story blocks as well as individual houses.[26] Gropius, along with four other architects, built the vast Siemens estate in Berlin for the employees of that electrical firm. These modernist dwellings all featured flat roofs, low rents, communal washhouses, and plenty of light, air, and open space.[27]

But neither Dessau nor Berlin matched the housing campaign of Frankfurt. There, in one of western Germany's largest cities, the building effort rehoused 9 percent of the entire population. Under the energetic direction of Ernst May, the city's official architect, no fewer than 15,174 dwelling units were completed between 1926 and 1930 (fig. 6.5). In 1925 May drew up a ten-year program that called for the city itself to build new housing and for building societies and foundations to do likewise following the city's plans and standards. May did much of the designing himself for the largest estates at the outskirts of the city. (Gropius did a block of 198 flats, while Mart Stam, whom Gropius had once asked to head the Bauhaus architecture effort, completed an 800-flat complex.) May's office established standards for the new building projects; these standards specified the size and placement of doors and windows, ground plans of different sizes, and the famous space-saving kitchens described next. May and his colleagues at the Municipal Building Department carried out research into special building techniques. Cost savings were a paramount concern; by 1927 a factory was turning out precast concrete wall slabs that permitted the walls of a flat to be put up in less than a day and a half. The modern Frankfurt "houses are as much alike as Ford cars," noted one American. "They are all built in units. The large house has more units than the small one, that is all. With the introduction of machine-made houses, the architect becomes an engineer."[28]

A bit south of Frankfurt, at Stuttgart, the emerging modern style had its first highbrow showcase in 1927. Although the city commissioned only sixty dwellings, the Weissenhof housing exposition had immense influence on international modernism. Mies van der Rohe was designated as the director, and he invited fifteen of the best-known modern architects—including Oud and Stam from the Netherlands, Corbusier from Paris, Josef Frank from Vienna, and many of the notable Germans, including Behrens, Poelzig, Taut, and Gropius. The model housing estate coincided with a major Werkbund exhibition and was on public view for an entire year. When the exhibition opened, up to 20,000 people a day saw the new architecture. What is more, May brought numerous exhibition-goers to view his projects in nearby Frankfurt. As Mies put it, the new architecture reflected "the struggle for a new way of habitation, together with a rational use of new materials and new structures."[29]

The Stuttgart exposition of 1927 was the first salvo in a wide-ranging campaign to frame modernism as rational, technological, and progressive. In 1932, the Museum of Modern Art in New York gave top billing to its "International Style" exhibition, which displayed and canonized the preponderantly European works representing this strain of modernist architecture. Later homegrown American contributions to the modern style included world's fair expositions at Chicago (1933) and New York (1939), especially the General Motors "World of Tomorrow" pavilion, which linked science, rationalization, and progress through technology.[30] The Congrès Internationaux d'Architecture Moderne, known as CIAM (1928–56), with its noisy conferences and its edgy and polemical "charters," also shaped the contours of the modern style; it did so in such an assertive way that it became known as a kind of international modernist mafia. As mentioned earlier many leading modernists came to the United States after fleeing Hitler, who mandated "authentic German" styles in architecture and brutally suppressed the left-wing social movements that had supported many of the modernists.[31] The influential teaching of Bauhaus exiles Gropius, Moholy-Nagy, and Mies van der Rohe in Boston and Chicago raised a generation of US-trained architects and designers who imbibed the modern movement directly from these masters. In the 1950s, in architecture at least, the International Style, or Modern Movement, became a well-entrenched orthodoxy.

While the public campaign to enshrine modernism in architecture is well known—one cannot overlook the thousands of modernist office buildings, apartments, hospitals, government buildings, and schools built worldwide in the decades since the 1920s as well as the "car friendly" cities that emerged from Stockholm to Los Angeles under the sway of modernist urban planning—an equally influential set of developments brought modernism to the home. Here, modernism's rationalizing and scientizing impulses interacted with established notions about the household and about women's roles as homemakers. The "Frankfurt kitchen," designed in 1926 by Margarete Schütte-Lihotzky (Grete Lihotzky), became a classic and well-regarded modernist icon.

The outlines of Lihotzky's dramatic life story make her an irresistible heroic figure. Trained in Vienna as an architect, she was one of very few women to thrive in that male-dominated field. She worked energetically to bring workers' perspectives to her projects for housing, schools, and hospitals in Vienna (1921–25), Frankfurt (1926–29), Moscow (1930–37), and Istanbul (1938–40). Her Frankfurt kitchen became so well known that when chief city architect Ernst May moved to Moscow, she agreed to continue working with him only if she *not* do any more kitchens. (In Moscow, she worked mostly on children's nurseries, clubs, and schools.) In 1940 she returned to Austria to join the resistance movement fighting fascism, but within weeks she was arrested, narrowly escaped a death sentence, and spent the war in a Bavarian prison. After the war, she helped with the reconstruction of Vienna, was active in CIAM, and kept up her architectural practice, engaging in many international study trips, publications, and projects through the 1970s.[32]

Household reform in Germany during the 1920s, as Lihotzky discovered, was crowded with diverse actors. The national government, while formally committed to equal rights for women and men, enacted a policy of "female redomestication." This policy encouraged young women to embrace traditional women's roles of homemaker, mother, and supporter for her husband. The government hoped to end the "drudgery" of housework and to reconceive it as modern, scientific, and professional. Modernizing housework through the use of Tayloristic "scientific management" principles was precisely the point of Christine Frederick's *The New Housekeeping: Efficiency Studies in Home Management* (published in the United States in 1913, translated into German in 1922 as *Die rationelle Haushaltführung*) and was a central message of Elisabeth Lüders and Erna Meyer, both prolific authors on women's reform issues and advisors to government and industry. In a leading German engineering journal, Meyer wrote that "the household, exactly like the workshop and the factory, must be understood as a manufacturing enterprise."[33]

The German government agency charged with rationalizing workshops and factories also worked closely with several women's groups to rationalize the household. The Federation of German Women's Associations (Bund Deutscher Frauenvereine), with its one million total members, was the country's largest association of women. With the federation's support the national government enacted compulsory home economics courses for girls and sponsored vocational secondary schools where women learned to be "professional" seamstresses, laundresses, and day-care attendants. Within the federation, the conservative Federal Union of German Housewives Associations (Reichverband Deutscher Hausfrauenvereine), originally founded to address the "servant problem," became a formal advisory body with special expertise on housewifery to the Reich Research Organization. The union's effort to modernize housekeeping resulted in numerous conferences, publications, and exhibitions, including one in Berlin in 1928 that featured a set of model kitchens.[34]

Lihotzhy's work on rational kitchen designs, then, emerged in the context of substantial governmental, industrial, and associational interest in the topic. Ernst May himself initiated a research program on "domestic culture" to shape his Frankfurt housing designs. The program's investigations involved psychology, evaluations of

FIG. 6.6. LIHOTZKY'S FRANKFURT KITCHEN.

"This kitchen was not only designed to save time but also to create an attractive room in which it was pleasant to be," wrote Grete Lihotzky of her space-saving kitchen. Her compact design eliminated "unneeded" steps that made a housewife's work inefficient. More than 10,000 of these factory-built kitchen units were installed in Frankfurt's large-scale housing program.

Peter Noever, ed., *Die Frankfurter Küche* (Berlin: Ernst & Sohn, n.d.), 45.

materials and products, and scientific management principles; researchers studied such diverse areas as household products, consumer markets, appliances, and home economics classrooms. May's chosen household products became an officially recommended line and were publicized in his journal, *The New Frankfurt*. The Frankfurt houses aimed at "air, light and sun in every room of the dwellings . . . sufficient bedrooms" for children, and "extensive lightening of the work of the housewife to free her for the education of the children and to take part in the interests of the husband and work in the garden," noted one of May's assistants. Lihotzhy developed her kitchen design using the principles of Frederick Taylor's time-and-motion studies (for example, reducing the "unneeded" steps a housewife made within her kitchen and in taking food to the nearby eating room), as well as giving careful attention to materials.

Lihotzky's kitchen combined diverse colors and an effective design into a compact and photogenic whole (fig. 6.6). (By comparison a contemporaneous kitchen designed by J. J. P. Oud and Erna Meyer for a Weissenhof house appears ugly, spare, and stark.) Summarizing Lihotzky's own description, one can tell that "some man" was not the designer. The gas range featured an enameled surface for easy cleaning and a "cooking box" (*Kochkiste*) where food that had been precooked in the morning could be left to stew all day, saving the working woman time and energy. The flour drawer, made of oak containing tannic acid, kept worms out. Fully illuminated by an ample window, the worktable, made of beech, featured an easily cleaned metal channel for vegetable waste. Cupboards for crockery were enclosed by glass windows and sealed against dust. Other features integrated into the compact plan of 1.9 by 3.44 meters were a fold-down ironing board, an insulated twin sink, a unit above the sink for drying and storing plates, and a moveable electric light. The kitchen fairly bristled with storage drawers. Most extant photographs of Lihotzky's Frankfurt kitchen are black-and-white, so they fail to reveal that Lihotzky featured color: "The combination of ultramarine blue wooden components (flies avoid the colour blue) with light grey–ochre tiles, the aluminum and white-metal parts together with the black, horizontal areas such as the flooring, work surfaces and cooker ensured that this kitchen was not only designed to save time but also to create an attractive room in which it was pleasant to be."[35]

During the peak years of the Frankfurt building campaign in the late 1920s, Lihotzky's kitchen was installed in 10,000 Frankfurt apartments. In fact, she had worked closely with the manufacturer, Georg Grumbach, to achieve an easily manufactured design. Grumbach's company assembled the kitchens as factory-built units and shipped them whole to the construction site, where they were lifted into place by cranes. Her kitchen also went into production in Sweden, after it was extensively praised in a Stockholm exhibition.

Looking at the modernist movement as a whole, then, a rich set of ironies pervades the history of aesthetic modernism and modern materials. While many of the key figures were, at least in the 1920s, activists committed to achieving better housing for workers, modernism in the 1950s became a corporate style associated with avowedly non-socialist IBM, Sears, and a multitude of cash-rich oil and insurance corporations. Modernism as an overarching style professed itself to be a natural and inevitable development, reflecting the necessary logic of modern technological society; and yet the campaign to enthrone modernism was intensely proactive and political: in the 1920s it was furthered by left-wing city housing projects, in the 1930s modernist architects were banned by the National Socialists, and in the 1940s onward into the Cold War years a certain interpretation of aesthetic modernism—with its socialist origins carefully trimmed away—was the object of intense promotional efforts by CIAM, the Museum of Modern Art, and other highbrow tastemakers. Finally, what can we make of Grete Lihotzky, a committed communist, negotiating with a private manufacturer to mass-produce her kitchen designs? (The Grumbach factories also took orders directly from private homeowners who wanted the Frankfurt kitchen.) The Frankfurt housing developments themselves remained too expensive for the working-class families for whom they were designed. Instead, the Frankfurt apartments filled up with families from the middle class and well-paid skilled workers.

Modern designers and architects, motivated by a variety of impulses, actively took up the possibilities of mass-produced machine-age glass and steel. These materials—typically inexpensive, available in large quantities, and factory-manufactured—made it economically possible to dream of building housing for the masses, in a way that was difficult to achieve with hand-cut stone or custom wood construction. The modern materials were also something of a nucleation point for technological fundamentalists asserting the imperative to change society in the name of technology. In examining how "technology changes society" we see that social actors, frequently asserting a technological fundamentalism that resonates deeply, actively work to create aesthetic theories, exemplary artifacts, supportive educational ventures, and broader cultural and political movements that embed their views in the wider society. If these techno-cultural actors fail to achieve their visions, we largely forget them. If they succeed, we believe that technology itself has changed society.

CHAPTER 7

1936-1990

The Means of Destruction

No force in the twentieth century had a greater influence in defining and shaping technology than the military. In the earlier eras of modernism and systems, independent inventors, corporations, designers, and architects played leading roles in developing such culture-shaping technologies as electricity, synthetic chemicals, skyscrapers, and household technologies, while governments and trading companies had dominated the era of imperialism. But not since the era of industry had a single force stamped a stronger imprint on technology. One might imagine a world in 1890 without the technologies of empire, but it is difficult to envision the world in 1990 absent such military-spawned technologies as nuclear power, computer chips, computer graphics, and the internet. Likewise, it was military-derived rockets that boosted the Apollo missions to the moon and military-deployed satellites that paid for the Space Shuttle missions. From the 1950s through the 1980s such stalwarts of the "high technology" (military-funded) economy as IBM, Boeing, Lockheed, Raytheon, General Dynamics, and MIT numbered among the leading US military contractors. Lamenting the decline of classic profit-maximizing capitalism, industrial engineer Seymour Melman termed the new economic arrangement as contract-maximizing "Pentagon capitalism." During these years of two world wars and the Cold War, the technology priorities of the United States, the Soviet Union, and France, and to a lesser extent England, China, and Germany, were in varied ways oriented to the "means of destruction."

Merely the use of military technologies is not what distinguishes this era, of course. Fearsome attack chariots and huge cannon figured in Renaissance-era warfare, while steam gunboats and Enfield rifles typified the imperialist era. What was new in the twentieth century was the pervasiveness of technological innovation and its centrality to military planning. This was true during World War II between 1939 and 1945, as military officers, scientists, engineers, and technology-minded businessmen forged new relationships that led to the "military-industrial complex" (named by President Dwight Eisenhower, an ex-general himself), and also during the prolonged Cold War (1947–90) that followed. The great powers' universities, technology companies, government institutes, and military services committed themselves to finding and funding new technologies in the hope of gaining advantage on the battlefield or in the Cold War's convoluted diplomacy. Above all, for military officers no less than researchers, military technology funding was a way of advancing one's vision of the future—and sometimes one's own career.

The swift pace of military-driven technical innovation did not come without cost. "The concentration of research on particular tasks greatly accelerated their achievement," writes Alan Milward of the Second World War, "but this was always at the expense of other lines of development." During the Cold War, "the increasing predominance of one patron, the military," writes Stuart Leslie, "indelibly imprint[ed] academic and industrial science with a distinct set of priorities" that set the agenda for decades. Such promising technologies as solar power, analog computers, and operator-controlled computer machine tools languished when (for various reasons) the military backed rival technical options—nuclear power, digital computers, and computer-controlled devices of many types—that consequently became the dominant designs in their fields.[1]

So pervasive was military funding for science and technology during these decades that perhaps only in our own (post–Cold War) era is a measured assessment possible. This is still no easy task, since for every instance of productive military-civilian "spin-off," such as Teflon, microchips, and nuclear power, there is a budgetary black hole from which no useful device, military or civilian, ever came. Among these ill-considered schemes in the United States must number the little-publicized Lexington, Mohole, and Plowshare projects. These were lavishly funded schemes designed, respectively, to build nuclear-powered airplanes (1946–61), to drill through the earth's crust (1957–66), and to use atomic explosions in making huge earthen dams and harbors (1959–73). For Project Lexington the Air Force spent $1 billion—when a billion dollars really meant something—on the patently absurd nuclear plane. The Soviet nuclear empire, even less constrained by common sense and safety concerns than its Western counterparts, spent untold sums on nuclear ships, nuclear explosions for earthmoving, and nuclear power that led to the Chernobyl disaster in 1986. Assessing the technology-laden "puzzle palace" of the National Security Agency, whose budget and technologies remain veiled in secrecy, may never be possible. Its critics chalk off as a budgetary black hole the $35 billion Strategic Defense Initiative (1985–94), the latter-day "technological dream" popularly known as Star Wars.[2]

A WAR OF INNOVATION

It may seem odd to distinguish between the two world wars, linked as they were by unstable geopolitics and debt economics, but in technology the First World War was not so much a war of innovation as one of mass production. The *existing* technologies of the time largely accounted for the outbreak, character, and outcome of this "war to end all wars." Telegraphs and railroads, no novelty at the time, figured prominently in the breakdown of diplomacy and outbreak of war in July–August 1914. Indeed, the fever pitch of telegraph negotiations, with their drumbeat of deadlines, simply overwhelmed diplomats on both sides. Telegraphs carried Austria's declaration of war on Serbia and later revealed to the United States Germany's designs on Mexico (in the intercepted Zimmerman telegram), while railroads carried German troops in a precise sequence of steps designed to defeat France and Russia. For German planners the problem was that the French held their ground at the Marne, just short of Paris. The two sides dug in. It would be a long war.

World War I's gruesome trench warfare was the direct result of the machine gun and mass-produced explosives. Invented in 1885, the machine gun had been used by imperialist armies with devastating effect against lightly armed native armies. In a few scant hours in the Sudan during 1896, the Anglo-Egyptian army with machine guns slaughtered 20,000 charging Dervishes. When both sides possessed machine guns, however, each side could readily mow down the other's charging soldiers. Taking territory from an enemy defending itself with machine guns and barbed wire was practically impossible. The stalemate of trench warfare persisted just so long as both sides successfully mass produced gunpowder and bullets. Even the horrific poison gases used by both sides had little strategic effect (see chapter 5). Tanks were one novel technology that saw wartime action, and the British devised a war-winning campaign using tanks called Plan 1919. But the war ended in November 1918 before the plan could be tried. The defeat of Germany did not depend much on novel technologies such as Dreadnought-class battleships, radios, automobiles, or even aircraft. Germany lost the war when its industrial and agricultural economy collapsed.[3]

As a madman bent on world domination, Adolf Hitler learned the right lessons from World War I. Wounded twice while serving in the army and partially blinded by mustard gas, Hitler concluded that Germany would certainly lose another protracted war of production. Cut off from essential raw materials during the First World War, Germany, he resolved, would be self-sufficient for the next war. Just three years after seizing power Hitler enacted the Four Year Plan (1936–40), his blueprint for self-sufficiency and rapid rearmament. Under his ideal model of autarky, Germany would produce everything it needed to fight a major war, including synthesizing gasoline and rubber from its plentiful domestic supplies of coal. Because of Germany's shortfalls of copper and tin needed for munitions as well as nickel and chromium needed for armor plate, that autarky would never be complete. As early as November 1937, Hitler grasped the point that Germany's early rearmament gave it only a short-lived technological and military advantage; the longer Germany delayed going to war, the more this advantage would fade. The inescapable conclusion, Hitler told his top advisors at the Hossbach Conference, was that war must come soon. Meantime the German air force practiced bombing and strafing in Spain culminating with the Condor Legion's destruction of Guernica (1937).

Hitler's invasion of Poland in September 1939 revealed the tactic he had chosen to avoid trench warfare. His *blitzkrieg*, or "lightning war," loosely resembled the British Plan 1919. A blitzkrieg began with massive aerial bombardment followed by a column of tanks, supported overhead by fighter aircraft, breaking through enemy lines and cutting off supply routes. Not merely a military tactic, blitzkrieg was more fundamentally a "strategic synthesis" that played to the strength of Germany's superior mobility technologies, especially aircraft and tanks, while avoiding the economic strain of a sustained mobilization. Back home Germany's industrial economy proved remarkably productive, using flexible general-purpose machine tools to alternate between military and civilian production. The intentionally short blitzkrieg attacks allowed for ample civilian production and substantial stockpiling of strategic raw materials between bouts of military action. Consequently, the extreme wartime deprivation visited upon Japanese civilians at an early stage in the war—when the Japanese military took over the economy and directed all production toward military ends, literally starving the civilian economy—did not arrive in Germany until the last months of the war. Across Poland, France, Belgium, and far into Russia, until the Battle of Stalingrad in the winter of 1941–42, the German army was undefeated and appeared invincible. But the moment Germany was thrown back on the defensive, the war was lost. Hitler's strategic synthesis was undone.[4]

Fortunately, Hitler did not foresee all the needed lessons. To begin with, Germany never had the unified, efficient, single-minded "war machine" pictured in Allied wartime propaganda and never totally mobilized its economy, as the British did. Battles between rival factions within the Nazi party, not to mention creative "delays" by the non-Nazi civil service, often made a mockery of announcements by Hitler and his advisors. For most of the war, German technologists matched their Allied counterparts in the innovation race. As military historian Martin van Creveld points out, both sides maintained a torrid pace of technological innovation that prevented either side from developing a war-winning weapon. While the Allies pioneered radar-based air defense, built the most powerful internal combustion aircraft engines, and developed the most sophisticated electronics, the Germans created superior navigational aids for bombing, led the world in jet and rocket engines, and consistently had superior tanks, artillery, and machine guns.[5]

Germany's scientists and engineers kept pace with their Allied counterparts until April 1944. In that month the Nazis, facing an economic crisis from the continual changes to factory technology and reeling under Allied bombardment, effectively halted research-and-development work and froze technical designs. With this drastic move they hoped to produce vastly larger quantities of the existing munitions, aircraft, and weaponry: this was a return to mass production World War I style. This decision ended the contribution to the war of Germany's jet-powered aircraft. Only select research projects, with promise of achieving a military result and soon, were continued. One of these was the German atomic program.

The National Socialist atomic program, facing shortages of people, materials, and funding, was a long shot at the outset. German physicists Otto Hahn and Fritz Strassmann had in 1938 split the uranium nucleus. Between 1940 and 1942 several German physicists, Werner Heisenberg most prominently, tried without success to sell an atom bomb project to the German Army and Reich Research Council. From 1942 onward Heisenberg and his colleagues had low-level funding for work on an atomic power reactor. Soon the country's economic turmoil and deteriorating military situation made it impossible to mount the industrial-scale effort necessary even for a reactor, let alone to build an atomic bomb. The German atomic effort was further hampered by Hitler's anti-Semitic ravings, which had driven out Germany's Jewish scientists, among them many of the country's leading atomic physicists. Germany had neither the enriched uranium, the atomic physicists, nor the governmental resources to manufacture an atomic bomb. In early March 1945, with a cobbled-together "uranium machine" located in an underground bunker south of Berlin, and with his country crumbling around him, Werner Heisenberg still pursued a self-sustaining atomic chain reaction.[6]

"TURNING THE WHOLE COUNTRY INTO A FACTORY"

A revealing measure of just how much distance separated the German and Anglo-American atomic programs came on 6 August 1945. The war in Europe had wound down in early May when Soviet tanks entered Berlin. Quickly the Allies rounded up ten top-ranking German physicists, who were brought to a country estate in England. There, while the war continued in Asia, Allied intelligence agents closely questioned them about their atomic research. Mostly, the German physicists spoke with a haughty pride about their wartime achievements—until the evening when word arrived that the Americans had dropped an atom bomb on the Japanese city Hiroshima. The existence of an atom bomb was surprise enough. The Germans were simply astonished to learn that the explosion at Hiroshima had equaled 20,000 tons of high explosives, and that the Allied atom project had employed 125,000 people to construct the necessary factories while another 65,000 people worked in them to produce the bomb material. Their own effort, scarcely one-thousandth that size in personnel and immeasurably less in effect, paled in comparison. The German atom project had been nothing more than a laboratory exercise, and an unsuccessful one at that.[7]

If the First World War is known as the chemists' war owing to military use of synthetic explosives and poison gases, it was the Manhattan Project that denominated the Second World War as the physicists' war. Physicists first began wartime atomic work at Columbia University, University of Chicago, and University of California–Berkeley. As early as December 1942 Enrico Fermi, at Chicago, had built a self-sustaining uranium reactor surpassing anything the Germans achieved during the war. Fermi's success paved the way for Robert Oppenheimer's Los Alamos laboratory. In the spring of 1943 physicists recruited by Oppenheimer began arriving at the lab's remote desert location in New Mexico. At its peak of activity, Los Alamos had 2,000 technical staff, among them eight Nobel Prize winners, who worked out the esoteric physics and subtle technology necessary to build a bomb. When the Los Alamos scientists, bright, articulate, and self-conscious about having made history, later told their stories, they in effect grabbed sole credit for the bomb project. Their tale was a wild exaggeration. In reality, Los Alamos served as the R&D center and assembly site for the bombs. The far greater part of the Manhattan Project was elsewhere, at two mammoth, top-secret factory complexes in Tennessee and Washington State. Indeed, these two factories were where most of those nearly 200,000 people worked and where virtually the project's entire whopping budget was spent (less than 4 percent of the project's plant and operating costs went to Los Alamos). Most important, it was the structure of these two factories, I think it can be fairly argued, that shaped the deployment of atomic weapons on Japan. One can also see, inscribed into the technical details of these factories, a certain vision for atomic weaponry that lasted far beyond the war.

In the physicists' fable, it was none other than the world's most famous physicist who convinced President Franklin Roosevelt to back a bomb project. In autumn 1939, with the European war much in their minds, Albert Einstein worked with Leo Szilard, one of the many émigré European physicists active in atomic research, setting down in a letter to the president the scientific possibilities and political dangers of atomic fission. Actually, it was Alexander Sachs, an economist, former New Deal staffer, and Roosevelt insider, carrying the physicists' letter, who made the pitch in person on 11 October 1939. Roosevelt saw immediately, before even reading the Einstein-Szilard letter, that a German atom project was menacing and that an Allied atom project needed prompt action. Roosevelt gave the go-ahead for a high-priority secret program. The project came to rest in the Office of Scientific Research and Development, or OSRD, a new government agency headed by the MIT engineer Vannevar Bush. Bush was charged with mobilizing the country's research and engineering talents in service to the war. From the first, then, the atom bomb project had the president's strong backing, an ample budget, smart and ambitious backers, and a brand of "top secret."

The industrial-factory aspect of the Manhattan Project took shape in the months after June 1942. Until that point the atom project had been simply, if impressively, big science. It was funded and coordinated by Bush's OSRD in Washington; the agency directed research money to several universities with active atomic research around the country. Earlier that year at the University of Chicago, physicist Arthur Holly Compton brought together several research groups, including Fermi's from Columbia, in an effort to centralize atomic research. The result was the evasively named Metallurgical Laboratory. That June marked a decision point on whether the United States should expand the promising laboratory-scale work to a modest level of funding, aiming for an atomic power reactor, or mount a massive, full-scale effort in pursuit of a bomb. Roosevelt, acting on Bush's recommendation, approved a large-scale project. A bomb it would be. Bush understood that this vast effort would overwhelm the capabilities of OSRD, whose mission was conducting military-relevant research and developing weapon prototypes but not industrial-scale production. It was entirely fitting that Roosevelt assigned the construction phase of the bomb project to the Army Corps of Engineers and that the Army assigned command over the Manhattan Engineering District to Brigadier General Leslie Groves, who took charge of building the Pentagon complex (1941–43).

Described as supremely egotistical, excessively blunt, and aggressively confident (after their first meeting Bush expressed doubt that the general had "sufficient tact for the job"), Groves went to work with dispatch.[8] In September, within days of tak-

FIG. 7.1. OAK RIDGE URANIUM FACTORY

The Oak Ridge complex in eastern Tennessee was the principal site for producing bomb-grade enriched uranium. The electromagnetic process, housed in the portion of the Oak Ridge complex shown here, promised a one-step separation of fissionable U-235 using atomic "racetracks" and huge magnets. Enriched uranium came also from Oak Ridge's equally massive gaseous diffusion plant and thermal diffusion plant.

Vincent C. Jones, *Manhattan: The Army and the Atomic Bomb* (Washington, D.C.: Government Printing Office, 1985), 147.

ing command, he secured for the Manhattan Project a top-priority AAA procurement rating, purchased a 1,250-ton lot of high-grade uranium ore from the Belgian Congo, and initialed the Army's acquisition of 54,000 acres in eastern Tennessee. In December, shortly after Fermi's Chicago pile went critical, he secured Roosevelt's approval of $400 million for two huge factories: one for uranium separation at Oak Ridge, Tennessee, and one for plutonium production at Hanford, Washington. From then on, Groves' most substantial task was mobilizing the legion of contractors and subcontractors. The Army had already tapped Stone & Webster, the massive engineering, management, and financial consulting firm, as principal contractor. Groves himself let subcontracts to divisions of Eastman Kodak, Westinghouse, General Electric, Chrysler, Allis-Chalmers, M. W. Kellogg, DuPont, and Union Carbide.

Groves needed 54,000 acres in Tennessee, because Oak Ridge would be a mammoth undertaking. The crucial task at Oak Ridge was to produce enough enriched uranium, somewhere between 2 and 100 kilograms, no one knew precisely how much, to make a bomb (fig. 7.1). Common uranium had less than 1 percent of the fissionable 235-isotope. Since no one had ever isolated uranium-235 in such quantities, Groves quickly authorized large-scale work on two distinct processes, electromagnetic separation and gaseous diffusion, while a third process, thermal diffusion, was later drafted to the cause. Electromagnetic separation was the baby of Ernest Lawrence, already accomplished at big science, who offered up his Berkeley laboratory's atom smashers. The faster atomic particles could be propelled, everyone realized, the more interesting was the physics that resulted when opposing streams of particles collided. Lawrence's particular genius was in seeing that a charged stream of particles could be accelerated to unbelievably high energies by a simple trick. While precisely positioned magnets bent the stream of charged particles to follow an oval "racetrack," a carefully tuned oscillating electric field sped them up, faster and faster, each time they went around. It was a bit like pushing a child on a swing, a bit higher with each pass. Armed with Lawrence's formidable skills at fund raising and machine building, his laboratory had, by 1932, accelerated hydrogen atoms to 1 million volts. By 1940, just after winning the Nobel Prize for the subatomic discoveries made by an impressive 16-million-volt machine, Lawrence lined up Rockefeller Foundation funds to build a really big accelerator in the Berkeley Hills. Its 184-inch-diameter magnet was the world's largest. The cost of the machine was a heady $1.5 million.

Lawrence, ever the hard-driving optimist, promised Groves that his cyclotron principle could make a clean separation between the two uranium isotopes. Lawrence knew that his powerful magnets would bend the path of a charged particle, including a charged uranium atom. Passing through a constant magnetic field, he explained, the lighter, fissionable uranium-235 would be bent slightly more than the heavier, stable uranium-238, a manageable fraction of an inch. He invented a "calutron," a C-shaped device something like a cyclotron cut in half, to do the work of separating the isotopes. While supremely elegant in theory, the electromagnetic-separation process in reality had been effective only at the laboratory scale; in 1941, after a month of intensive effort using a smaller Berkeley machine, Lawrence had produced 100 *micro*grams of *partially* enriched uranium. A uranium bomb would need at least 100 million times more, perhaps 40 *kilo*grams of *highly* enriched uranium.

At Groves' direction Stone & Webster broke ground in February 1943 on the Tennessee complex. By August 20,000 construction workers were building the atomic racetracks and calutrons on a site that stretched over an area the size of twenty football fields. When all the necessary supporting infrastructure was completed, there would be 268 buildings. (On an inspection trip that spring, even

Lawrence was dazzled by the size of the undertaking.) Obtaining enough copper wire for the countless magnet windings looked hopeless. Coming to the rescue, the Treasury Department offered its stores of silver. Groves borrowed 13,540 tons of silver—a precious metal generally accounted for in ounces—valued at $300 million, and used it in place of copper to wind magnets and to build solid silver bus bars to conduct the heavy electric currents. All the while, troublesome vacuum leaks, dangerous sparks, and innumerable practical problems kept Lawrence and his assistants problem solving at a frenetic pace. Facing intractable difficulties with the magnetic windings, Groves shipped the entire first batch of 48 huge silver-wound magnets back to the Allis-Chalmers factory in Milwaukee, idling the 4,800 Tennessee Eastman production workers for a full month. Production of uranium-235 resumed in January 1944.

The second uranium enrichment process tried at Oak Ridge required even larger machines. Gaseous diffusion used small effects to achieve a big result. This method exploited the established fact that lighter molecules move more readily through a porous barrier than do heavier ones. In this case, atoms of fissionable uranium-235 made into uranium hexafluoride would move through a porous barrier a bit more quickly than would the heavier uranium-238. Initial work by researchers at Columbia University and Bell Telephone Laboratories suggested that a nickel barrier would separate the isotopes effectively while withstanding the highly corrosive uranium hexafluoride gas. Groves authorized the Kellex Corporation to build acres of diffusion tubes at Oak Ridge, but the early results were somewhat disappointing. In autumn of 1943 one of Kellex's engineers, Clarence Johnson, hit upon the idea of sintering the nickel metal to form a barrier; sintering involved pressing powdered nickel into a semiporous block, and it would require the company to handmake the new barriers, at staggering cost. Groves' willingness to scrap the existing nickel barriers, to junk the Decatur, Illinois, factory that had been constructed to make them, and to push on with the revamped gaseous diffusion plant convinced physicist Harold Urey, among others, that Groves was building for the ages. Oak Ridge was to be not merely a temporary wartime factory but a permanent bomb-producing complex.

As the original goal of building a bomb by the end of 1944 slipped, the needed amounts of enriched uranium nowhere in sight, Groves brought a third separation process to Oak Ridge. Thermal diffusion was the simplest. Its chief backer, Philip Abelson, had worked three years for the Navy—cut off from the Army-led atom bomb project—on enriching uranium with an eye to making atomic reactors for submarines. In 1941 Abelson had perfected a method to make liquid uranium hexa-fluoride. He then lined up thermal diffusion tubes to enrich it. Trapped between hot and cold surfaces, the lighter uranium-235 floated up while the heavier uranium-238 drifted down.

To exploit this separating effect Abelson built, at the Naval Research Laboratory, a single experimental 36-foot-tall column in which the liquid uranium hexafluoride was piped between 400°F steam and 130°F water. Workers simply skimmed enriched ura-nium-235 off the top. In January 1944, with the Navy's go-ahead, Abelson began build-ing a 100-column pilot plant next to the steam plant of the Philadelphia Navy Yard. Abelson never promised the one-stop separation of Lawrence's cyclotrons; on the other hand, his columns worked reliably, yielding sizable quantities of moderately enriched uranium. In mid-June Groves contracted with the H. K. Ferguson engineering firm to build at Oak Ridge a factory-scale complex consisting of twenty-one exact copies of the entire Philadelphia plant, or 2,100 columns in all. By 1945 the three processes were brought together: thermal diffusion provided an enriched input to gaseous diffusion and electromagnetic separation. Concerning the immense effort of making enough uranium-235 for a bomb, the Danish physicist Niels Bohr told Edward Teller, another physics colleague on the Manhattan Project, "I told you it couldn't be done without turning the whole country into a factory. You have done just that."[9]

The Oak Ridge complex with its network of designers, contractors, fabrica-tors, and operators was just one division of this far-flung national factory. The site at Hanford, with 400,000 acres, was nearly eight times larger (fig. 7.2). Groves had chosen this remote site in south-central Washington State for several reasons. The Columbia River provided water for the atomic reactors, while electric power was readily available from the Grand Coulee and Bonneville hydroelectric dams. Most important of all, since it did not take an atomic physicist to understand that a large-scale plutonium factory would be a grave safety concern, Hanford was distant from major population centers. Conceptually, making plutonium was a snap. One needed only to bombard common uranium-238 (atomic number 92) with a stream of neu-trons to make neptunium (atomic number 93). Neptunium soon decayed into pluto-nium (atomic number 94). Berkeley physical chemist Glenn Seaborg had discovered plutonium in a targeted search for a new fissionable element for a bomb; he named it plutonium to continue the pattern of naming elements after the outermost planets, Uranus, Neptune, and Pluto. After lengthy negotiations, Groves by December 1942 had convinced the DuPont company to become prime contractor for the Hanford Works and take charge of designing, building, and operating a full-scale plutonium

FIG. 7.2. HANFORD WORKS PLUTONIUM FACTORY

The Hanford Works consisted of seven large complexes, separated for safety by 6 miles or more, built on 400,000 acres in south central Washington State. The 100 B Pile, shown here, was one of three plutonium production reactors. The production pile has a single smokestack; the electricity plant has two smokestacks, and to the right beyond it is the pump house on the Columbia River.

Vincent C. Jones, *Manhattan: The Army and the Atomic Bomb* (Washington, D.C.: Government Printing Office, 1985), 216.

production and separation facility. A pilot plant at the Oak Ridge complex was also part of the plutonium project. At the time, DuPont had no experience in the atomic field, but it did have extensive experience in scaling up chemical plants. And separating plutonium from irradiated uranium entailed building one very large chemical-separation factory, involving a multistep bismuth-phosphate process.

At Hanford the most pressing problems were engineering ones. Whereas Seaborg's laboratory at Berkeley had produced a few micrograms of plutonium, DuPont's engineers at Hanford needed to scale up production not merely a million-fold to produce a few grams of plutonium but a thousandfold beyond that to make the kilograms of plutonium believed necessary for a bomb. Intermediate between the laboratory and the factory was the "pilot plant," where theoretical calculations could be tested and design data collected. Conducting research at the pilot-plant stage was usually a task for industrial scientists or engineers, but Chicago's academic Metallurgical Lab scientists had insisted that they direct its reactor. Groves, assenting reluctantly to their demand, directed them to conduct experiments to see if plutonium could be made in quantity. At the same time, and in advance of the pilot-plant data, DuPont engineers designed a full-scale plutonium facility at Hanford. Relations between the scientists and the engineers were strained from the start. The DuPont project manager quieted the Chicago scientists' criticism when he asked them to sign off on the thousands of blueprints. Only then did the scientists see the magnitude of the plutonium factory. It would consist of three huge uranium reactors to produce the plutonium and four gigantic chemical separation plants to isolate and purify the deadly element. Construction at its peak in June 1944 employed 42,000 skilled welders, pipefitters, electricians, and carpenters.

By mid-September 1944 the Hanford plant was ready for its first test run. The uranium pile went critical as expected—then petered out. After some hours the reactor could be restarted, but no chain reaction lasted for long. Morale plunged when Hanford appeared to be a vastly expensive failure. Some trouble-shooting revealed that the element xenon-135, a powerful neutron absorber with a brief half-life, was poisoning the reactor. When xenon-135 decayed, the reactor restarted but it could not run for long. Groves had ordered the Met Lab scientists to run their pilot-plant reactor at full power, and had they done so the xenon effect would have been revealed. But they had not run at full power. Fortunately, the DuPont engineers had designed room for extra slugs of uranium (the Chicago scientists had denounced this seeming folly). With the additional uranium in place, 500 slugs added to the original 1,500, the increase in neutrons overwhelmed the xenon poisoning. The DuPont engineers' cautious design had saved the plant. It successfully went critical in mid-December.[10]

In the physicists' fable the next chapter is Los Alamos, where the best and brightest physicists created a workable atom bomb. The narrative climaxes with the incredible tension surrounding the Trinity test explosion on 16 July 1945 and the relief and wonderment at its success. The standard epilogue to the story acknowledges the actual atom bombing of Hiroshima on 6 August and, sometimes, the atom bombing of Nagasaki on 9 August. Various interpretations of these events exist. Our public memory has enshrined the most comforting version, President Truman's, that the atom bomb was dropped on Japan to end the Second World War, even though there were dissenting voices at the time. General Eisenhower no less told Secretary of the War Henry Stimson, "The Japanese were ready to surrender and it wasn't necessary to hit them with that awful thing."[11] Debate on these momentous events has continued ever since. Here I would like to explore the irresistible machine-like momentum that had built up behind the atomic bomb project. Winston Churchill, no foe of the bomb, wrote that "there never was a moment's discussion as to whether the bomb should be used or not."[12]

Many commentators, even Eisenhower and Churchill, obscure the crucial point that the two bombs dropped on Japan were technologically distinct: the Hiroshima bomb used Oak Ridge's uranium while the Nagasaki bomb used Hanford's plutonium. The Hiroshima bomb, a simple gun-type device code-named Little Boy, worked by firing one subcritical mass of uranium-235 into another, creating a single larger critical mass; so confident were the designers that no test of this bomb was ever attempted. Hiroshima was the world's first uranium explosion. Entirely different was the Nagasaki bomb dropped three days later, code-named Fat Man, a plutonium device detonated by implosion. (Two supposedly subcritical masses of plutonium, as in the uranium bomb, were deemed dangerously unstable.) Implosion relied on high explosives, properly detonated, to generate a spherical shock wave that compressed a hollow sphere of plutonium inward with great speed and precise force. When the bowling ball of plutonium was squeezed to grapefruit size, and a hazelnut-sized core released its initiating neutrons, the plutonium would go critical and explode. This was the bomb tested at Trinity in mid-July and dropped on Nagasaki in August (fig. 7.3).

That spring, a committee of top-level army officers drafted a targeting plan for dropping atomic bombs on Japan. Hiroshima, Kokura, Niigata, and Nagasaki were the preferred targets. The order approving the plan was drafted by Leslie Groves and approved by Henry Stimson and George Marshall, respectively secretary of war and chairman of the joint chiefs of staff. Truman signed nothing. "I didn't have to have the President press the button on this affair," recalled Groves.[13] The Army air force planned to use whatever bombs it had at hand. Two bombs were en route to the forward base at Tinian Island in the Pacific. A third bomb, still under final assembly at Los Alamos, would have been ready for shipping to Tinian by 12 or 13 August, but before it could be dispatched, Truman, rattled by the thought of "all those kids" who had been killed in Hiroshima, ordered a halt in the atomic bombing. He gave the order on 10 August, the same day the Japanese emperor's surrender terms reached Washington. The third bomb stayed at Los Alamos. While Truman and his advisors dickered over the precise terms of Japan's surrender, the fire bombing of Japan continued. General Hap Arnold ordered out 800 of his B-29s, escorted by 200 fighters, for a final all-out incendiary bombing of Honshu on 14 August. Even as the 12 million pounds of incendiaries and high explosives were falling, the emperor's acceptance of Allied surrender terms was on its way to Washington. The war ended, officially, the next day.

The rush to drop the two atom bombs on Japan must be an enduring puzzle. President Truman's rationale, while comforting, does not account satisfactorily for

FIG. 7.3. NAGASAKI MEDICAL COLLEGE HOSPITAL, OCTOBER 1945

The buildings of Nagasaki Medical College Hospital, about 700 yards from the hypocenter, just outside the zone of total destruction, withstood the blast of the second bomb dropped on Japan (the plutonium one). The interiors were burned out, and ruined; wooden structures in this area were instantly destroyed. The first bomb (the uranium one) had destroyed the center of Hiroshima out to approximately 2,000 yards.

Photograph by Hayashi Shigeo, used by permission of his wife Hayashi Tsuneko. Published in Hiroshima-Nagasaki Publishing Committee, *Hiroshima-Nagasaki* (Tokyo: Hiroshima-Nagasaki Publishing Committee, 1978), 220–21.

a number of facts. As vice president until FDR's death on 12 April Truman had not even been told of the bomb project, and, as Groves plainly stated, it is unclear how much of the decision was his anyway. An alternative view suggests that the bombs were dropped on Japan but "aimed" at the Soviet Union, then mobilizing its army for a long-promised invasion of Japanese-controlled Manchuria. In this view, propounded by physicist P. M. S. Blackett and historian Gar Alperovitz, the atom bombs did not so much end the Second World War as begin the Cold War. There is no doubt that the Allies wanted to keep the Soviets out of the Far East (the Red Army's presence in Eastern Europe was already troublesome). Yet, with the Soviets moving troops on 8 August, the quickest end to the war would have been to accept Japan's proffered surrender on the tenth. (Possibly even earlier: on 11 July Japan's foreign minister directed a Japanese ambassador in Moscow to seek the Soviets' assistance in brokering a surrender agreement, a message that US intelligence had intercepted and decoded.) And, in any event, these geopolitical considerations do not really explain the second bomb.

We know that concerns about budgetary accountability pressed mercilessly on Groves. "If this weapon fizzles," Groves told his staff on Christmas Eve 1944, "each of you can look forward to a lifetime of testifying before congressional investigating committees." Jimmy Byrnes—FDR's "assistant president" and Truman's secretary of state—told Szilard, in the scientist's words, that "we had spent two billion dollars on developing the bomb, and Congress would want to know what we had got for the money spent."[14] The Manhattan Project's total cost of $2 billion was 2 percent of the war-swollen 1945 US federal budget, perhaps $50 billion in current gross domestic product terms.

One hesitates to put it this way, but the two bombs dropped on Japan were partly "aimed" also at the US Congress. "The bomb was also to be used to pay for itself, to justify to Congress the investment of $2 billion, to keep Groves and Stimson out of Leavenworth prison," writes historian Richard Rhodes. After all, there were *two* hugely expensive atomic factories. Indeed the pressure to quickly drop the plutonium bomb *increased* after the uranium bomb was dropped on Hiroshima. "With the success of the Hiroshima weapon, the pressure to be ready with the much more complex implosion device became excruciating," recalled one of Fat Man's assembly crew.[15] The plutonium bomb's testing schedule was slashed to move its target date up a full two days (to 9 August from 11 August). It does not seem accidental that Fat Man was dropped just before Truman's cut-off on the tenth. Thus, in various measures, was Japan delivered a surrender-making deathblow, while simultaneously the Soviets were shut out of the Far East and the Manhattan Project's two massive atomic factories were justified by the logic of war. Bohr's observation that the atomic project would transform "the whole country into a factory," true enough in the obvious physical and organizational sense, may also be insightful in a moral sense as well.

The *structure* of wartime atomic research had long-term consequences, far beyond the Hiroshima–Nagasaki debates. The Army-dominated Manhattan Project had effectively shut out the US Navy. While the Navy had developed the promising thermal diffusion method of enriching uranium and funded its early trial at the Philadelphia Navy Yard, the Navy was sidelined once this project too was moved to Oak Ridge. The Army not only controlled the atom bomb project during the war but also, as the result of recruiting the redoubtable Wernher von Braun and his German rocket team to its facility in Huntsville, Alabama, had the most advanced rocket project after the war. The newly organized Air Force meanwhile was laying its own claim to the atomic age with its long-range bombers.

These interservice blows shattered the Navy's prestige. While the Army and Air Force stepped confidently into the brave new atomic era, the Navy seemed stuck with its outmoded battleships and oversized aircraft carriers. A little-known Navy captain took up the suggestion that the Navy develop its own atomic reactors for submarines. During the war Captain Hyman Rickover, who had a master's degree in electrical engineering and a solid record in managing electronics procurement, had been part of a low-level Navy delegation at Oak Ridge. Rickover, soon-to-be Admiral Rickover, would become the "father of the nuclear Navy."

By 1947 the hard-driving Rickover—something of a cross between the bureaucratically brilliant Groves and the technically talented Lawrence—took charge of the Navy's nuclear power branch as well as the (nominally civilian) Atomic Energy Commission's naval reactors branch. He worked most closely and successfully with Westinghouse and Electric Boat. While Westinghouse worked on a novel pressurized light-water-cooled nuclear reactor, Electric Boat prepared to build the submarine itself at its Groton, Connecticut, construction yard. (Meanwhile, a parallel project by General Electric to build a sodium-cooled reactor fell hopelessly behind schedule.) By early 1955, the Navy's first nuclear submarine, *Nautilus*, passed its initial sea trials and clocked unprecedented speeds. It also completed a submerged run of 1,300 miles to San Juan, Puerto Rico, ten times farther than the range of any existing diesel-powered submarine.

Nautilus, it turned out, was a precedent for more than just the US Navy, which in time fully matched the other military branches with its nuclear-powered submarines capable of launching nuclear-tipped missiles. In 1953, while *Nautilus* was still in development, President Eisenhower announced his "Atoms for Peace" initiative. He aimed to dispel the heavy military cast of the country's nuclear program by highlighting civilian applications. Electricity production topped his list. Eisenhower's insistence on speed and economy left the AEC little option but to draw heavily on the reactor options that were already on Rickover's drawing board. In fact, among

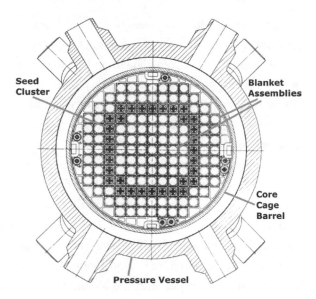

Seed
Cluster

Blanket
Assemblies

Core
Cage
Barrel

Pressure Vessel

FIG. 7.4. SHIPPINGPORT NUCLEAR REACTOR

Commercial nuclear power in the U.S. grew up in the shadow of the US Navy's work on submarine reactors. The pioneering Shippingport reactor's distinctive "seed and blanket" design was invented in 1953 at the Naval Reactors Branch of the Atomic Energy Commission. This cross-sectional diagram of the cylindrical reactor vessel shows the "seed" of highly enriched uranium (90 percent fissionable U-235) surrounded by "blanket" assemblies of natural uranium.

Redrawn from Naval Reactors Branch, *The Shippingport Pressurized Water Reactor* (Washington, D.C.: Atomic Energy Commission, 1958), 6, 43, 61.

his several duties, Rickover headed up the civilian reactor project. Working with Westinghouse once again, Rickover quickly scaled up the submarine reactor designs to produce a working civilian reactor. The reactor was installed at an AEC-owned site at Shippingport, Pennsylvania, and its steam was piped over and sold to the Duquesne Light Company to spin its generating turbines. Ten of the first twelve civilian reactors closely followed the Shippingport design, being knock-offs of submarine reactors.

The design decisions that made perfect sense in a military context, however, were not always well suited to a civilian, commercial context. Safety was a paramount concern in both domains, but the submarine reactor designers had not worried much about either installation or operating costs, which were of obvious importance to a profit-making company. Shippingport cost about five times what Westinghouse and General Electric considered commercially viable. Shippingport's "seed and blanket" submarine reactor (fig. 7.4) had been designed for maximum duration between refueling—of clear importance to submarines—but not for ease or speed or economy of maintenance. This mismatch between civilian and military imperatives further inflated the cost of the atomic power program.[16]

The enduring legacy of the Manhattan Project above and beyond its contribution to the atomic power effort was its creation of a nuclear weapons complex that framed years of bitter competition between the United States and the Soviet Union. The atomic fission bombs were small fry compared to the nuclear fusion bombs vigorously promoted by physicist Edward Teller. Where fission bombs split atoms, fusion bombs joined them. Fusion bombs, whose explosive yields were measured in multiple megatons of TNT, were a thousand-fold more powerful than the Hiroshima and Nagasaki bombs. (By comparison, the US Army air force dropped approximately 3 megatons of explosives during the entire Second World War.) A cycle of recurrent technology gaps and arms races began with the atomic fission bombs (America's in 1945 and the Soviets' in 1949) and accelerated with the nuclear fusion bombs (in 1952 and 1953, respectively). The resulting nuclear arms race pitted the high-technology industries of each superpower against the other in a campaign to invent, develop, produce, and deploy a succession of fusion bombs with ever-larger "throw weights" atop quicker and more accurate delivery vehicles.

In the nuclear age technology race, the advantage inclined to the offense. Each side's long-range bombers, intercontinental nuclear missiles, and finally multi-warhead MIRV missiles (multiple, independently targeted reentry vehicles) were truly fearsome weapons of mass destruction. The defensive technologies arrayed against them on both sides—the early-warning systems and the antiballistic missile, or ABM, systems—appeared hopelessly inadequate to the task. In fact, because its stupendous cost promised little additional safety, the United States shut down its $21 billion Safeguard ABM system located in North Dakota (1969–76). The fantastical cost of remaining a player in the nuclear arms race strained the finances of both rivals. The cost from 1940 to 1986 of the US nuclear arsenal is estimated at $5.5 trillion.[17] No one knows the fair dollar cost of the former Soviet Union's nuclear arsenal, but its currently crumbling state—nuclear technicians have in effect been told to find work elsewhere, while security over uranium and plutonium stocks is appallingly lax—constitutes a severe and present danger.

COMMAND AND CONTROL: SOLID-STATE ELECTRONICS

The Manhattan Project was by far the best known of the wartime technology projects. Yet, together, the massive wartime efforts on radar, proximity fuzes, and solid-fuel rockets rivaled the atom bomb in cost. A definite rival to the bomb project in breadth of influence on technology was the $1.5 billion effort to develop high-frequency radar. Centered at the Radiation Laboratory at the Massachusetts Institute of Technology, the radar project at its peak employed one in five of the country's physicists. The leading industrial contractors for the project included Bell Laboratories, Westinghouse, General Electric, Sylvania, and DuPont, as well as researchers at the University of Pennsylvania and Purdue University. Even as its radar aided the Allied war effort, the Rad Lab sowed the seeds for three classic elements of the Cold War military-industrial-university complex: digital electronic computing, high-performance solid-state electronics, and mission-oriented contract research. The US military's influence expanded after the promise of digital computing and electronics opened up in the 1950s, just when the military services were enjoying open-ended research budgets and when the Cold War placed new urgency on quick-acting command and control systems. Either by funding new technologies directly or by incorporating new electronics technologies in missile guidance, antimissile systems, nuclear weapons, and fire control, the military services imprinted a broad swath of technology with a distinct set of priorities. At the core was the military's pursuit of command and control—beginning with the bomber-alert systems of the 1950s through the electronic battlefield schemes tested in Vietnam down to the latter-day Strategic Defense Initiative (1984–93).

At the Radiation Laboratory during World War II, researchers focused on three technical steps. Radar worked by, first, sending out a sharp and powerful burst of radio waves that bounced off objects in their path; receivers then detected the returning signals; and finally the results were displayed on video screens. While the first and third of these steps provided researchers with valuable experience in designing high-frequency electronics (which would be useful in building digital computers) the second step resulted in a great surge of research on semiconductors. Vacuum tubes were sensitive only to lower frequency signals, so when the wartime radar project concentrated on the higher microwave frequency (3,000 to 30,000 megahertz), it needed a new type of electronic detector. The radar researchers found that detecting devices, similar to the "cat's whisker" crystals used by ham-radio buffs (who had used galena, silicon, or carborundum) could handle the large power and high frequencies needed for radar. Much of the solid-state physics done during the war focused on understanding the semiconductor materials germanium and silicon and devising ways to purify them. "The significance of the wartime semiconductor developments in setting the stage for the invention of the transistor cannot be overemphasized," writes historian Charles Weiner.[18] Between 1941 and 1945 Bell Laboratories turned over three-fourths of its staff and facilities to some 1,500 military projects, including a large effort in radar research.

In July 1945, just as the war was ending, Bell launched a far-reaching research effort in solid-state physics. Bell's director of research, Mervin Kelly, organized a new Solid State Department to obtain "new knowledge that can be used in the development of completely new and improved components and apparatus elements of communications systems."[19] By January 1946, this department's semiconductor group had brought together the three physicists who would co-invent the transistor—William Shockley, Walter Brattain, and John Bardeen. Ironically, no one in this group had had direct experience in the radar project's semiconductor work (though other Bell researchers had). The group focused on germanium and silicon, familiar because of the wartime work. Across the next year and a half, the group's experimentalists carefully explored the surface properties of germanium—while the group's theorists struggled to explain the often-baffling experimental results. Unlike metals whose negatively charged electrons carry electrical current, in silicon and germanium something like a positively charged electron seemed to carry current. With this insight the group built and demonstrated the first transistor in mid-December 1947.

Bell's transistor group explored several different devices. In December 1947 came the first transistor, a point-contact device that was descended from a ham-radio crystal. The junction transistor, invented two years later, was the type of transistor most widely used during the 1950s. It was a three-layered semiconductor sandwich in which the thin middle layer acted like a valve that controlled the flow of electrical current. The third device, a field-effect transistor, sketched in a laboratory notebook in April 1945, was first built only in the late 1950s. It exploited the electronic interactions between a plate of metal and a semiconductor. Beginning in the 1960s, such field-effect, or MOS (metal-oxide silicon), transistors formed the elements of integrated circuits. Their descendants (see chapter 10) are used in virtually all computer chips today.

Several well-known companies entered the new field by making junction transistors. Raytheon, aided by Army and Navy research contracts to improve its production, was manufacturing 10,000 junction transistors each month by March 1953, for use in radios, military applications, and hearing aids. That same year Texas Instruments established its research laboratory, headed by a former Bell researcher,

Gordon Teal. The first of the three major developments TI would contribute was a silicon transistor, which the military bought in large numbers because of its ability to operate at high temperatures and in high radiation fluxes (the Air Force's planned nuclear plane was the occasion for this unusual design criterion). The second project resulted in the first pocket-sized, mass-market transistor radio, on the market by Christmas 1954 and selling for $49.95. The third TI project, purifying large amounts of silicon, was achieved in 1956. What became Silicon Valley, named much later, also took form during these years. Bell's William Shockley moved back to his hometown of Palo Alto, California, in 1955. While a brilliant theoretical physicist, Shockley was a lousy manager, and his start-up company's employees were unhappy from the first. Eight of them left Shockley's company two years later to found Fairchild Semiconductor. Fairchild not only mastered the "planar" process for manufacturing transistors (described later), which in the 1960s allowed for mass-producing integrated circuits, but also served as the seedbed for Silicon Valley. By the early 1970s various Fairchild employees had moved on again to found forty or more companies in that region, including Intel, the firm responsible for commercializing the microprocessor—described as a "computer on a single chip."

While Bell Laboratories had internally funded the researchers who invented the transistor, the development of the device in the 1950s resulted from a close and sometimes strained partnership between Bell and the military services. In the transistor story, much like the Shippingport nuclear reactor, we see how the tension between military and commercial imperatives shaped the emergence of a widely used technology. While Bell Laboratories saw the transistor as a replacement for mechanical relays in the telephone system, the military services became greatly interested in its potential for miniaturizing equipment, reducing power requirements, and serving in high-speed data transmission and in computing. Bell Laboratories briefed the military services on the transistor a week before the invention's public unveiling on 30 June 1948. The military significance of the transistor was so large that the military services considered imposing a blanket secrecy ban on the technology, but, while many military applications were kept secret, no wide-ranging classification was placed on the transistor field as a whole.

Indeed, instead of classifying transistors, the armed services assertively publicized military uses for them. As a specified task of its first military contract, signed in 1949, Bell Laboratories jointly sponsored with the military a five-day symposium in 1951. Attended by 300 representatives of universities, industrial companies, and the armed services, this symposium advertised transistors' advantages—size, weight, and power requirements—in military equipment and disseminated the circuit concepts needed to use the new devices. The symposium material formed a 792-page textbook. In April 1952, the Bell System sponsored a second weeklong symposium for its patent licensees. For the hefty $25,000 entrance fee, credited as advance payment on the transistor patents, the licensees received a clear view of Bell's physics and fabrication technology. Each licensee brought home a two-volume textbook incorporating material from the first symposium. The two volumes composing *Transistor Technology* became known as the bible of the industry. They were originally classified by the government as "restricted" but were declassified in 1953. Even four years later, an internal Bell report could describe these volumes as "the first and still only comprehensive detailed treatment of the complete material, technique and structure technology" of transistors and added that they "enabled all licensees to get into the military contracting business quickly and soundly."[20] A third volume in the textbook series *Transistor Technology* resulted from a Bell symposium held in January 1956 to publicize its newly invented diffused base transistor. Diffused transistors used oxide masking and gaseous diffusion to create very thin layers of semiconductor that became transistors (the basis for the planar technique exploited by Fairchild). For several years Bell sold these high-performance diffused transistors only to the military services.

Beyond these publicity efforts, the military services directly financed research, development, and production technology. The military support at Bell for transistor research and development rose to fully 50 percent of the company's total transistor funding for the years 1953 through 1955. While Bell's first military contract (1949–51) had focused on application and circuit studies, its second military contract (1951–58) specified that it was to devote services, facilities, and material to studies of military interest. Bell's military systems division, conducting large projects in antiballistic missile systems, specified the applications for which customized transistors were developed. From 1948 to 1957, the military funded 38 percent of Bell's transistor development expenses of $22.3 million.

In 1951 the Army Signal Corps was assigned the lead-agency responsibility for promoting military transistor development. In the late 1930s, the Signal Corps Engineering Laboratory at Fort Monmouth, New Jersey, had designed the portable radios used during the war. Just after the war, the Signal Corps had announced a major electronics development effort to apply some of the wartime lessons learned. Miniaturizing its equipment was a major objective of this effort, and the transistor amply fit the bill. In 1953, the Army Signal Corps underwrote the construction costs of a huge transistor plant at Laureldale, Pennsylvania, for Western Electric (the manufacturing arm of the Bell System). In addition to Western Electric, other companies benefiting from this mode of military support were General Electric, Raytheon, Radio Corporation of America, and Sylvania. Altogether, the Army spent $13 million building pilot plants and production facilities.

The Army Signal Corps also steered the transistor field through its "engineering development" program, which carried prototypes to the point where they could be manufactured. For this costly process the Army let contracts from 1952 to 1964 totaling $50 million (averaging $4 million per year). Engineering development funds fostered the specific types of transistor technologies the military wanted. In 1956, when Bell released its diffusion process to industry, the Signal Corps let $15 million in engineering development contracts. The program, according to a Signal Corps historian, was to "make available to military users new devices capable of operating in the very high frequency (VHF) range which was of particular interest to the Signal Corps communications program."[21]

Bell Laboratories had not forgotten its telephone system, but its commercial applications of transistors were squeezed out by several large high-priority military projects. In 1954 Bell had planned a large effort in transistorizing the telephone network. Two large telephone projects alone, Rural Carrier and Line Concentrator, were supposed to receive 500,000 transistors in 1955 and a million the year after; all together Bell System telephone projects were to use ten times the number of transistors as Bell's military projects. But these rosy forecasts fell victim to the Cold War. In February 1955 the Army Ordnance Corps asked Bell to begin design work on the Nike II antiballistic missile system, a twenty-year-long effort that Bell historians describe as "the largest and most extensive program in depth and breadth of technology carried out by the Bell system for the military services."[22] Many of Bell's transistor development engineers, as well as much of its transistor production facilities, were redirected from telephone projects to a set of designated "preferred" military projects. These included Bell's own work on the Ballistic Missile Early Warning System (BMEWS) as well as the Nike-Hercules and Nike-Zeus missiles; work by other companies on the Atlas, Titan, and Polaris missiles; and several state-of-the-art military computer projects, including Stretch and Lightning. Given the press of this military work, Bell had little choice but to delay transistorizing its telephone network until the 1960s. It took two decades to realize Bell's 1945 plan for "completely new and improved components . . . [in] communications systems."

The integrated circuit was also to a large degree a military creation. The integrated

TABLE 7.1. US INTEGRATED CIRCUIT PRODUCTION AND PRICES, 1962–1968

YEAR	TOTAL PRODUCTION ($ MILLION)	AVERAGE PRICE ($)	MILITARY PRODUCTION AS PERCENTAGE OF TOTAL PRODUCTION
1962	4	50.00	100
1963	16	31.60	94
1964	41	18.50	85
1965	79	8.33	72
1966	148	5.05	53
1967	228	3.32	43
1968	312	2.33	37

Source: John Tilton, *International Diffusion of Technology: The Case of Semiconductors* (Washington, DC: Brookings Institution, 1971), p. 91.

circuit relied on the Bell-Fairchild planar technique for fabricating transistors described previously. Normally production workers made a number of transistors on one side of a semiconductor wafer, then broke up the wafer and attached wire leads to the separate devices. In 1959 both Jack Kilby at Texas Instruments and Robert Noyce at Fairchild instead attached transistors together on a single wafer. Military applications consumed nearly all the early integrated circuits (table 7.1). Texas Instruments' first integrated circuit was a custom circuit for the military; in October 1961 the firm delivered a small computer to the Air Force. In 1962 Texas Instruments received a large military contract to build a family of twenty-two special circuits for the Minuteman missile program. Fairchild capitalized on a large contract to supply integrated circuits to NASA. By 1965 it emerged as the leader among the twenty-five firms manufacturing the new devices. Although Silicon Valley's latter-day promoters prefer to stress civilian electronics, the importance of the military market to many of the pioneering firms is clear.

Across the 1950s and 1960s, then, the military not only accelerated development in solid-state electronics but also gave structure to the industry, in part by encouraging a wide dissemination of (certain types of) transistor technology and also by helping set industrywide standards. The focus on military applications encouraged the development of several variants of transistors, chosen because their performance characteristics suited military applications and despite their higher cost. Some transistor technologies initially of use only to the military subsequently found wide application (silicon transistors, the diffusion-planar process) while other military-specific

technologies came to dead ends (gold-bonded diodes and several schemes for miniaturization that were rendered obsolete by integrated circuits). The large demand
generated by the guided-missile and space projects of the 1950s and 1960s caused
a shortage of transistor engineers in Bell Laboratories' commercial (telephone)
projects. These competing demands probably delayed the large-scale application of
transistors to the telephone system at least a half-dozen years (from 1955 to the early
1960s). As Alan Milward noted at the beginning of the chapter, "The concentration
of research on particular tasks greatly accelerated their achievement . . . but this was
always at the expense of other lines of development."

COMMAND AND CONTROL: DIGITAL COMPUTING

Compared to solid-state electronics, a field in which the principal military influence
came in the demands of top-priority applications, military sponsorship in the field
of digital computing came more directly, in the funding of research and engineering
work. Most of the pioneers in digital computing either worked on military-funded
contracts with the goal to develop digital computers or sold their first machines
to the military services. Code-breaking, artillery range-finding, nuclear weapons
designing, aircraft and missile controlling, and antimissile warning were among the
leading military projects that shaped digital computing in its formative years, from
the 1940s through the 1960s. Kenneth Flamm found that no fewer than seventeen
of twenty-five major developments in computer technology during these years—
including hardware components like transistors and integrated circuits as well as
design elements like stored program code, index register, interrupt mechanism,
graphics display, and virtual memory—not only received government research and
development funding but also were first sold to the military services, the National
Security Agency, or the AEC's Livermore or Los Alamos weapons-design labs. In
addition, micro-programming, floating-point hardware, and data channels benefited from one or the other mode of government support. Many specific developments in memory technology, computer hardware, and computer software (both
time-sharing and batch operating systems) were also heavily backed by government
R&D funds or direct sales, often through the network of military contractors that
were using computers for missile and aviation projects.[23]

Most of the legendary names in early computing did significant work with military sponsorship in one form or another. The fundamental "stored program" concept
emerged from work done during and immediately after World War II on code-
breaking, artillery range-finding, and nuclear weapons designing. At Harvard,

Howard Aiken's Mark I relay computer, a joint project with IBM before the war, became a Navy-funded project during the war. At the University of Pennsylvania's Moore School, J. Presper Eckert gained experience with memory techniques and high-speed electronics through a subcontract from MIT's Radiation Laboratory. Soon he and John W. Mauchly built their ENIAC and EDVAC machines with funds from the Army Ordnance Department, which had employed 200 women as "computers" laboring at the task of computing artillery range tables by hand. Eckert and Mauchly, struggling to commercialize their work after the war, paid their bills by building a digital computer to control the Air Force's Snark missile through a subcontract with Northrop Aircraft. At the Institute for Advanced Study (IAS) at Princeton, John von Neumann started a major computer development project. During the war, von Neumann had worked with Los Alamos physicists doing tedious hand calculations for the atomic bomb designs. Immediately after the war, he collaborated with RCA (much as Aiken had done with IBM) on a joint Army and Navy–funded project to build a general-purpose digital computer. Copies of the IAS machine were built for government use at six designated sites: Aberdeen Proving Ground in Maryland, Los Alamos, Argonne outside Chicago, Oak Ridge, RAND (an Air Force think tank), and an Army machine (code-named ILLIAC) at the University of Illinois.[24]

The Harvard, Pennsylvania, and IAS computer projects are justly remembered for developing the basic concepts of digital computing. Project Whirlwind at MIT, by comparison, was notable for developing computer hardware—particularly, fast magnetic-core memory—and notorious for consuming an enormous amount of money. It also showed that computers could do more than calculate equations, providing an influential paradigm of command and control. Whirlwind initially took shape during the war as a Navy-sponsored project to build a flight trainer. At MIT's Servomechanisms Laboratory, Jay Forrester used a successor of Vannevar Bush's analog differential analyzer (see chapter 5) in an attempt to do real-time airplane flight simulation. Forrester, an advanced graduate student at the time, used an array of electromechanical devices to solve numerous differential equations that modeled an airplane's flight. But in October 1945, at an MIT conference that discussed the Moore School's ENIAC and EDVAC machines, Forrester was bitten by the digital bug. Within months he and the MIT researchers dropped their analog project and persuaded the new Office of Naval Research to fund a digital computer, a thinly disguised general-purpose computer carrying a price tag of $2.4 million. Forrester's ambitious and expensive plans strained the bank, even though the ONR was at the time the principal research branch for the entire US military (separate research offices for the Navy,

FIG. 7.5. WHIRLWIND COMPUTER CONTROL CENTER

The Whirlwind computer in 1950, at a moment of transition. Gone by this time was the original Navy-funded plan to use Whirlwind as a flight trainer. This image shows Forrester's vision of Whirlwind, then funded by the Air Force, as a command-and-control center. An array of computers built to the Whirlwind design formed the core of the bomber defense system known as SAGE (Semi-Automated Ground Environment). Left to right are Stephen Dodd, Jr., Jay Forrester, Robert Everett, and Ramona Ferenz.

Courtesy of Massachusetts Institute of Technology Museum.

Army, and Air Force, as well as the National Science Foundation, were all still on the horizon). Whirlwind's annual funding requests were five times larger than those for the ONR's other major computer effort, von Neumann's IAS project.

Faced with mounting skepticism from the Navy that it would ever see a working flight simulator, Forrester articulated instead an all-embracing vision of Whirlwind as a "Universal Computer" at the heart of a "coordinated CIC" (Combat Information Center) featuring "automatic defensive" capabilities as required in "rocket and guided missile warfare." Computers, he prophesied, would facilitate military research on such "dynamic systems" as aircraft control, radar tracking and fire control, guided missile stability, torpedoes, servomechanisms, and surface ships.[25] Forrester hoped Whirlwind might become another mega-project like the Radiation Laboratory or Manhattan Project. The Navy blanched at the prospect and dramatically scaled back its Whirlwind funding, although the ONR continued substantial funding across the 1950s for MIT's Center for Machine Computation. It was the Air Force, increasingly concerned with defense against Soviet bombers, that took up Forrester's grand vision for command and control (fig. 7.5). In 1950, with Whirlwind amply funded at just under $1 million that year—the bulk of the money now coming from the Air Force—

other MIT faculty began a series of Air Force–commissioned studies of air defense. These advocacy studies would culminate in the fantastically complex bomber defense system known as SAGE.

Operationally obsolete before it was deployed, SAGE (Semi-Automated Ground Environment) nonetheless was an important link between the advanced work on computing and systems management at MIT and the wider world. (SAGE was first operated a year after the Soviets launched the Sputnik satellite in 1957, by which time missiles flying much too fast to intercept had replaced slow-flying bombers as the principal air defense concern.) SAGE aimed to gather a continuous flow of real-time data from radar sites arrayed across North America, with the goal of directing the response of the Air Force in the case of a Soviet bomber attack. Such a huge effort was beyond even Forrester's grasp, and, just as Bush had spun off the Manhattan Project from his Office of Scientific Research and Development, the MIT administration created the Lincoln Laboratory as a separate branch of the university to manage the massive SAGE project. Forrester's Whirlwind project thus became Lincoln Laboratory's Division 6. Most of the country's leading computer and electronics manufacturers were SAGE contractors, including RCA, Burroughs, Bendix, IBM, General Electric, Bell Telephone Laboratories, and Western Electric. The newly founded Systems Development Corporation wrote the several hundred thousand lines of computer code that the complex project required. Estimates for the total cost of SAGE range surprisingly widely, from $4 billion to $12 billion (1950s dollars).[26]

At the center of this fantastic scheme was a successor to Forrester's Whirlwind, or more precisely fifty-six of these machines. To produce the Whirlwind follow-ons for SAGE, Lincoln Laboratory in 1952 selected a medium-size business machines manufacturer previously distinguished for supplying tabulating machines to the new Social Security Administration and punch cards by the billions to the Army. The firm was named, rather grandly, International Business Machines. At the time it began its work on SAGE, IBM ranked squarely among its computing rivals RCA, Sperry Rand, and Burroughs. IBM would make the most of this military-spawned opportunity, launching itself into the front ranks of computing. IBM was already working on an experimental research computer for the Naval Ordnance Laboratories and had begun work on its first electronic computer, the IBM 701 Defense Calculator. With its participation in SAGE, IBM gained a healthy stream of revenues totaling $500 million across the project's duration. Fully half of IBM's domestic electronic data-processing revenues in the 1950s came from just two military projects: SAGE and an analog computer for the B-52 bomber. These handsome revenues in turn convinced IBM's president, Thomas J. Watson, Jr., to push the company into the computer age and bankrolled the company's major developmental effort for the System solidus 360 computer (1962–64).

Besides the $500 million in revenues, IBM gained unparalleled exposure to state-of-the-art computing concepts. With its participation in SAGE, IBM gained firsthand knowledge of MIT's magnetic-core memories and invaluable experience with mass-producing printed-circuit boards for computers. It incorporated the computer-networking and data-handling concepts of SAGE into its SABRE airline reservation system (the name is revealing: *Semiautomatic* Business-Research *Environment*), which became operational in 1964. Another military undertaking of IBM's was Project Stretch. Beginning in 1956 Stretch involved IBM with the National Security Agency and the Atomic Energy Commission's Los Alamos nuclear weapons laboratory. The two agencies had radically different computing needs. The AEC lab needed to do numerical computations with high precision, while the NSA needed to match strings of text for code-breaking. The NSA's Stretch computer, delivered in 1962 as the core of a data-processing complex, served the NSA until 1976, when it was replaced by the renowned Cray I supercomputer.[27]

IBM was hardly alone in tapping the lucrative military market in computing. In his study of Silicon Valley, Christophe Lécuyer found the region's original bellwether firms sold microwave electronics, high-power radar, and early semiconductor devices to the military and space agencies, while Margaret O'Mara points out that the immense Lockheed Missile and Space complex in Sunnyvale was "the Valley's largest high-tech employer for decades" (working on Polaris ballistic missiles, Corona spy satellites, and other projects) and that Stanford University scooped up its share of military R&D funding. Steve Wozniak's father, among many thousands, worked as a Lockheed engineer. Anchoring Boston's dynamic high-tech region were MIT and Raytheon, respectively, the country's top university military contractor and a leading high-tech military contractor originally founded by MIT's Vannevar Bush. The Project SAGE–derived Lincoln Labs, its spin-off Mitre Corporation, the Air Force Cambridge Research Laboratories, and Hanscom Air Force Base added additional military heft to the region. In Minnesota, the dominant computing firms— Engineering Research Associates, Univac, Control Data, and Seymour Cray's several supercomputing ventures—built their most impressive computers for the military and intelligence agencies—while giant Honeywell specialized in control systems to guide missiles, rockets, aircraft, and more.[28]

Even though the commercial success of IBM's System solidus 360 in the 1960s and beyond made computing a much more mainstream activity, the military retained its pronounced presence in computing and computer science. This was the result once again of open-ended research budgets, connections to urgent Cold War applications, and a cadre of technically savvy officers. The center of the military's computing research efforts was a small office known by its acronym IPTO, Information Processing Techniques Office, located within the Pentagon's Advanced Research Project Agency. Founded in 1959, in the wake of the Sputnik crisis, ARPA itself funded research on a wide range of military-related topics, ranging from ballistic missile defense, nuclear test detection, and counter-guerrilla warfare to fields such as behavioral sciences, advanced materials, and computing. The IPTO was the nation's largest funder of advanced computer science from its founding in 1962 through the early 1980s. Its projects strengthened existing computing centers at MIT and Carnegie-Mellon and built new computing "centers of excellence" at Stanford, UCLA, Illinois, Utah, Caltech, USC, Berkeley, and Rutgers. Among the fundamental advances in computer science funded by the IPTO were timesharing, interactive computer graphics, and artificial intelligence. Time sharing prompted some of the first computer security concepts, also under military auspices. J. C. R. Licklider, head of the IPTO program in the early 1960s, also initiated work on computer networking that led, after many twists and turns, to the internet.

In the mid-1960s IPTO saw a looming budget crunch. Each of its sponsored research centers wanted a suite of expensive computers, and not even the Pentagon could keep up with their insatiable budget requests. The solution, pretty clearly, was to enable researchers at one site to remotely use the specialized computing resources at distant sites. Ever canny, the IPTO's program managers, from the 1960s through the 1980s, sold a succession of internet-spawning techniques to the military brass as the necessary means to a reliable military communications network. One result of the 1962 Cuban missile crisis was the military's realization that it needed the ability to communicate quickly and reliably across its worldwide field of operations. A break in the fragile communications network might isolate a battlefield commander from headquarters and (as satirized in the film *Dr. Strangelove* [1964]) lead to all sorts of dangerous mishaps. Alternatively, a single Soviet bomb strike on a centralized switching facility might knock out the entire communication system. A 1964 RAND Corporation report, "On Distributed Communications," proposed the theoretical grounds for a rugged, bombproof network using "message blocks"—later known as "packet switching"—to build a distributed communications system. RAND endorsed this novel technique because it seemed to ensure that military communication networks could survive a nuclear attack. Even if part of the distributed network were destroyed, the network itself could identify alternative paths and route a message through to its destination.[29] These concepts became the conceptual core of the internet (see chapters 8 and 10).

Throughout the decades of world war and cold war, the various arenas of military technology (East and West) appeared to many outsiders to be a world unto themselves, walled off from the broader society. This image was reinforced by the esoteric nature of the science involved in military technologies—dealing with the smallest wavelengths and the largest numbers. The pervasive security regimes swallowed up entire fields of science and technology (including some not-so-obviously military ones, such as deep-water oceanography). Western critics faulted the "military-industrial complex" or "national-security state" for being antithetical and even dangerous to a democratic society, with its ideals of openness and transparency.

Through the military-dominated era there was an unsettling tension between the West's individual-centered ideology and its state-driven technologies. Visions of free markets and open democracy were casualties of the secrecy regime of the Manhattan Project, the Oppenheimer-Teller conflict about the hydrogen bomb, President Eisenhower's "military-industrial complex," the studied deceptions of the Vietnam War (fanned by the Kennedy, Johnson, and Nixon administrations), and the pitched techno-political battles in the Reagan years over the Strategic Defense Initiative. Throughout the Cold War "national security" was a trump card to silence troublesome critics.

On balance, in my view, the West was lucky to be rid of the Cold War, because (as in many conflicts) we took on some of the worst characteristics of our enemies. Faced with the image of a maniacal enemy driven by cold logic unencumbered by the niceties of democratic process, Western governments were all too often tempted to "suspend" open processes and crack down on troublemakers. By comparison, dissidents in the Soviet Union found that criticizing the government, especially on military and nuclear matters, was hazardous in the extreme, although criticisms pertaining to the environment permitted. The Soviet Union's state-centered ideology and state-centered technologies "fit" the imperatives of the Cold War better than the West's did. What the Soviet system failed to do was generate economic surpluses large enough to pay for the heavy military expenditures. "One of the reasons that the Soviet Union collapsed," according to Seymour Melman's critical analysis, "is that the

Soviets were so focused on making military equipment that they let their industrial machinery literally fall apart."[30]

What is striking to me is the wide reach and pervasiveness of military technologies. Scientist Niels Bohr remarked that the atom bomb project was "turning the whole country into a factory." Developing the bomb involved, in addition to the innumerable military officers and government officials, the managers and workers of the nation's leading corporations; a similar list of dependent enterprises and enrolled people might be compiled for each of the large Cold War projects. Together, these military endeavors were not so much an "outside influence" on technology as an all-pervading environment that defined what the technical problems were, how they were to be addressed, and who would pay the bills. While closed-world, command-and-control technologies typified the military era, the post–Cold War era of globalization generated more open-ended, consumer-oriented, and networked technologies.

CHAPTER *8*

1970 – 2001

Promises of Global Culture

"Globalization . . . is the most important phenomenon of our time. It is a contentious, complicated subject, and each particular element of it—from the alleged Americanization of world culture to the supposed end of the nation-state to the triumph of global companies—is itself at the heart of a huge debate."[1] A helpful way of thinking about globalization, it seems to me, is as a recent-day industrial revolution, a set of unsettling shifts in society, economics, and culture in which technology appears to be in the driver's seat. We'll see on closer inspection that technology in the era of global culture, no less than in earlier eras, was both a powerful shaper of political, social, cultural, and economic developments as well as decisively shaped by these developments.

The arguments about globalization, like many arguments involving technology, were largely about contending visions for the future. Debates about technology are seldom about wires and boxes, but rather are about the alternative forms of society and culture that these technical details might promote or discourage. As we will see, a positive vision of globalization was inscribed in core technical details of fax machines and the internet, even as the commercialization of these technologies, among others, promoted globalization in economics and culture.

Globalization has been a contested concept for fifty years. A set of events points to a striking upsurge in globalization, as described later in further detail, in the 1970s. An "American Century" ended that decade with the economic crisis of stagflation, the Arab oil embargo, the protracted agony of the Vietnam War, and the political turmoil of Richard Nixon's Watergate scandal. Even though the Soviet Union did not dissolve, it too faced internal economic problems that led its leadership to embrace "détente" with the West and temporarily eased the Cold War superpower tensions that had defined and focused the military era. China emerged from an isolated economy when its leadership declared in 1978 that capitalistic practices could be tolerated. Agriculture was loosened up, foreign investment streamed in, and China began a long march that lifted many of its billion-plus residents out of poverty. A bookend was China's joining the globalizers' World Trade Organization in 2001. Also in the

1970s, Europe saw the end of its last dictatorships (in Portugal and Spain) as well as enlargement of the European Union as an economic and political bloc. So-called Asian "tigers" like Hong Kong, Singapore, South Korea, and Taiwan participated strongly in globalization, with Indonesia, Malaysia, the Philippines, Thailand, and Vietnam also joining the race for technology-driven, export-oriented economic growth. The decade of the 1970s through the disruptions following 11 September 2001 form a meaningful historical period with intense globalization accompanied by a spirited debate about its character.

Advocates of globalization point out that the rise in world trade, especially since the 1970s, worked to increase average incomes across the world and in many nations. Countries that were open to globalization typically did better economically than those that were not. In this view, globalization rescued roughly a *billion* people in Asia and Latin America from poverty (while shortfalls remain in Africa). Further, advocates maintain that the glare of global publicity improved working conditions and environmental regulation in countries lacking proper labor and environmental laws. In the end, advocates sometimes claim that—in a latter-day version of tech-nological fundamentalism—the global economy was the only game in town. "The global economy in which both managers and policy-makers must now operate . . . has its own reality, its own rules, and its own logic. The leaders of most nation states have a difficult time accepting that reality, following those rules, or accommodating that logic," wrote business guru Kenichi Ohmae.[2]

Critics counter that globalization failed to ease the crushing burdens on the world's poorest citizens and that globalization increased pollution. For instance, an industrial revolution in coal-rich China, with its filthy cement factories, widespread heavy-metal contamination, and rampant automobile use produced air pollution as deadly as any in industrial-era Britain. With energy-intensive economic growth, globalization hastened global warming. Equally important, the profile of freewheel-ing multinational corporations revived arguments against the excesses of capitalism. Critics blamed globalization for the gap between global "haves" and "have-nots," for the cultural imperialism manifest in American media and fast food such as McDonald's, and the antidemocratic power of international entities such as the World Bank and the World Trade Organization. "Satellites, cables, walkmans, video-cassette recorders, CDs, and other marvels of entertainment technology have created the arteries through which modern entertainment conglomerates are homogenizing global culture," wrote Richard Barnet and John Cavanagh. Today, the argument about cultural imperialism remains much the same, if perhaps less America-centric, with global attention on the mass-media stars of India's famous movie industry, which makes twice the number of movies and sells twice the number of cinema seats as Hollywood's, or the multimedia phenomenon known as the Korean Wave. The first YouTube video to pass one *billion* views was the K-pop confection "Gangnam Style." According to *Forbes* no less, "Bollywood's Shah Rukh Khan is the biggest movie star in the world." And, the three most-popular video games of all time—Minecraft, Grand Theft Auto V, and Tetris—hail from Sweden, Scotland, and Russia.[3]

The persisting debate on globalization is a healthy one. The debate itself modestly altered the "nature" of globalization. The gap between globalization and development was spotlighted in the dramatic street protests at the World Trade Organization's Seattle summit meeting in 1999. In hindsight, the clouds of tear gas aimed at anti-globalization protesters—in Seattle, Washington, Prague, and Genoa—ended the globalizers' pie-in-the-sky optimism and led to modest changes in institutional practices. For decades, World Bank had enforced a narrow measure of economic development. Its sole concern was increasing national GDP levels (an aggregate measure for economic development), and it typically required a recipient country to undergo painful "structural adjustment" of its macroeconomic policies. Because the World Bank's structural adjustments usually required cutting government expenditures and/or raising taxes, it earned a frightful reputation in many recipient countries. In the wake of globalization debates, however, the World Bank embraced broader indexes of social development, giving greater attention to education, public health, local infrastructure, and the environment.[4]

Whatever the economic and political consequences of globalization, the threat of cultural homogenization concerned many observers. Seemingly everyone, during the era of globalization, watched Fox News or Disney, drank Coca-Cola, and ate at McDonald's (fig. 8.1). Perhaps the citizens of the world were becoming more and more alike? There is some support for such "convergence." Look at alcohol and industry. Alcohol consumption during the first industrial revolution was a much-discussed "problem": factory masters wanted their workers sober and efficient, while many workers preferred to spend their spare income on beer (see chapter 3). As recently as the mid-1960s the citizens of France, Italy, and Portugal each drank an astounding 120 liters of wine each year, with Spaniards not far behind; but the citizens of thoroughly industrial Germany drank one-*sixth* that amount, while producing wine for export. By the late 1990s, however, wine consumption had plummeted—by half or more—in each of the four southern European countries, while Germans' wine drinking remained steady. It appeared that northern habits of industry and efficiency

FIG. 8.1. COCA-COLONIZATION OF EUROPE?

A typical Dutch mobile snack bar, serving French fries with mayonnaise, at Zaanse Schans, 2001.

Photograph by Marco Okhuizen, used by permission of Bureau Hollandse Hoogte.

eroded southern habits of wine drinking. If southern Europe gained industrial efficiency, Europe as a whole lost cultural diversity.

While mindful of convergence, I believe there is substantial evidence for a contrary hypothesis, let's say "persisting diversity." The fervently pro-globalization *Economist* magazine knocked down the "convergence" hypothesis by citing strong evidence of persisting cultural diversity in mass media. In most countries during the 1990s, American television programming filled in the gaps, but the top-rated shows were *local* productions. In South Africa the most-watched show was *Generations*, a locally produced multiracial soap opera; in France, *Julie Lescaut*, a French police series; and in Brazil, *O Clone*, a drama produced by TV Globo. For its part MTV, which once boasted "one world, one image, one channel," created diverse productions across thirty-five different channels (fifteen in Europe alone) in response to local tastes and languages. Even McDonald's, a symbol of globalization, was more complex than it might appear. McDonald's is best understood as a case of active appropriation by local cultures of a "global" phenomenon. The internet is another construct that appears to be the same across the world only so long as you don't look closely at European electronic commerce or China's burgeoning social media.[5]

"Persisting diversity" is also consistent with earlier eras. As we saw in chapters 3 and 5, the paths taken by countries during the first and second industrial revolutions varied according to the countries' political and institutional structures, social and cultural preferences, and the availability of raw materials. It is worth adding here that Japan successfully industrialized and modernized during the twentieth century, but it did not become "Western," as is attested by the many distinctive political, industrial, and financial structures and practices in Japan that to this day baffle visiting Westerners. A case in point is Japan's singular affection, through today, for the paper-based fax machine.[6] This chapter first surveys the origins of the global economy and then turns to case studies of the facsimile machine, McDonald's, and the internet.

THE THIRD GLOBAL ECONOMY

The late twentieth-century's global economy is not the first or second global economy we have examined in this book, but the third. The first was in the era of commerce (see chapter 2). The Dutch in the seventeenth century created the first substantially "global" economy through multicentered trading, transport, and "traffic" industries that stitched together cottons from India, spices from Indonesia, enslaved people from Africa, sugarcane in the Caribbean with tulip speculators, stock traders, boat builders, sugar refineries, and skilled artisans in the home country. This early global economy lasted until France and England overwhelmed it in the early eighteenth century. Early industrialization in England—even though it imported raw materials, such as Swedish iron and American cotton, and exported cotton textiles, machine tools, and locomotives—did not create a multicentered Dutch-style trading system.

A second global economy developed in the 1860s and lasted until around the First World War, overlapping the era of imperialism (see chapter 4). This second global economy involved huge international flows of money, people, and goods. Britain, with its large overseas empire, was the chief player. A key enabling technology was the oceangoing steamship, which by the 1870s cheaply and efficiently carried people and cargo. In compound steam engines, "waste" steam exiting one bank of cylinders was fed into a second bank of lower-pressure cylinders to extract additional mechanical power. By the 1890s oceangoing steamships employed state-of-the-art triple-stage compound steam engines; some builders even tried four stages. In these super-efficient engines, the heat from burning two sheets of paper moved a ton of cargo a full mile. Linking efficient steamships with nationwide railroad networks dramatically cut transport costs for many goods. Even decidedly middling-value products found a ready global market. Butter from New Zealand, sheep from Australia, jute from the Far East, and wheat from North America all flowed in huge volumes to European ports.[7]

The complex tasks of trading, warehousing, shipping, insuring, and marketing these global products led to a sizable and specialized financial superstructure. During the second global economy Chicago became the world's grain-trading center and New York specialized in cotton and railroad securities, while London became the world's investment banker. London, as noted in chapters 3 and 4, had the world's most sophisticated financial markets and substantial exportable wealth from the first industrial revolution. So active were its international investors that Britain's net outflows of capital averaged 5 percent of gross domestic product (GDP) from 1880 to 1913, peaking at an astounding 10 percent (while by comparison Japan's net outflows of capital were in the 1980s considered "excessive" at 2–3 percent). In the decade prior to the First World War, Britain's wealthy made foreign direct investments (in factories and other tangible assets) that nearly equaled their domestic direct investments, and Europe's investors displayed similar behavior. In fact, foreign direct investments from Britain, France, Germany, and the Netherlands were substantially *larger* in GDP terms in 1914 than in the mid-1990s. The United States, Canada, Argentina, and Australia were the chief recipients of European investment.

Widespread use of the gold standard assisted the international flow of money, while a world without passports facilitated the international flow of people. The industrialization of North American cities and the extension of agriculture across the Great Plains tugged on peasants, smallholders, and artisans in Europe facing economic difficulties, religious persecution, or family difficulties. Some 60 million people left Europe for North and South America in the second half of the nineteenth century. This huge transatlantic traffic came to a screeching halt with the First World War, as the Atlantic filled up with hostile ships and deadly submarines. In the early 1920s, a set of blatantly discriminatory laws in the United States finished off immigration from southern and eastern Europe, where the most recent generation of immigrants had come from. Immigration to the United States did not reach its pre-1914 levels until the 1990s.

While wartime pressures and American nativism in the 1920s weakened the second global economy, it was brought low by economic nationalism during the 1930s. Hitler's Germany offered an extreme case of going it alone (see chapter 7). Across the industrialized countries, increased tariffs, decreased credits, and various trading restrictions effectively ended the international circulation of money and goods. Import restrictions in the United States culminated in the notorious 1931 Smoot-Hawley Tariff, which hiked average US import tariffs to just under 60 percent. This was a deathblow to world trade because the United States was, at the time, the world's leading creditor nation. High US tariffs made it impossible for European debtor nations to trade their goods for US dollars and thus to repay their loans and credits (a hangover from the First World War). Trade among the seventy-five leading countries, already weak in January 1929, during the next four years spiraled down to less than one-third that volume. Country after country defaulted on its loans and abandoned the gold standard. Global movements of people, capital, and goods had looked strong in 1910, but this trend was a dead letter within two decades. Even before the Nazis' brutally nationalistic Four Year Plan of 1936, the second global economy was dead.

Since around 1970 global forces in the economy and in society resurged, but like all runs the third global economy is a creature of history. The institutions responsible for the global economy, including the World Bank, International Monetary Fund, and World Trade Organization, are human creations and not facts of nature. And even the floodtide of the third global economy had odd countercurrents. In the 1990s, when such countries as Singapore and the Philippines became the globalizers' darlings for their openness to trade and investment, some countries actually became *less* globalized. According to an index drawn up by A. T. Kearney, globalization actually *decreased* in Malaysia, Mexico, Indonesia, Turkey, and Egypt. McDonald's, after leaving Bermuda and Barbados in the 1990s, also left Bolivia, Iceland, Jamaica, Macedonia, and Montenegro in subsequent years. The recent rise of nationalism in many countries, the re-emergence of sharp trade wars, and the emergence of nation-specific internets (see chapter 10) suggest that the third global economy, under stress after 2001, is fighting for its life. The following sections of this chapter indicate clearly how fax machines, the internet, and McDonald's concretely expressed and perceptibly entrenched the promises of global culture.

FAX MACHINES AND GLOBAL STANDARDS

The very first electromechanical facsimile machines were something of a global technology in the mid-nineteenth century, and the rapid commercialization of their electronic descendants in the 1970s and 1980s was due in large measure to international standards for technology. Something like the geographical shifts during the era of science and systems, the geography of consumer electronics shifted with globalization. Fax machines were one of numerous consumer electronics products—transistor radios, televisions, videocassette recorders, and computer-game consoles—that Japanese manufacturers soundly dominated.[8] While many US firms accepted the imperatives of military electronics, Japanese firms embraced the possibilities of consumer electronics. One might say that in the United States the military market displaced the consumer market, while in postwar Japan it was the other way around. Consumer electronics radiated outward from Japan to South Korea, Taiwan, and other so-called Asian "tigers" that seized on export-driven globalization.

In one way, Japanese manufacturers had a cultural incentive for facsimile communication. Image-friendly fax machines are especially compatible with pictographic languages like Japanese. Indeed it was just after a Chinese telegraph commission visited Boston in 1868 that a young Thomas Edison, impressed by the difficulties of fitting the Chinese language into the digital structure of Morse code telegraphy, began work on an early electromechanical fax machine. While English has only twenty-six characters and was easily translated into a series of telegraphic dots-and-dashes (or digital ones-and-zeros), the standard Japanese system of *kanji* has more than two thousand characters plus forty-six *katakana* phonetic characters used for foreign words. Chinese has even more characters. Edison saw that transmitting pictures by telegraph would be ideal for such ideographic languages.

While other facsimile-transmitting machines typically used some form of raised-type (including the first fax machine patented by Scotsman Alexander Bain, in 1843), and often used up to a dozen wires, Edison's fax machine used a single telegraph wire. His backers in Gold and Stock Telegraph Company thought it promising enough to invest $3,000. A message written on paper with a standard graphite pencil was "read" by his ingenious machine. The electricity-conducting graphite completed a circuit between two closely spaced wires on a pendulum, tracing an arc over the message, and sent bursts of electricity encoding the message from the transmitting machine to the receiving machine. In the 1930s William G. H. Finch's radio facsimile deployed an identical conceptual scheme (including synchronized pendulums and chemically treated paper) to successfully transmit text and illustrations. Finch's machine also deployed selenium photocells—available since around 1900—to read the image for transmission. Stations in Sacramento, Fresno, Des Moines, and nearly twenty other US cities experimented with radio-broadcasting small newspapers direct to subscribers' homes, while Fitch's greatest triumph was at the 1939 New York World's Fair with 2,000 recipients.

From the 1920s through the 1960s facsimile remained a niche product for specialized users. Facsimile schemes of RCA, Western Union, and AT&T targeted mass-market newspapers seeking circulation-boosting news photographs. The best-known venture in this area was the Associated Press's Wirephoto service, established in 1934, competing with rival services from the Hearst and Scripps-Howard syndicates as well as the *New York Times*, *New York Daily News*, and *Chicago Tribune*. Another specialized user unconstrained by facsimile's high costs was the military. During the Second World War, both the German and the Allied Forces transmitted maps, photographs, and weather data by facsimile. Even after the war the Times Facsimile Corporation, a subsidiary of the New York Times Company, found its largest market in the military services. By 1960 Western Union, seeking to preserve its aging telegraph network against the onslaught of telephones, had sold 40,000 Desk-Fax units as "a sort of self-contained telegraph office" directly connecting medium-sized companies with the nearest Western Union central office. It stopped manufacturing Desk-Faxes that year.

In the 1960s Magnavox and Xerox began selling general-purpose, analog facsimile machines to compete with market leader Stewart–Warner. Their modest sales and meager profits can be traced to their slow transmission and corresponding high telephone costs; furthermore, in the absence of commonly accepted standards, most fax machines operated only with machines of their own brand. A further barrier was that Bell Telephone, a regulated monopoly, prevented users connecting non-Bell equipment to its phone lines. (Bell maintained that such a restriction ensured the "integrity" of the telephone system, even as it locked in a lucrative monopoly for Western Electric, its manufacturing arm.) To sell one of its Telecopier machines for use over the public phone system, Xerox needed to add a special box, sold by Western Electric, that connected the Telecopier safely to the phone system. A regulatory change in 1968—the Carterphone decision—eased this restriction, allowing fax machines and computer modems (in 1976) to be directly connected to the telephone system. Parallel changes occurred in Japan.

The explosive growth of facsimile was made possible by the creation of worldwide technology standards. The adoption of global standards made it possible for fax machines from different manufacturers to "interoperate." At the time, in the late 1960s, the leading fax manufacturers—Magnavox, Xerox, Stewart–Warner, and Graphic Sciences (a spin-off from Xerox)—each had proprietary standards that only vaguely conformed to the prevailing US standards. Accordingly, sending a message between machines of two different brands was "fraught with error." Companies clung to their own standards for fear of helping a rival or hurting themselves, even though it was clear that incompatible standards constrained the industry. Enter the principal standards-setting body for facsimile, the CCITT, a branch of the United Nations' International Telecommunication Union, whose headquarters are in Geneva, Switzerland.

The CCITT, or Comité Consultatif International Télégraphique et Téléphonique, was the leading international standards-setting body for all of telecommunications beginning in the 1950s. Its special strength was and remains standards setting by committee. In some instances, when there is a single producer, a single market, or a single country large enough to pull others along, such a dominant player can set standards. For example, with the stunning success of IBM's System solidus 360 line of computers in the 1960s, that company's proprietary computer standards became de facto international ones. Governments, too, often legislate standards, such as North America's 525-line televisions contrasted with Europe's 625-line ones. But for many communications technologies, including facsimile, no single firm or single country could set durable international standards. This fragmentation became increasingly problematic; the collective benefit of common standards in a "network technology" (increased certainty, economies of scale, lowered transaction and legal costs) were reasonably clear and certainly attractive to manufacturers and users alike. It was here that the CCITT—with its quadrennial plenary meetings, ongoing formal study groups, scores of working committees, legions of subcommittees, not to mention informal coffee breaks—shaped the standards-setting process for fax technology.

Standards setting for facsimile started slow with two generations of weak standards but gained force with the emerging G3 standard, which, according to fax historian Jonathan Coopersmith, became "one of the most successful standards in telecommunications history." It can hardly be an accident that the G3 standard was created (1977 to 1980) during the heyday of globalization when international cooperation, transnational technology, and global standards were on everyone's mind. Japanese fax manufactures, like American ones, offered initially a chaos of incompatible standards with no less than six different algorithms. The Japanese state intervened when the Ministry of Posts and Telecommunications brought together the varied players in 1977 and insisted on a uniform national standard. Japan's compromise national standard, submitted by British Telecom, was adopted by CCITT's plenary assembly in 1980.

It was CCITT's success with the G3 standards in 1980 that made facsimile into a global technology—and relocated the industry to Japan. The pre-1980 standards had specified a worldwide standard using amplitude-modulated (AM) signal processing, and transmission times at three minutes per page were close to the theoretical maximum. The 1980 G3 standard effected a shift to digital processing that offered a way to "compress" the scanned image; for instance, instead of coding individually thirty-six times for a blank space, a compression routine would code a short message to repeat a blank space thirty-six times. The fax machine's internal modem transformed the compressed digital information into an analog signal and sent it over the phone system. Machines conforming to the digital standard could send an entire page in one minute. The achievement of worldwide standards, digital compression, and flexible handshaking, in combination with open access to public telephone systems, created a huge potential market for facsimile. By the late 1990s facsimile accounted for fully half the telephone traffic between Japan and the United States.

With Japanese electronics manufacturers poised to manufacture the digital fax machines, American manufacturers in effect declined to make the switch. Already by the late 1960s Matsushita had pioneered flat-head scanning technology, using glass fibers, as well as electronic-tube scanning; in 1971, when the firm started work on solid-state scanning and high-speed (digital) scanning, it had made 296 fax patent applications and had 110 engineers working on various aspects of fax development. An early signal of Matsushita's prominence was that the US firm Litton, once a leading exporter of fax machines, stopped developing its own machines in 1968 and started importing Matsushita's. Ricoh also began its high-speed fax development in 1971, with approximately fifty engineers organized into three teams. In 1973 Ricoh introduced onto the market a digital fax machine capable of 60-second-per-page transmission, while Matsushita answered two years later with its 60-second machine (and by 1986 had achieved a 12-second machine). The distinctive Japanese mixture of technological cooperation with market competition meant that Matsushita and Ricoh, followed closely by Canon, NEC, and Toshiba, emerged as world leaders. By the late 1980s Japanese firms produced over 90 percent of the world's fax machines.[9]

Japanese people quickly entrenched fax machines into their working and living routines. Japan became hooked on fax. By 2011, when old-style paper fax machines in the West were mostly long retired or being donated to museums, in Japan nearly all offices and 45 percent of all households still had them in regular use. A mass-market bento lunchbox delivery business failed to switch over to online ordering yet positively flourished with 30,000 daily orders coming in via fax. "There is still something in Japanese culture that demands the warm, personal feelings that you get with a handwritten fax," noted its owner. Remarkably, faxes remained popular with government workers, banks, their customers, and even organized-crime gangs. Japan's singular love affair with fax hit a hard stop in 2020 with the coronavirus pandemic, however. Telework was difficult when contracts and government paperwork were made "official" by personally applied *hanko* stamps, and when hospital orders and medical records were handwritten and submitted by fax. The paperwork to gain government subsidies for remote telework reportedly required printing out 100 pages of paper documents and delivering them in person. In May 2020, public health clinics in Japan gained permission to report coronavirus cases by email.[10]

While fax machines in the late 1980s were mostly used for traditional business activities, such as sending out price lists and product information (not to mention lunch orders), a group of inspired secondary-school teachers and students in Europe spotted an exciting possibility in using fax machines for broader political and cultural purposes. This network of students and teachers, along with some journalists and government officials, is notable not only for creatively using fax technology but also for explicitly theorizing about their culture-making use of technology. Secondary-school students working in the "Fax! Programme" collaborated with student-colleagues across Europe to produce a series of newspapers that were each assembled and distributed, in a single day, by fax.[11]

The Fax! Programme embodied and expressed a "shared desire to transform a European ideal into reality" (39). The idea of using fax machines for building European identity and youth culture originated with the Education and Media Liaison Center of France's Ministry of Education, which was in the middle of a four-year project to boost public awareness of telematics and videotext. (France's famous Minitel networked personal-computing system came out of this same context of state support for information technology.) The French media center was particularly impressed that fax technology was then "accessible to a very wide public" at modest cost, that the fax format made it "easy for those participating to do their [own] text and image layout," and that links between schools in different countries were possible because current fax machines could be "found anywhere" and had "maximum compatibility" (18). The vision was to have groups of students, across Europe, compose individual newspaper pages, and fax them in to a central editorial team. "The fax—the 'magical little machine'—is at the symbolic centre of the project" (45). The same day, the editorial team would assemble the contributions into a multilingual student newspaper, compose an editorial on the chosen theme, and arrange for the paper's distribution by fax and in traditional printed form. The first issue, with the theme "a day in the life of a secondary school student," was assembled and published on 3 November 1989 by thirty teams of students from twelve countries, coordinated by an editorial team in Renne, France.

The Fax! Programme quickly developed a life of its own. The fifth issue (with the theme "aged 20 and living in Europe") was the first created entirely by students and on a European theme, while the sixth issue, assembled by students in Bilbao, in the

Basque country of Spain, was an international milestone. The Bilbao issue, part of a major educational exhibition, was published in an exceptionally large print run of 80,000 as a supplement to a local daily newspaper. In its first three years (1989–92) the Fax! Programme produced thirty issues and involved 5,000 students from forty countries. An issue was typically composed by students from six countries, often drawing on resources from their schools, local governments, and professional journalists. While self-consciously centered on Europe—defined as the continent from Portugal to Russia and from the United Kingdom to Uzbekistan (and including the French West Indies)—the effort has included "guests" from sixteen countries in Asia, Africa, and the Americas.

Producing a multilingual, intentionally multicultural newspaper was the students' chief achievement. Each of the participating teams wrote its material in their home language and provided a brief summary in the organizing country's language. Students laid out their text and images, using a suggested template, and composed their page of the collected newspaper—before faxing it to the central editorial team. Besides the main European languages, the newspaper featured articles in Arabic, Chinese, Japanese, Korean, Nigerian dialects, Occitan, and Vietnamese. Interest on the part of primary school teachers led to four issues of a "junior" fax newspaper, focused on themes of interest to primary school students (e.g., the sports and games we play). While many of the chosen themes had a feel-good quality about them, the student organizers at times chose difficult themes, including environmental challenges to the coasts (issue 10), Antarctica after the extension of the Antarctic Treaty (24), and human rights (26). Issue 28, organized in Vienna, invited its sixty student contributors to ask themselves who and what was "foreign" to them. The Fax! Programme "is already an important and original vector for the European identity" (34); rather than parroting some bureaucrat's view, students instead "express their vision of Europe, their recognition of difference, their reciprocal curiosity" (50).

The idea of publishing an international student newspaper was certainly more groundbreaking in the 1980s than it might appear in our present internet-saturated world. What is still notable about the Fax! Programme was the students' ability to identify and develop new cultural formations in a technology then dominated by business activities. The global character of facsimile technology, and especially the possibility of linking students from several different countries, resonated with students intrigued by European integration. Like the internet visionaries examined later, they successfully translated their idealism about society and technology into tangible results. Fittingly, the European students drew on a technology that had been explicitly designed—through the CCITT's standards—as a global technology.

MCWORLD OR MCCURRY?

The promise of global culture has many faces, but none is better known than Ronald McDonald's, whose fame exceeds that of Santa Claus. McDonald's was and remains without peer as a supercharged symbol of globalization. Born in the 1950s Californian car culture and growing up alongside the baby-boom generation, McDonald's emerged as the 800-pound gorilla of the international fast-food jungle. McDonald's became the largest single consumer in the world of beef, potatoes, and chicken, and its massive presence has shaped the culture, eating habits, and food industries in roughly 120 countries. Sales outside the United States reached 50 percent of the corporation's revenues in 1995. Indeed, McDonald's was a creature of globalization; in 1970, it had restaurants in just *three* countries, but by 2001 it operated in 100 independent countries (not counting overseas territories and dependencies such as Guam, Curaçao, or Gibraltar). Its self-conscious corporate "system" of special fast-food technology, rigorous training, entrepreneurial franchising, aggressive advertising, and active public relations inspired an entire cottage industry to scrutinize the company's activities. The multiple "Mc" words testify to some uneasiness: "McJobs" are low-wage, nonunion, service-sector positions. "McWorld" points to cultural homogenization and rampant Americanization denounced by many critics of globalization. "McDonaldization" notes the spread of predictability, calculability, and control—with the fast-food restaurant as the present-day paradigm of Max Weber's famous theory of rationalization. And everywhere the discussion turns to globalization—among critics, advocates, or the genuinely puzzled—sooner or later McDonald's is sure to pop up.

Critics of globalization have seized on McDonald's as a potent symbol. José Bové, an otherwise retiring forty-six-year-old French sheep farmer, became an international celebrity in 1999 when he targeted a nearby McDonald's, pulling down half its roof with a tractor and scrawling on what remained, "McDo Dehors, Gardons le Roquefort" (roughly, McDonald's get out, let's keep the Roquefort). While focused on a single McDonald's, his protest had far wider aims. It embodied and expressed the resentment felt by many French people about American economic influence, at that moment focused on the punitive 100 percent American import duties levied against Bové's beloved Roquefort cheese and other upmarket French foods, as well as the way American popular culture challenged a more leisurely and traditional way of life. Big Macs and fancy cheeses were just the tip of a transatlantic trade tangle; Europeans had banned imports of hormone-treated American beef, a long-running trade dispute only distantly related to McDonald's. As he went off to jail, fist defiantly in the air, Bové promised to persist in "the battle against globalization and for the right of peoples to feed themselves as they choose." French President Jacques Chirac went on record that he "detests McDonald's food."[12]

When facing such protests McDonald's counters that its restaurants are owned and run by local people, employ local people, and sell locally grown food, but its well-honed communication strategy backfired in the "McLibel" case in Britain. In 1990 McDonald's charged two members of a London protest group with making false and misleading statements in a strongly worded anti-McDonald's pamphlet. As Naomi Klein writes in No Logo, "it was an early case study in using a single name brand to connect all the dots among every topic on the social agenda." The activists condemned McDonald's for its role in rain forest destruction and Third World poverty, animal cruelty, waste and litter, health problems, poor labor conditions, and exploitative advertising aimed at children. While the company eventually won the legal battle, in 1997, the anti McDonald's activists gained immense television and newspaper publicity. In Britain alone, they distributed some 3 million printed copies of the contested pamphlet, What's Wrong with McDonald's. And a support group created McSpotlight, a website featuring the original pamphlet, available in a dozen languages; some 2,500 files on the company, including videos, interviews, articles, and trial transcripts; chat rooms for discontented McDonald's employees; and the latest antiglobalization news. More than a quarter century later, McSpotlight keeps alive such courtroom howlers as the company's claim that Coca-Cola is a "nutritious" food and the assertion that "the dumping of waste [is] a benefit, otherwise you will end up with lots of vast empty gravel pits all over the country."[13]

Advocates of globalization, for their part, suggest the crowds lining up to get into McDonald's restaurants are voting with their feet. McDonald's extolled the 15,000 customers that lined up at the opening of its Kuwait City restaurant in 1994, following the Gulf War. The presence of McDonald's in the conflict-torn Middle East was good news to Thomas L. Friedman, the author of The Lexus and the Olive Tree (1999). In his spirited pro-globalization tract, Friedman framed the "golden arches theory of conflict prevention." As he phrased his theory, gained in a lifetime of munching Quarter Pounders and fries in his extensive international travels for the New York Times: "No two countries that both had McDonald's had fought a war against each other since each got its McDonald's." His happy thought was that

FIG. 8.2. CAPITALIST MCDONALD'S IN EASTERN EUROPE

McDonald's in the historic city center of Prague, Czech Republic, May 1999.

Photograph by Marleen Daniels, used by permission of Bureau Hollandse Hoogte.

a country's leaders must think twice about invading a neighboring country when they know that war making would endanger foreign investment and erode domestic support from the middle class. "People in McDonald's countries don't like to fight wars any more, they prefer to wait in line for burgers," he claimed. The actual record is mixed. Argentina got its McDonald's in 1986, four years after the Falklands war with Great Britain, but the theory does not account for the breakup of Yugoslavia and the India-Pakistan clash over Kashmir in the late 1990s, nor the 2008 Georgian war as well as Russia's 2014 invasion of Georgia—each with plenty of McDonald's on both sides of these conflicts. Friedman's unbounded enthusiasm about America's leading role in a globalized world led him to be openly dismissive of globalization critics: "In most societies people cannot distinguish anymore between American power, American exports, American cultural assaults, American cultural exports and plain vanilla globalization."[14]

For many reasons, then, McDonald's can reveal economic and cultural dimensions of globalization. The history of McDonald's is neatly packaged on the company's website. The legend began in 1954 when Ray Kroc discovered the McDonald brothers' hamburger stand in San Bernardino, California, where he saw eight of his Multimixers whipping up milk shakes for the waiting crowds. Quickly, he signed up the brothers and within a year opened the "first" franchised restaurant in Des Plaines, Illinois, "where it all began." The website emphasizes the human face of the company's growth—with cheery blurbs on the origins of Ronald McDonald, Big Macs, Happy Meals, and the company's charity work—saying little about the efforts to find financing, franchisees, and locations, and entirely passing over the real reason for its success: "the system" for turning out exactly standardized hamburgers and French fries. The company's overseas activities began in 1970 with a trial venture in the Netherlands, and within five years it had full initiatives in Japan, Australia, Germany, France, Sweden, England, and Hong Kong. To help focus our inquiry we can examine McDonald's in Russia and in the Far East and ask the following questions: What are the central elements of the McDonald's system, and how has it exported its system to other countries? How has McDonald's *changed* the eating habits, food industry, and culture of foreign countries? And in turn how have global operations *changed* McDonald's?

The multinational capitalist era in Russia began 31 January 1990 with the opening of the world's largest McDonald's in Moscow. From the start the corporation's efforts to expand into Russia were bound up with Cold War rivalries and complicated by a hundred practical questions about delivering a standard McDonald's hamburger in a country lacking anything like a Western-style food industry. At the center stood the colorful George Cohon, whom Pravda honored in a rare moment of humor as a "hero of capitalist labor." Cohon, as chief of McDonald's Canada, had had his entrepreneurial eye on Russia ever since hosting a delegation of Russian officials at the 1976 Montreal Olympics. In 1979, during his failed effort to secure for McDonald's the food-service contract for the 1980 Moscow Olympics, Cohon told a Canadian audience, "If companies like McDonald's . . . deal with the Russians, this, in a certain way, breaks down the ideological barriers that exist between communism and capitalism."[15] As it happened, Western countries boycotted the Moscow Olympic Games in protest of the Soviet invasion of Afghanistan; during the next decade Cohon's extensive personal contacts and some high-level diplomacy eventually landed McDonald's a go-ahead from Moscow's city government.

Cohon's quest had the good luck of accompanying Mikhail Gorbachev's famed *glasnost* and *perestroika* initiatives in the mid-1980s. Having a McDonald's in Moscow would signal that the Soviets were serious about "openness" and "restructuring." Cohon's team focused first on finding high-profile sites ("the Soviets' instincts seemed to be to put us behind the elevator shafts in hotels," he wrote) and settling on real-estate prices in a society without proper market mechanisms. They then turned to securing food sources that would meet McDonald's stringent standards, especially difficult since the late-Soviet agriculture system was in an advanced state of decay. Meat was frightful. "The existing Soviet abattoirs and processing facilities were . . . like something out of Upton Sinclair."[16] For a time they considered importing supplies from the huge McDonald's distribution site in Duisburg, West Germany, but decided instead to build from scratch a local complex in a Moscow suburb.

This 100,000-square-foot "McComplex," built at a cost of $21 million, replicated an entire Western-style food-industry infrastructure. It featured special lines for processing meat, dairy products, potatoes, and bakery goods, as well as dry storage, freezer storage, and a plant for making ketchup, mustard, and Big Mac sauce. Employing 400 workers, and processing 57 tons of food each day, McComplex met the company's policy of using local food supplies wherever possible. When it had trouble making proper French fries from the traditional small, round Russian potato, for instance, the company imported large Russet potatoes from the Netherlands and encouraged Russian farmers to grow them. McComplex also embodied the company's antiunion practices, further angering antiglobalization labor activists.

McDonald's arrived in Moscow when the Soviet economy was crumbling. It appeared that Muscovites in droves were voting against communism with their hands and feet. When McDonald's initially sought 630 workers, it received 27,000 written applications. And on opening day in Pushkin Square in 1990 McDonald's served a record 30,567 customers who lined up until midnight. The Russian manager of the restaurant, Khamzat Kasbulatov, declared to a friendly television crew: "Many people talk about *perestroika,* but for them *perestroika* is an abstraction. Now, me—I can touch my *perestroika.* I can taste my *perestroika.* Big Mac is *perestroika.*" And while cultural conservatives in Russia bemoaned McDonald's sway over young people, and some anti-capitalism leftists denounced its impact on the environment, working conditions, and scarce local food supplies, on balance McDonald's was not perceived as a corrosive foreign influence. In the unsettled days of August 1991, following the attempted coup by old-guard Soviet officers, when the streets of Moscow were filled with "anxious milling crowds," McDonald's was safe. "McDonald's?" the comment went, "They're a Russian company. They're okay."[17]

Yet even as McDonald's deployed its "system"—in operating Western-style restaurants, while managing and training workers and building an entire Western-style food supply—the challenges of operating in Russia perceptibly changed the company. The company fronted an enormous investment, perhaps $50 million, to get the Russian venture off the ground; and the decade-long groundwork had involved McDonald's executives directly in a tangle of high-level diplomacy, including the contentious issue of Soviet Jews' emigration to Israel. This depth of involvement changed the company. Elsewhere in the world, McDonald's relied on a decentralized network of independent contractors who supply everything from meat to milk shake mix. The fully integrated McComplex in Moscow stood in sharp distinction. "This was new for us; we had never been so vertically integrated," admitted Cohon.[18] And McDonald's Canada, anticipating licensing fees of 5 percent of total sales, was forced to reinvest the entirety of its accumulated revenues in Russia until the ruble eventually became convertible to Western currencies.

McDonald's operations in East Asia, beginning with Japan in 1971, lacked the drama of Pushkin Square and *perestroika*, but the company's presence there involved it in the region's complex politics and dynamic cultures. Anthropologist James Watson, in *Golden Arches East*, argues that understanding globalization requires seriously considering such pervasive everyday phenomena as McDonald's.[19] Based on colleagues' fieldwork in China, Taiwan, Korea, Japan, and his own in Hong Kong, Watson's findings do not fit the notion that McDonald's was corrupting local cultures or turning them into pale copies of America, as the critics of cultural imperialism often charge. Watson and his colleagues found instead a complex process of appropriation, in which McDonald's was one agent among many and in which McDonald's itself was transformed.

McDonald's certainly altered Asian eating habits. It arrived in Tokyo in 1971, Hong Kong in 1975, Taipei in 1984, Seoul in 1988, and Beijing in 1992. Except in Korea, where beef was already eaten, a hamburger was little known before McDonald's appeared on the scene. In Taiwan, owing much to McDonald's, young people enthusiastically took up French fries. In Japan, McDonald's diners break the taboo of *tachigui*, or "eating while standing." Through its promotions McDonald's transplanted to Asia the (Western) concept of an individual's annual birthday celebration. And McDonald's taught its Asian customers, and others around the world, about standing in line, seating yourself, and taking away your own garbage.

Yet McDonald's has not been an unmoved mover. Asian consumers subverted the concept of "fast food," in the United States understood to be food quickly delivered and speedily devoured within just eleven minutes. In dense-packed Hong Kong, McDonald's customers on average took a leisurely twenty to twenty-five minutes to eat their meals, while during off-peak times in Beijing the average mealtime stretched out to fifty-one minutes. From 3:00 to 6:00 in the afternoon, McDonald's restaurants in Hong Kong—including seven of the ten busiest McDonald's in the world—filled up with students seeking a safe place to relax with their friends, time away from teachers and parents, and a bit of welcome space. "The obvious strategy is to turn a potential liability into an asset: 'Students create a good atmosphere which is good for our business,' said one manager as he watched an army of teenagers—dressed in identical school uniforms—surge into his restaurant."[20] And in a sharp departure from the company's pervasive laborsaving practices (it's hard to believe in the days of frozen French fries, but long ago at each individual restaurant, fresh Russet potatoes were stored, washed, peeled, cut, soaked, dried, and fried), each McDonald's in Beijing hired ten extra women to help customers with seating.

McDonald's corporate strategy of localization not only accommodates local initiatives and sensibilities but also, as the company is well aware, disarms its critics. McDonald's International takes a 50 percent ownership stake in each overseas restaurant, while its local partner owns 50 percent. Accusations of economic or cultural imperialism are less compelling when a local partner pockets half the profits, has substantial operating autonomy (setting wage rates and choosing menu items), and navigates the shoals of the local business environment. A nationalistic Korean marketing manager maintained that his local McDonald's is a *Korean* business, not an American multinational, since it was conducted entirely by Koreans with the necessary savvy to deal with the country's complex banking system and government bureaucracies. Localization is a matter, then, not of opportunism or chance but of long-standing corporate policy. McDonald's strives to "become as much a part of the local culture as possible," according to the president of McDonald's International. "People call us a multinational. I like to call us *multilocal*." Much the same point was made by a troop of Japanese Boy Scouts, traveling abroad, who "were pleasantly surprised to find a McDonald's in Chicago."[21]

Not even the appearance of a standard "American" menu, down to the last perfectly placed pickle, should be automatically assumed to be an act of cultural hegemony. In Beijing the menu was self-consciously standard because the local franchisee aimed for an "authentic" American experience. (Across Asia, Watson's ethnographers report that what is achieved is Americana *as constructed by* the local culture, a far more complex and open-ended process than the one-way control claimed by many cultural-imperialism theorists.) In Japan, the legendary Den Fujita, a supremely flamboyant entrepreneur who established McDonald's Japan, positioned his new venture as a genuine Japanese company while offering a standard American menu. His tactic was to label the hamburger not as an American import, which might have irritated Japanese sensibilities, but as a "revolutionary" food. This led to some goofy publicity. "The reason Japanese people are so short and have yellow skins is because they have eaten nothing but fish and rice for two thousand years," he told reporters. "If we eat McDonald's hamburgers and potatoes for a thousand years, we will become taller, our skin will become white, and our hair blond."

In his localizing campaign, Fujita brazenly flouted McDonald's corporate formula for starting restaurants. To begin, instead of opening restaurants in McDonald's tried-and-true suburban locations, his first one was in Tokyo's hyperexpensive Ginza shopping district. His previous career in importing luxury handbags gave Fujita the contacts to secure 500 square feet of precious space from a long-established department store in a prime location. Fujita squeezed an entire restaurant into the tiny site, one-fifth the standard size, by custom designing a miniature kitchen and offering only stand-up counters. His real challenge was erecting his restaurant, a complex job typically taking three months, in the thirty-nine hours mandated by his agreement with the department store. The contract required Fujita to start and complete his restaurant's construction between the department store's closing on a Sunday evening and its reopening on Tuesday morning. Seventy workers practiced setting up the custom-made restaurant three times at an off-site warehouse before shaving the time down to an acceptable thirty-six hours. ("But where's the store?" asked a mystified McDonald's manager just two days before Fujita's grand opening on 20 July 1971.) In the succeeding eighteen months Fujita opened eighteen additional restaurants as well as the Tokyo branch of Hamburger University. In fact, the Japanese mastery of McDonald's notoriously detailed "system" actually rattled McDonald's president Fred Turner. "In Japan, you tell a grill man only once how to lay the patties, and he puts them there every time. I'd been looking for that one-hundred-percent compliance for thirty years, and now that I finally found it in Japan, it made me very nervous."[22]

Whether we are more likely, then, to get a homogenized "McWorld" or a hybrid "McCurry" depends somewhat on the trajectory of McDonald's and its resident managers and much on consumers' appropriating or even transforming its offerings.

We may well end up with a culturally bland "McWorld" if the company's localized menu and offerings are only a temporary gambit. Australia fits this pattern. When launching McDonald's in the early 1970s, the local franchisee initially tailored to Australian tastes. His menu offerings included English-style fish-and-chips, a custom fried chicken product that was unknown in the United States yet made up 30 percent of Australian sales, and hamburgers in line with Aussies' decided preference for lettuce, tomato, and mayonnaise, but no perfectly placed pickles. (When customers were sold the standard hamburger, the franchisee recalls, "we had pickles all over the store—sticking on the ceilings and walls.") Yet over time McDonald's was able to woo Aussies to its standard American menu—and these localized offerings were put aside.[23]

On the other hand, we might hope for a zesty "McCurry" world. In Japan, for instance, a locally inspired teriyaki burger became a permanent menu item. Vegetarian burgers are popular at certain upscale McDonald's in the Netherlands, while the company requires alternatives to beef hamburgers and pork breakfast sausages in countries where religious traditions forbid the eating of cows or pigs or meat altogether. In India, McDonald's served a mutton-based Maharaja Mac as well as Vegetable McNuggets, while in Malaysia and Singapore the company satisfied Muslim authorities' stiff criteria for a *halal* certificate, requiring the total absence of pork. My own fieldwork confirms that beer is commonly available in German, Danish, Swedish, and Dutch McDonald's, while the delicious icy vegetable soup called *gazpacho* appeared in Spain. In a true "McCurry" world, these multilocal innovations would circulate to the United States. American-as-apple-pie Disney, facing crippling financial losses at Disneyland–Paris, was forced to sell wine, beer, and champagne to its patrons. Walt Disney World in Florida, entirely alcohol-free for years after its opening in 1971, followed Disneyland–Paris in offering wine and beer to the public in 2012 while Disneyland in California followed suit in 2019. Is it too much to hope in the United States for a vegetarian burger, with tamari-flavored fries, and a glass of cold beer?

INTERNET CULTURE

As McDonald's found to its distress with McLibel, much of our public culture has moved onto the internet. A long list of multinational companies, including McDonald's, Nike, Shell, and Monsanto, have found that taking out expensive advertisements in newspapers or on television no longer secures the hearts and minds of the public. McSpotlight represents a wave of activist websites that challenge once-dominant commercial institutions. "The most comprehensive collection of alternative material ever assembled on a multinational corporation is gaining global publicity itself, and is totally beyond McDonald's control," noted John Vidal.[24] Unraveling the many threads of the internet and assessing its interactions with global cultures is complex; "the internet" consists of many different components—email, chat rooms, websites, intranets, e-commerce, and social media.

Understanding the influence of the internet requires us to grapple with two rival viewpoints on its origins and nature. Many popular accounts offer a "civilian origins" storyline. Katie Hafner and Matthew Lyon's pioneering *Where Wizards Stay Up Late: The Origins of the Internet* (1996) locates its roots in the civilian computer-science subculture of the 1960s and emphatically denies significant military influence. "The project had embodied the most peaceful intentions . . . [and] had nothing to do with supporting or surviving war—never did," they write.[25]

Despite Hafner and Lyon's dismissal, the internet's military origins are pronounced. Many of the important technical milestones—the RAND concept of packet switching (1964), the Navy-funded Alohanet that led to the Ethernet (1970–72), the Defense Department's ARPANET (1972), the diverse military communication needs that led to the "*inter*network" concept (1973–74), and the rapid adoption of the now-ubiquitous TCP/IP internetworking protocols (1983)—were funded or promoted or even mandated by the military services. "Let us consider the synthesis of a communications network which will allow several hundred major communications stations to talk with one another after an enemy attack," is the *first sentence* of Paul Baran's celebrated 1964 RAND report. "Every time I wrote a proposal I had to show the relevance to the military's applications . . . but it was not at all imposed on us," recalled another top researcher. Vinton Cerf, a leading computer scientist who is widely credited as "co-inventor" of the internet, reflected on the two faces of military funding. Historian Janet Abbate summarizes his view: "Although Principal Investigators at universities acted as buffers between their graduate students and the Department of Defense, thus allowing the students to focus on the research without necessarily having to confront its military implications, this only disguised and did not negate the fact that military imperatives drove the research." One senior ARPA official put it this way, "We knew exactly what we were doing. We were building a survivable command system for nuclear war."[26]

Overall, we can discern three phases in the internet story: (1) the early origins, from the 1960s to mid-1980s, when the military services were prominent; (2) a transitional decade beginning in the 1980s, when the National Science Foundation

(NSF) became the principal government agency supporting the internet; and (3) the commercialization of the internet in the 1990s, when the network itself was privatized and the World Wide Web came into being. During each of these stages, its promoters intended the internet to be a "global" technology, not merely with worldwide sweep but also connecting diverse networks and communities and thereby linking people around the world.

In the phase of military prominence, the heady aim of being global was sold to the military services as a way to build a robust, worldwide communications system. In their "civilian origins" interpretation, Hafner and Lyon skip right over a 1960 technical paper by Paul Baran, the RAND researcher who invented the concept of packet switching, who describes a "survivable" communication system: "The cloud-of-doom attitude that nuclear war spells the end of the earth is slowly lifting from the minds of the many. . . . It follows that we should . . . do all those things necessary to permit the survivors of the holocaust to shuck their ashes and reconstruct the economy swiftly."[27] (Not everyone appreciates such vintage RAND-speak.) As noted in chapter 7, the Defense Department's Advanced Research Projects Agency, or ARPA, used Baran's concepts to build a distributed network linking the fifteen ARPA-funded computer science research centers across the United States. The resulting ARPANET was the conceptual and practical prototype of the internet. While an emerging networking community saw ARPANET as an exciting experiment in computer science, ARPA itself carefully justified it to the Pentagon and Congress in economic and military terms. For example in 1969 the ARPA director told Congress that the ARPANET "could make a factor of 10 to 100 difference in effective computer capacity per dollar," while two years later the message to Congress was that the military would benefit enormously from the ability to transmit over the ARPANET "logistics data bases, force levels, and various sorts of personnel files."[28]

The groundwork for interconnecting different networks also took place under ARPA's guidance and owed much to military goals. By the mid-1970s, ARPA was running, in addition to ARPANET, two other packet-based networks: PRNET and SATNET. PRNET evolved from ARPA's project to link the University of Hawaii's seven campuses using "packet radio," the sending of discrete data packets by radio broadcast; ARPA then built a packet-radio network, PRNET, in the San Francisco region. (PRNET was the conceptual basis of the foundational Ethernet concept. Ethernets transmit packets of data that are "broadcast" between computers at first connected by physical cables and now more commonly by local wireless networks.[29]) The third significant ARPA network was SATNET, which used satellite links to rapidly transmit seismic data collected in Norway to analysis sites in Maryland and Virginia. With SATNET speed was of the essence, since the data might indicate Soviet nuclear explosions. ARPA's dilemma was that ARPANET was a point-to-point network built with near-perfect reliability and guaranteed sequencing, while PRNET was a broadcast network with lower reliability, and SATNET experienced unpredictable delays as data packets bounced between ground stations and satellites.

The *inter*networking concept responded to the military's desire to connect its technically diverse communication networks. The internet conception resulted from an intense collaboration between Vinton Cerf, a Stanford computer scientist who had helped devise the ARPANET protocols, and Robert Kahn, a program manager at ARPA. In 1973 they hit upon the key concepts—common host protocols within a network, special gateways between networks, and a common address space across the whole—and the following year published a now-classic paper, "A Protocol for Packet Network Intercommunication." Although this paper is sometimes held up as embodying a singular Edisonian "eureka moment," Cerf and Kahn worked very closely for years with an international networking group to develop and refine their ideas. In 1976 Cerf became ARPA's manager for network activities, and a year later Kahn and Cerf successfully tested an internet between PRNET in California, the Norway-London-Virginia SATNET, and the multinode ARPANET. "All of the demonstrations that we did had military counterparts," Cerf later commented. "They involved the Strategic Air Command at one point, where we put airborne packet radios in the field communicating with each other and to the ground, using the airborne systems to sew together fragments of Internet that had been segregated by a simulated nuclear attack."[30]

Not only did Kahn and Cerf advance the internet as a solution to a military problem; the military services forcefully promoted the internetworking concepts they devised for operating these networks. At the time, there were at least three rival protocols for linking different computers into a network, but Kahn and Cerf's internet protocols—the now-ubiquitous transmission control protocol (TCP) and Internet Protocol (IP), combined as TCP/IP—won out owing to the strong support it received from the Defense Department. The ARPANET, a single network, had been built around a protocol named NWP. The switchover to TCP/IP internet protocols was mandated in 1982 by the Defense Communication Agency (DCA), which had taken over operating the ARPANET and was concerned with creating an integrated military communications system. It was a classic case of a forced technology choice. As of 1 January 1983, DCA directed the network operators to reject any transmissions that did not conform to the new protocols. By 1983, IP incorporated the "datagram" concept of French researcher Louis Pouzin; datagrams were individual data packets forming an email, image, or any other information that could take several different paths from sender to recipient in a "connectionless" mode, which helped reliably link disparate individual networks into a single "global" network. ARPA also directly funded implementing the TCP/IP protocols on Unix-based computers, and it spent an additional $20 million to fund other computer manufacturers to do the same. "The US military had a lot of leverage because it could withhold funding, and such coercive tactics proved rather effective in getting TCP/IP adopted," notes historian Shane Greenstein.[31]

The direct military impetus that had built the ARPANET and fostered the internet concept did not persist long into the 1980s. ARPA for its part, having already by 1975 given up the day-to-day operation of the ARPANET to the DCA, took up the Reagan-era Strategic Defense Initiative. The military services split off their secret and sensitive activities to the newly created MILNET in 1983, which was reconnected to ARPANET, by then dominated by universities. ARPA's concern with interconnecting diverse computer networks, embodied in TCP/IP, turned out to cope brilliantly with two massive developments in computing during the 1980s. The personal-computer revolution expanded computing from the restricted province of a few thousand researchers into a mainstream activity of the millions, while the deliberate civilianization of the internet by the NSF created a mass network experience. Both of these developments powerfully shaped the culture of the internet.

While computing in the 1960s had meant room-sized "mainframe" computers, followed by refrigerator-sized "minicomputers" in the 1970s, computing by the 1980s increasingly meant "microcomputers," or personal computers, small enough and cheap enough for individuals to own. Whereas small computers in the era of Radio Shack's TRS-80s had required luck and patience, IBM's entry into personal computing in 1981, and the many clones that followed, turned computing into a mainstream activity. Many universities, government offices, and businesses small and large found that their computing needs could be met by purchasing dozens or hundreds or even thousands of personal computers—rather than renting computing power from a networked mainframe, ARPANET's original inspiration. The ready availability of local-area networks, or LANs, such as Ethernet, meant that personal computers could be locally networked and thus enabled to share software and files. Ethernets fairly exploded. By the mid-1990s there were five million Ethernet LANs, and each could be connected to the internet. The parallel expansion of proprietary networks—such as the IBM-based BITNET or the Unix-derived USENET—added even more potential internet users.

Closely related to the massive spread of computing was the invention of electronic mail, for years the predominant use of the internet. Email is an excellent example of how computer users transformed ARPANET into something that its designers had never foreseen. ARPANET, one recalls, was conceived to share data files and to efficiently use expensive computers. An early ARPA blueprint explicitly dismissed email: sending messages between users was "not an important motivation for a network of scientific computers." Yet some anomalous data from MIT suggested that ARPA's vision had missed something. The small switching computer that linked MIT to the ARPANET was always busy; yet surprisingly few files actually went out over the ARPANET. It turned out that MIT users, via the ARPANET switching computer, were sending emails to one another. Soon enough ARPA converted to email, owing to the enthusiasm of its director, Stephen Lukasik; he made email his preferred means of communicating with his aides, who quickly got the message that if they wanted to talk to the boss they needed email, too. Email was initially piggybacked onto preexisting protocols for sending data files (e.g., the file transfer protocol, or ftp). Meantime, the sprouting of special-subject email lists, a practice in which a single computer kept a file of users' addresses, created the notion of a "virtual community" of like-minded individuals interacting solely through computers. All of this unforeseen activity was quietly tolerated by ARPA, since ARPANET was in fact underused and the heavy network traffic sent out by large membership lists like "sf-lovers" provided useful experiments. Beginning in the early 1980s, the internet community specified a simple mail transfer protocol, or SMTP, widely used today.[32]

The second broad development of the 1980s was the "civilianization" of the internet. While ARPA's contractors at the leading computer research centers were the only authorized users of the ARPANET (there were around 70 nodes besides MILNET), the NSF-funded expansion of computing and computer science in the 1970s had created a large community of potential network users. In 1980 the NSF approved a $5 million proposal by computer scientists outside ARPA's orbit to create CSNET. CSNET was initially conceived with the rival X.25 protocol popular in Europe, but Vinton Cerf offered to set up connections between CSNET and ARPANET *if* CSNET would adopt TCP/IP instead. When it began in 1982, CSNET consisted of 18 full-time hosts plus 128 hosts on PhoneNet (which used telephones to make network connections as needed), all of which were linked to the ARPANET. CSNET soon also connected to newly formed computer-science networks in Japan, Korea, Australia, Germany, France, England, Finland, Sweden, and Israel. Commercial use of the network was strictly forbidden, however, which reinforced the idea that "cyberspace" was a free collection of individuals rather than an advertisement-strewn shopping mall.[33]

CSNET was just the start of the NSF-led phase of the internet. The earlier ARPA vision of networking large-scale computing facilities guided the NSF's supercomputing program of the mid-1980s. The NSF planned and built six supercomputing centers that were linked, along with nine additional regional centers, by a super-fast internet "backbone." Altogether the NSF spent $200 million creating this NSFNET, which became the hub of an "information infrastructure." Soon, NSFNET was joined to smaller regional networks. For example, forty universities in New England formed one regional consortium; universities in the southeast, northwest, Rocky Mountain states, and New York State area, among others, formed independent regional consortia, often with NSF assistance; in turn, all of them were linked to the NSFNET. NSFNET initially used a non-ARPA protocol, but it too was converted to TCP/IP during a five-year upgrade that began in 1987. As a creation of the federal government, NSFNET further reinforced the noncommercial appearance of cyberspace.

Another example of how the internet gained its "global" character is the domain-name system, or DNS. Most users experience this system, which emerged in the mid-1980s, as the two halves of email addresses ("user" and "host.domain") joined by the ubiquitous "at" sign (@). The earliest users of email remember that it often took several tries with different addresses, through different network gateways, to reach an email recipient on a network other than yours. In this era, a central "host file" was kept of *all* approved host names, correlated with their numerical addresses. As the number of hosts grew, the "host file" became unmanageably large; just updating each of the individual host computers could overwhelm the network. With the spread of the DNS, any user could be addressed with one address that worked worldwide. More important, the DNS established an address space that was massively expandable and yet could be effectively managed *without* central control.

To retire the default ".arpa" used for ARPANET, the internet community created six defined top-level domains (.edu, .gov, .mil, .net, .org, and .com). A consortium managed each domain, and it assigned names within its domain. In turn, for example, within the top-level .edu domain, each educational institution was assigned a subdomain and could assign subnames as needed. Subdomains, like suburbs, sprawled across the prairie; in 1997 the longest official host name was challenger.med.synapse. uah.ualberta.ca. With DNS there was no need for a central "host table" connecting all host names with numerical addresses; rather, each domain, and each subdomain, was responsible for keeping track of the host names within its purview. And, even though it irritated some non-US users, the DNS system effortlessly added "foreign" countries, by establishing a two-letter country domain (for instance, the .su domain was claimed in 1990 by a computer-users group in the Soviet Union while four years later the .ru domain for Russia was claimed by a network of research institutes). Absent some decentralized addressing scheme, it is difficult to see how the internet could have become "global" in scope.

Looking forward, far beyond the 1990s, the adoption of TCP/IP and DNS had two consequences that strengthened the internet's "global" scope but also entailed significant but subtle problems. No one predicted just how big the internet would be; the original addressing scheme in 1983 had room for around four billion users, an immense number at the time. (The address space in IPv4 was 32 binary "bits" in length, or 2^{32}). Since sizable address blocks were set aside for administrative uses, and since big companies secured huge blocks of addresses, whether or not they needed them, the four billion addresses went quickly; so by 1998 the successor IPv6 was first introduced and both versions of IP operate today (IPv6 has space for 2^{128} addresses). The short length of individual internet packets used with IPv4 left room for only a limited number of "root servers": the physical servers that handled all queries for any particular website's numerical address, something like a phone book for the internet where your browser looked up a website's *name*, such as www.tjmisa. com, and found its numerical *address*, such as 23.229.129.165 Ten "root servers" were in the United States with three others in Stockholm, Amsterdam, and Tokyo. For a time, this US-centric system brought an unusual degree of centralization to the global internet. Even today—after the adoption of "anycasting" effectively distributed the "root servers" so that there are more than 1,000 worldwide—attacks on the DNS system are how hackers block access to websites they wish to damage (see chapter 10).

The entirely civilian, resolutely noncommercial internet lasted five years and two months. When the twenty-year-old ARPANET was closed down in February 1990, NSFNET took over the internet as a government-owned and -operated entity. In turn, in April 1995 the NSFNET was replaced by today's for-profit "Internet Service Providers" (or ISPs). All along, NSF was interested in the internet mostly as an infrastructure for research. Enforcing the "acceptable use" policy required by the US Congress, which strictly forbad commercial activities, became ever more troublesome as more businesses and commercial ventures came online. The path toward privatization actually began in 1987, when the NSF awarded a five-year contract for upgrades and operations to MERIT, a regional consortium based in Michigan, and its private-sector partners, IBM and phone company MCI. In 1990, MERIT spun off a nonprofit corporation named Advanced Network Services, which in turn spun off a *for*-profit company that began selling commercial network services. Several other regional networks hatched similar privatization plans; three of these networks formed Commercial Internet Exchange in 1991. Finally, the large telecommunications companies, including MCI, Sprint, and AT&T, began offering commercial internet services.[34]

What evolved through 1995 were two parallel internets. One was linked by the NSFNET backbone, with no commercial activities; while the other was formed by the for-profit companies, and onto it rushed all manner of money-making (and money-losing) ventures. After 1995, a truly "internetwork" internet emerged, with private companies (ISPs) not only wiring homes and schools and businesses but also providing the high-speed backbones that carried long-distance data traffic. In turn, gateways and exchanges connected the networks of the private telecommunications companies, as well as overseas networks. This array of networks, backbones, exchanges, and gateways creates the internet writ large (see also chapter 10).

The "global" internet took off with the World Wide Web, the multimedia extravaganza that created a mass internet experience, paved the way to e-commerce, and laid the groundwork for social media. The Web is, at least conceptually, nothing more than a sophisticated way of sending and receiving data files (text, image, sound, or video). It consists of a protocol (hypertext transfer protocol, or http), allowing networked computers to send and receive these files; an address space (uniform resource locator, or URL), which permits remote computers to access all files; the language of the Web (e.g., hypertext markup language, or html), which combines text, images, formatting, and other elements onto a single page; along with a Web browser able to interpret and display the files. The Web relied on the existing infrastructure of the internet, and following its public release in 1991 became the internet's chief source of traffic, displacing telnet (remote login) in 1994 and ftp transfers in 1995 (and, by 2010, streaming video with fully half of US internet traffic eclipsed the Web, with "peer-to-peer" file sharing a close third). Its "global" character owes much to the vision and forceful guidance of its creator, Tim Berners-Lee.

Tim Berners-Lee had the good fortune to turn a brilliant idea into a full-time career without ever going through the trouble of starting a company and to hold prestigious professorships at MIT and Oxford without ever writing a PhD disserta-

FIG. 8.3. INTERNET CAFÉ

Cybercafé in Amsterdam's historic Waag building, constructed in 1488 as a city gate. It was rebuilt in 1617 as the city's customs house and for 200 years also provided rooms for the city's guilds. More recently, it has been a museum and a new-media center. It became one of Europe's leading experimental sites for wireless internet access.

Photograph courtesy of Bureau Hollandse Hoogte.

tion. (He has innumerable honorary doctorates and received the 2016 Turing Award, often called computer science's Nobel Prize, "for inventing the World Wide Web, the first web browser, and the fundamental protocols and algorithms allowing the Web to scale.") Born in 1955, Berners-Lee grew up with computers at home. His mother and father were both mathematicians who worked on programming Britain's state-of-the-art Ferranti Mark 1, one of the very earliest commercially available computers. From them, he says, he first learned about humans' abilities to create and control computers. After earning an honors degree in physics from Queen's College, Oxford, in 1976 and at the same time building his first computer, he took a series of software and consulting jobs. One of these was a six-month stay in 1980 at CERN, the European high-energy physics laboratory in Geneva, Switzerland. While there he wrote a trial program called Enquire to store information and easily make hypertext links. This was the conceptual core of the World Wide Web. He returned to CERN in 1984 and five years later formally proposed an ambitious global hypertext project that became the World Wide Web. Within a year CERN had a version of the WWW program running on its network; CERN released it to the world in 1991. Berners-Lee came to MIT in 1994 and has since then directed the World Wide Web Consortium, which sets standards for the Web's user and developer communities.

From the start, Berners-Lee built into the Web a set of global and universal values. These values were incorporated into the design at a very deep level. "The principles of universality of access irrespective of hardware or software platform, network infrastructure, language, culture, geographical location, or physical or mental impairment are core values in Web design," he wrote describing the architecture of the Web. And it is not merely that all computers should be able to access the Web. "The original driving force was collaboration at home and at work," he writes of the philosophy behind the Web. "The idea was, that by building together a hypertext Web, a group of whatever size would force itself to use a common vocabulary, to overcome its misunderstandings, and at any time to have a running model—in the Web—of their plans and reasons." While earlier hypertext systems, such as Apple's HyperCard, had allowed users to create links between image, text, and sound documents, the space was limited to documents on a local filing system. Often they used a central database to keep track of all links. The result was a reliable but *closed* system to which outsiders could not establish links. As Berners-Lee commented, "there was no way to scale up such a system to allow outsiders to easily contribute information to it. . . . I sacrificed that consistency requirement [of centrally stored and reliable links] to allow the Web to work globally."[35]

Two technical developments and one massive political subsidy made e-commerce into a worldwide phenomenon, further extending the sway of globalization. The US Internet Tax Freedom Act of 1998 supported the Clinton–Gore administration's vision of a high-tech future. For three years, initially, the act forbade states collecting sales taxes on internet sales of goods or services (sold outside the state's border) as well as any taxes whatsoever on internet access. So-called brick-and-mortar retailers were of course subject to sales taxes as before, so the act and its successors distinctly tilted the playing field toward early e-commerce vendors such as Amazon, eBay, and Priceline. Extended several times, the act was made permanent with its inclusion in the Trade Facilitation and Trade Enforcement Act of 2015.

In the early 1990s webpages were "static" with any content hand-coded into an HTML document, and so any changes also required additional rounds of hand-coding. The Perl programming language, hailed as "the indispensable duct tape, or glue, that holds the entire Web together," permitted online computers to achieve basic functionality with doing computations, printing results, automating tasks, connecting disparate programs, linking otherwise incompatible computers, and soon enough *(important:)* making database queries. Perl was used with Common Gateway Interface (CGI), which specified how web *servers* should respond to requests from web *clients* or browsers, to create short programs, or "scripts," that could be called by any web browser visiting your site. Such "cgi.bin" scripts permitted a web browser to make database lookups or queries, do calculations, reformat data, and to create "dynamic" webpages with the results. It might be real-time temperatures in San Francisco, stockmarket quotes from Frankfurt, or the inner recesses of database transactions on prices, orders, invoices, and shipping—the basic information infrastructure that made e-commerce take off. The late 1990s "dot-com" boom was one direct result. The enduring legacy was the worldwide spread of "dynamic" webpages. It's no accident that China's e-commerce titans Alibaba and JD.com were likewise founded in 1998–99.[36]

These notable examples—worldwide financial flows, fax machines, McDonald's, and the internet—taken together indicate that globalization was both a fact of contemporary life and a historical construction that emerged over time. For years, globalization was a strong force that oriented people's thinking about technology and culture. To the extent that people believed it to be real, it *was* real.

Globalization was and is materialized through technologies. A vision of a "global" world was consciously and explicitly designed into fax machines, McDonald's, and the internet. Fax technology's open standards, international standards setting, and Japanese manufacturers all created in fax machines an impressive ability to operate between different phone systems and across different cultural systems. Fax machines were not always this way: for nearly a half-century (1930s–70s) electronic facsimile was a closed, proprietary technology. Similarly, McDonald's is explicit about its corporate capitalist strategies of localization and cultural responsiveness. And finally "global values" were built into the internet from its origins in the 1960s onward through the emergence of the World Wide Web and e-commerce in the 1990s.

This situation, of globalization as both solid fact and emergent creation, presents both a paradox and a challenge. The paradox is that at the same moment when consumers' values are actively being solicited by businesses, governments, and nongovernmental organizations, and in theory the "consumer is king" everywhere, citizens in many countries feel powerless to deal with their changing world. Globalization, in the form of multinational corporations and international agencies and disruptive, impersonal flows of goods, labor, and capital often takes the rap.

Indeed, the enthusiasm during the 1990s that globalization would continue and expand, seemingly without borders, ended with the attacks on 11 September 2001 and the resulting political turmoil. With vast expansion in nation-states' effective power during the "war on terrorism," the nation-state is, contrary to the globalizers' utopian dreams, thriving as never before. McDonald's cannot deliver security in subways and airports. More to the point, just as the reduction of military spending after the Cold War in the 1990s forced technology companies to turn toward civilian markets, including global media and entertainment and manufacturing, the sharp military buildup that began in the United States during 2002 abruptly reversed this trend.[37] What an irony that the September 11th attackers—acting in the name of antimodern ideologies—because of the Western nations' national security–minded and state-centered reactions, brought an end to this phase of global modernity.

Viewed now from beyond the year 2016, the "promises of global culture" appear more distant and yet, perhaps, slightly more attractive. One success of neoliberal capitalism was significant economic gains including "since 1990 [global] trade has helped to halve the number of people living in extreme poverty." As recently as the mid-1980s China's per-capita national income was less than $1 per day, with pervasive extreme poverty; in 2020, the same figure was just over $11,000 per year with sharply reduced poverty and a burgeoning middle class among its 1.4 billion people. In India and Indonesia, too, the reduction in extreme poverty was impressive. But a certain failure of this era of "globalization" was generating unhealthy inequality in too many countries, which led to a resurgence of nationalism with pronounced authoritarianism and dangerous xenophobia. Such populism and nationalism brought authoritarian and antiglobalization leaders to power in the United States and the United Kingdom in 2016, while strengthening authoritarian and nationalist governments in China, Hungary, Russia, Turkey, and many additional countries. It's not that globalization "caused" these developments, yet a reasonable question asked in 2016 by a former trade minister of Costa Rica is: "Can the open global economy be saved from populist challengers?"[38] Assertions that globalization is the "only answer" will not go far; but it's difficult to muster enthusiasm for the present world condition with debilitating trade wars, inhumane immigration restrictions, authoritarian curtailment of dissent, and persisting disparities between global "haves" and "have-nots." Whereas technologies in the era of globalization were consciously designed to connect people across countries and cultures, too many technologies in the present are built to divide, distract, and discipline.

CHAPTER 9

2001–2010

Paths to Insecurity

Who knows when some slight shock, disturbing the delicate
balance between social order and thirsty aspiration, shall
send the skyscrapers in our cities toppling?
—Richard Wright, *Native Son* (1940)

Richard Wright was born in Mississippi, a generation removed from the slavery of his grandparents, but he came of age as a young man in Chicago during the 1920s and 1930s. There, "free of the daily pressure of the Dixie environment," he sculpted the sting of cultural oppression into his celebrated first novel *Native Son*. At the time, skyscrapers were everywhere the preeminent symbol of modern society, strong and muscular, seemingly expressing unbounded optimism and unlimited power. It was Wright's genius to discern their fragility. Skyscrapers, as Wright knew firsthand from working as a hospital research technician in Chicago, depended on a legion of ill-paid cleaning staff, ash-cart men, newspaper boys, elevator attendants, even cooks and restaurant waiters.[1]

While construction crews were typically white, skyscrapers required the labor of black hands in steel mills and excavation gangs as well. Skyscrapers in downtown Chicago and elsewhere depended also on extended technological networks. Rapid transportation systems—trolleys, commuter trains, subways, a rare automobile—brought secretaries, clerks, accountants, lawyers, bankers, and managers to work in the morning and carried these mostly white people back home in the evening. Information systems made skyscrapers into nodes for finance, wholesaling, transportation, retailing, and mass media. Energy systems delivered coal, gas, or electricity, and the buildings expelled vast amounts of smoke, ashes, and heat to the urban environment.[2]

Yet in this passage Wright does more than evoke a strained moment in urban history. With the unsettling contrast between "thirsty aspiration" and "social order," Wright suggests an allegory with a collision of the opposing forces. The novel's central character, Bigger Thomas, is a composite of aspiring young black men whom

Wright had personally known and closely observed. They were troublemakers; many met bad ends. None of them fit the subservient roles offered by white-dominated society. Cracking under the pressure of conformity, Bigger commits murder and rape and goes on the lam. For a brief moment, he experiences an unfamiliar and thrilling sense of freedom. But he is soon tracked down and trapped like an animal. The "social order" depicted in *Native Son* was one of stark racism, mob violence, and grinding injustice, with the civil rights era far distant. Lacking a means for reconciling order and aspiration, Wright adopted the skyscraper as a fragile symbol of modern society: suspending it in a "delicate balance" between the forces of modern capitalism that erected the lofty buildings and the wider society that sustained them. Bigger's lawyer points to the Loop's skyscrapers: "You know what keeps them in their place, keeps them from tumbling down? It's the belief of men. . . . Those buildings sprang up out of the hearts of men, Bigger. Men like you" (426).

The twentieth century delivered recurrent gales of economic disruption, social dislocation, and world wars, cold wars, and postcolonial wars, among other shocks. Nevertheless, as we know, skyscrapers did not topple. Originally expressing corporate pride and civic ambition, skyscrapers came to embody the global social and economic order. Today's roster of tallest office buildings spotlights the cities and countries striving for a top spot in the "global society." Of the world's twenty-five tallest skyscrapers in 2020, three are in Malaysia, twelve in China, plus one in Hong Kong and one in Taipei, and one each in Dubai, South Korea, Saudi Arabia, Russia, and Vietnam, as well as three in the former skyscraper capitals New York and Chicago. Wuhan Center, at 1,312 feet the twenty-fifth tallest, is roughly three miles from the Huanan Seafood Wholesale Market believed to be the epicenter of the 2020 global coronavirus pandemic.

As the perceived link between global trade and American empire grew stronger, as described in chapter 8, it's not terribly surprising that "thirsty aspiration" took aim at the World Trade Center towers. It's even possible the terrorists who directed attacks against them in 1993 and 2001 knew that the WTC towers were shockingly dangerous structures. The Port Authority of New York and New Jersey, the government agency that had built the tower complex in the 1960s, ignored many provisions of the New York City building code and consequently constructed two 110-story towers with inadequate stairwells, unsafe elevators, ineffective fireproofing, and a risky experimental steel structure. In effect, a century's hard-won experience in designing tall buildings to be strong and safe was thrown out. The terrorist attacks on the WTC towers in September 2001 in a way proved Wright's point.

Compounding the 2001 disaster was the Port Authority's desire for "the world's tallest building." To achieve this lofty ambition the Port Authority required the architect to add twenty additional stories above his original design. For years, entire massive floors of the Twin Towers stood empty. Of course, the race for the world's tallest building is rarely about proper economics. Securing superlative bragging rights motivated the entire twentieth-century series, beginning with the Woolworth Building (1913), the Chrysler Building (1930), the Empire State Building (1931), the World Trade Center (briefly tallest in 1972–73), the Sears Tower (1974), the twin Petronas Towers (1998), Taipei 101 (2004), and through to the Burj Dubai, the world's tallest today. When it was officially opened on 4 January 2010, the 160-story spire was renamed the Burj Khalifa to honor the monarch of oil-rich Abu Dhabi, which had extended $25 billion in emergency financing. One could hardly imagine a worse time, in the aftermath of the 2008 global recession triggered by rampant property speculation, to bring a super-luxury hotel, 900 apartments, and thirty-seven floors of pricey office space onto the real-estate market. The Burj Khalifa was in 2010 "mostly empty, and likely to stay that way for the foreseeable future." One resident wrote, a couple years later, "there's never a wait for the elevator, gyms and pools are typically empty or have only a couple people there . . . —maybe you can see me waving from the Jacuzzi on the balcony of the 76th floor. Don't worry, you can't miss me—I'll be the only one there."[3]

Might there be equivalents today of Wright's allegorical structures "delicately balanced" and at some risk of toppling? This chapter examines technologies that have generated "systemic risks" that need our active engagement. Some systemic risks such as nuclear war have largely passed into history, although nuclear proliferation remains worrisome;[4] while others such as oil spills, air pollution, and climate change are already vividly on the public agenda. My concern here is the systemic risks created and propagated by technological systems, where multiple systems or networks are interconnected and where failure in one can ripple throughout the whole, often spreading over sizable geographic areas.[5]

For example, a wide-ranging blackout in August 2003 knocked out power to 55 million people in Canada and the United States, the results of power overloads and automatic shutdowns in the Eastern Interconnection grid.[6] Since 2003 similar large-scale electricity blackouts have affected an additional 215 million people in Brazil, Paraguay, Italy, and Java, as well as 15 million households in the Europe-wide system blackout from Portugal to Croatia of 4 November 2006. In July 2012 two massive blackouts across the north of India, the product of "major technical faults in India's

FIG. 9.1. SYSTEMIC RISK IN PARIS, 1895

Granville-Paris Express at Montparnasse Station killed a woman newspaper vendor
(22 October 1895), the result of excess train speed, inadequate training, a malfunctioning
Westinghouse air brake, and crowded urban space.

grid system," knocked out power to 670 million people, fully half of the nation's population, with catastrophic results for railways, hospitals, water treatment, and even underground coal mining.[7] Amory Lovins labeled the large, tightly coupled electric power systems, aptly enough, "brittle power."

Looking closely, contemporary society very much depends on innumerable technological structures, systems, and networks. Especially fragile—in Wright's precise sense of merely appearing strong and robust—are the technical infrastructures that create and channel the global flows of energy, information, and goods. More frequently than not, these infrastructures were designed, built, implemented, and put into use without any great attention to their security in the face of slight or serious shocks. The technologies examined in this chapter—energy systems, information networks, global shipping—are not ones that consumers ever had much choice about. Largely, these technologies were selected, promoted, and shaped by mixed networks of corporate and governmental institutions. In fact, these technological systems and networks emerged and became durable structures without anyone really thinking much at all about their long-term security. Today, our only "choice" is to rely on them or opt out of mainstream society.

A new discourse about technology seems necessary to perceptively examine and constructively engage the systemic risks of concern here. Without intending it, we have constructed a global social and economic order dependent on technical systems and networks. I believe these are "fragile" in the sense that Richard Wright suggested. Devising ways to address and correct this unbidden state of affairs needs the attention of the next generation of citizens and technologists.

UNSUSTAINABLE ENERGY

Few of the 40 million happy shoppers that visit Minnesota's Mall of America each year can be aware that it is not the most important shopping mall in the state. People fly in from neighboring states to make a long weekend with its 520 retail establishments, though most locals simply leave their cars in one of the 12,550 on-site parking spaces. Shopping, entertainment, recreation, and bargains galore are readily at hand. One road map labels the Mall of America as a "major attraction," and an estimated 40 percent of visitors travel at least 150 miles, but in fact it is merely the largest shopping mall in the United States. The West Edmonton Mall in Alberta is the largest in North America. The largest in the world are in Asia—Kuala Lumpur, China, Taiwan, and four in the Philippines—and the most important shopping mall is 5 miles away, in Edina, Minnesota.[8] Built in the mid-1950s, Southdale Center is the world's prototype of the indoor enclosed shopping mall, the dominant space of mass consumption. (The Census Bureau's most-recent Statistical Abstract indicates e-commerce accounted for just 4.0 percent of total US retail sales, while the Federal Reserve Bank of St. Louis estimates a January 2020 e-commerce figure of 11.8 percent, with subsequent coronavirus restrictions likely to boost e-commerce.)[9] Southdale is moreover a perfect example of how technologies encourage and entrench certain social and cultural formations. In this instance, what emerged was an energy-intensive paradigm of low-density housing and far-flung suburban development. One could say that Southdale was an early step on a path toward unsustainable energy.

The designer of Southdale was Victor Gruen, an Austrian-born architect and planner fully under the sway of high modernism (see chapter 6).[10] At the time, architects and planners believed in "functional" urban planning that separated the characteristic urban activities, such as living, working, and shopping, while spatially connecting them with purpose-built transportation corridors. Traditional cities with their dense historical cores typically prevented full-blown modern urban planning, where a sizable swath of existing urban fabric was torn up and reengineered. Compact modern districts exist in many cities around the world today, and occasionally there is an exceptional complete modern capital city like Brasília or Chandigarh. Modernist functional planning had its heyday in the newly developing areas outside the major cities.[11] With factories around the urban periphery and commercial activities still at the urban core, postwar suburbs featured immense residential areas punctuated by defined oases of shopping, such as shopping malls. The decentralized way of life meant long distances from home to work or shopping, and the absence of viable public transit in many postwar suburban developments resulted in a near-total reliance on automobiles. If the nineteenth-century city had been the "walking city" or "streetcar suburb," then the twentieth-century city often became the "car friendly city." Many European cities imported American traffic engineering in the 1950s when they constructed ring roads and parking lots. In recent years, Shanghai and Beijing have built large-scale elevated freeways that are now jammed with cars. There is no remaining US monopoly, if there ever really was, on the automobile. In 2009, with 13.6 million in total vehicle sales, China handily surpassed the United States at 10.4 million as the world's largest automobile market.[12]

Gruen, the consummate high modernist, designed the Southdale shopping mall as a tightly controlled environment. Early photographs show crowds of all-white faces, seemingly safe from urban diversity and disorder. The formula for shopping malls pioneered by Southdale was that two or more large department stores would serve as "anchors" for the smaller retailers, who paid premium rents in exchange for proximity to the traffic-making anchors. Shopping centers before Southdale featured separate stores with separate external entrances, and so shoppers leaving one store might decide simply to return to their cars and drive off for a picnic. Gruen purposely created Southdale with an enclosed, all-indoor design. You can still enter Southdale today through its original large doorway, walk down a short nondescript entrance hall, and then gain an impressive panoramic view of the whole. Gruen intended arriving shoppers to be stunned by the view. The result, then and now, is that shoppers leaving one store immediately face a profusion of enticing choices: large department stores for serious shopping, smaller specialty shops, places to grab a bite to eat, or just an inviting stroll. Gruen's enclosed climate-controlled space gives shoppers every reason to stay at Southdale (fig. 9.2).

FIG. 9.2. INTERIOR OF SOUTHDALE MALL, 2009

Compare this recent "stunning view" with similar views at Southdale's 1956 opening.

web.archive.org/web/20101101172522/https://www.shorpy.com/node/5007. Wikimedia Commons image of Bobak Ha'Eri (11 June 2009); permission via CC BY 3.0.

It may be too pessimistic to dismiss shopping malls and the postwar suburbs that grew up around them as dreary sites of mass consumption. Historians of suburbs have uncovered a more complex picture than the one-dimensional conformity and mass culture that was targeted by social critics such as Jane Jacobs, John Keats, and Lewis Mumford. Not all suburbs were the same, and even seemingly identical houses, such as those in the prototypical Levittown development of 10,600 new homes on Long Island, New York, might be customized by their owners and appropriated to different lifestyles.[13] Also, while critics still blast Levittown's restrictive racial covenants that expressly forbade "any person other than members of the Caucasian race" (home contracts put these words in capital letters for emphasis) so that US suburbs remained 95 percent white in 1970, eventually white, black, and brown families did move to the suburbs, if not always the same ones.[14] Here, close attention to suburbs can uncover their variety and difference in experiences. Merely observing this heterogeneity, however, does not reveal systemic risk. Just as all users of internal-combustion engines depend on fossil fuels, all suburban dwellers—whether their house style is Cape Cod or Colonial, ranch or rambler—depend on far-flung road networks and affordable energy supplies.

Whatever their character, low-density suburbs relied on cheap and abundant energy. As Gruen himself noted, "the real shopping center will be the most profitable type of chain store location . . . it will include features to induce people to drive considerable distances to enjoy its advantages."[15] However families might "customize" their homes, this structural characteristic of suburban development is deeply entrenched and, needless to say, not only in the United States.[16] Individual families loved larger houses and water-hungry lawns, and the "culture of abundance" was fueled by cheap oil and cheap electricity. Until the energy crisis of the 1970s, houses large and small were kept warm in the winter months by turning up the heat and then kept cool in the summer by running fans and air conditioners. Then came the dizzying spread of car culture. As one critic saw it, "any suburban mother can state her role" quite simply: "it is to deliver children—obstetrically once and by car forever after."[17] And when Americans bought monstrous-sized cars, they needed monstrous-sized parking lots surrounding their schools, offices, factories, and shopping malls. All of this meant more driving. In *Sprawl: A Compact History*, Robert Bruegmann observes that "low-density living has undoubtedly been accompanied by more miles of driving and more energy use per capita and this in turn has led to the production of more greenhouse gases."[18]

Suburban sprawl was a complex product that resulted from national housing policies, tax laws, regulatory codes, cultural norms, and professional standards about the built infrastructure as well as individual family choices about where to live. Without exactly intending it, these choices, policies, and practices heightened the systemic risks connected to the international oil economy. In 1973 during the OPEC oil embargo, President Richard Nixon worried when around 30 percent of America's oil consumption was imported from overseas; while in 2008 Americans imported fully 62 percent of the 20 million barrels of oil they consumed each day. (In ranked order, the largest sources of America's imported crude oil in that latter year were Canada, Saudi Arabia, Mexico, Venezuela, Nigeria, and Iraq.)[19] Without much thought about yoking their destiny to regions of the world with substantial geopolitical instability, Americans continued a binge of fossil-fuel consumption. The rise of shale oil "fracking" in several US states since 2010 trimmed oil imports at the cost of disastrous environmental impacts.

Suburban lifestyles with gas-guzzling SUVs, and the other "light trucks" they are grouped with to circumvent gas-mileage standards, are only the tip of the ice-

FIG. 9.3. SUBURBAN SPRAWL OUTSIDE CALGARY, ALBERTA, 2007

New community of Arbour Lake (founded 1992), where "a freshwater lake . . . is currently under construction."

Wikimedia Commons image of Qyd (January 2007); permission via CC BY 2.5.

berg.[20] Existing industrial approaches to agriculture, too, are energy intensive and petroleum dependent. In the United States, the country's leading "crop" with greater acreage devoted to it than wheat, corn, or tobacco is lawn grass. Lawn care could be a low-tech affair of encouraging native species to grow naturally, but this environmentally friendly approach is not predominant. To sculpt their 40 million acres of lawns, Americans deploy a mechanical army of 200 million internal-combustion-engine lawnmowers as well as innumerable weed whackers, leaf blowers, and chain saws. Super high-tech golf courses stress arid climates with scarce water. Rising incomes and the desire for outdoor recreation have brought the proliferation of additional internal-combustion engines for boats, snowmobiles, all-terrain vehicles, motorcycles, and personal watercraft.[21] These consumer-oriented technologies add tailpipe emissions to global warming and increase dependence on imported oil and exposure to geopolitical instability.

Americans' thirst for low-priced petroleum derails seemingly all efforts to deal rationally with energy policy. Instead of raising taxes on fossil fuels to encourage energy efficiency and reduce dependency on imported oil—trimming systemic risk both ways—the United States instead put in place policies and subsidies that perversely encouraged *greater* fossil-fuel consumption. A patchwork of federal policies and state-by-state subsidies created an immense corn-based ethanol industry.

Together, the Energy Policy Act of 2005 and the Energy Independence and Security Act of 2007 mandated the use in the United States of 36 billion gallons per year of renewable fuels by 2022, neatly quadrupling the industry's volume. Ethanol derived from corn is the only feasible option, even though the 2007 law encouraged "advanced biofuels" such as ethanol derived from nonfood crops or biodiesel derived from oil-seed crops. Soybean-based biodiesel is roughly one-tenth the volume of corn-based ethanol, with around 12 percent of the soybean harvest used for biodiesel in 2016. Ethanol became popular in the Midwest "corn belt" states, since the huge demand for ethanol roughly doubled the price of a bushel of corn.[22]

Ethanol became popular also with automobile makers since it allowed them a cheap way to comply with federal mileage standards. Many of the largest pickup and SUV models offered a "flex-fuel" engine that could run on 85 percent ethanol as well as regular gasoline. For every flex-fuel vehicle a manufacturer sold, owing to the Alternative Motor Vehicle Fuels Act of 1988 (and subsequently extended), regardless whether it ever touched a drop of E-85 or not, "such vehicles get credit from the government for nearly double the gas mileage that they actually achieve." Here's the game. Each automobile manufacturer must meet average fleet standards of 27.5 miles per gallon (mpg) for cars (somewhat less for "light trucks"); with the Corporate Average Fuel Economy, or CAFE, created in 1975 in the wake of the OPEC oil embargo, lawmakers intended that raising fleet mileage standards would reduce oil consumption by increasing average fuel efficiency. The mileage standards were raised modestly over the years, and each low-mileage vehicle that was sold needed to be offset by a high-mileage vehicle. Over time, the average mileage of the nation's automobiles would go up and, it was hoped, the total gasoline consumed would go down.

But the flex-fuel "light trucks," which cost around $100 extra to manufacture, permit manufacturers to sell large gas-guzzlers while gaining credits that artificially inflate their fleet mileage. The low-mileage flex-fuel vehicles become, hey presto, high-mileage vehicles. The Alternative Motor Vehicle Fuels Act makes the strained assumption that all E-85 vehicles run half the time on gasoline and half the time on E-85; the E-85 mileage is "divided by 0.15," and the two figures are averaged together. A flex-fuel vehicle getting 20 mpg on gasoline and 15 mpg on ethanol is treated as though it gets 20 mpg half the time and 100 mpg the other half, for a made-up average of 60 mpg. Credits for "excess" mileage surpassing the CAFE standard in any particular year can be applied for up to three years before, or after, so manufacturers can use the ethanol credits to escape any fines for their fleet averages missing the CAFE targets. Some $750 million in fines have been paid, mostly by European makers of luxury cars such as BMW and Mercedes Benz. In recent years, domestic automobile manufacturers have not paid any such fines. "The credits are a motivating factor," noted one General Motors official.[23]

On a vacation trip in the upper Midwest in 2010, we found that while corn-based E-85 was widely available, it was no bargain. Just nine states in the "corn belt" then had most of the nation's 2,000 E-85 gas stations.[24] The cost at the pump for E-85 was lowered by generous federal blending credits, several federal tax credits, and state subsidies like Minnesota's 20-cent-per-gallon direct payment to ethanol producers. But our mileage with the rented "flex fuel" vehicle on E-85 was terrible, falling by around 25 percent over regular gasoline, in line with other flex-fuel vehicles.[25] Mile for mile, we actually paid more money at the E-85 pump. So, in a narrow economic sense, ethanol was a clear loser.

Examining the entire energy cycle makes corn-based ethanol look even worse. While it is marketed as a domestic substitute for imported petroleum, corn is hardly petroleum free. The corn-to-ethanol process requires considerable energy inputs: petroleum-based chemical fertilizer for the cornfields and fuel for plowing, seeding, harvesting, and trucking, not to mention the energy-intensive chemical processing of corn into ethanol. (Some analysts estimate that to manufacture 10 gallons of ethanol the equivalent of 7 gallons is required as an energy input for growing and processing the corn.)[26] At best, E-85 may have a tiny net positive effect on reducing imported oil. It certainly artificially boosts government-mandated fleet mileage figures, increases the number of oversize gas-guzzling vehicles permitted on the road, and encourages Americans to consume more gasoline rather than less.

Critics of US energy policies also observe that such policies unwittingly bring about a perceptible drift toward global insecurity. In sharp contrast to benefits for American corn farmers, Mexican consumers suffered a quadrupling in the price of corn flour for tortillas. An average Mexican family of four consumes 1 kilogram or 2.2 pounds of tortillas each day. As a result of the 1994 North American Free Trade Agreement, Mexico opened its borders to then-cheap US corn imports and ended reliance on its own domestic supplies. Food riots spread across Mexico in January 2007, as higher corn prices kicked in. The "huge increase in the demand for industrial corn for the production of ethanol . . . inevitably pushes up the price of food stuffs," observed one US economist. It's not "a very good idea to start using yet another vital and limited resource [corn] to wean ourselves off oil."[27]

Higher corn prices that bring prosperity to America's corn growers, then, harm Mexico's millions of corn consumers. In 2006, around 20 percent of the total US corn harvest went to ethanol production rather than food or animal feed. In 2010, after the ethanol subsidies were fully in place, 39 percent of the corn harvest went to ethanol. Two-fifths of the nation's corn seems destined for the E-85 pump, despite the Earth Policy Institute's calculation that "the amount of grain needed to fill the tank of an SUV with ethanol just once can feed one person for an entire year."[28] Continuing troubles in Mexico's domestic economy encouraged Mexicans to brave the hazards of the border crossing to the United States. For this American taxpayers paid the bills. With the multibillion-dollar Secure Border Initiative (2005–11), the United States spent an average of $7.5 million per mile in building pedestrian and vehicle fencing along 670 miles of the 1,969-mile-long US-Mexico border; after 2016 the new US presidential administration revived this wall-building effort despite more than a decade of "program challenges, management weaknesses, and cost, schedule, and performance risks."[29]

Diverting corn to ethanol brought measurably higher food prices to American consumers already by 2009, according to the Congressional Budget Office.[30] More is likely to come. To meet the federally mandated 2022 target of 36 billion gallons of ethanol, additional US farmland will need to be converted to corn for ethanol, reducing land available for food-crop production and increasing domestic food costs. There will be upward price pressures for corn chips, corn syrup, and corn-fed beef and pork. Beef and pork producers, among others, supported the 2009 Affordable Food and Fuel for America Act (H.R. 3187) in an effort to rein in the ethanol subsidies, and so lower the price of corn paid by feedlots, but the bill died in committee. "The price of grain is now tied to the price of oil," noted Lester Brown. "In this new situation, when the price of oil climbs, the world price of grain moves upward toward its oil-equivalent value. If the fuel value of grain exceeds its food value, the market will simply move the commodity into the energy economy."[31] Indeed, corn-based ethanol sadly connects energy consumption, farm subsidies, the international energy-and-food economy, and the unhappy consequence of heightened geopolitical insecurity.

INSECURE INFORMATION

In the 1970s John Draper was a man on the move. From a phone booth in the Bay Area of California, he could go nearly anywhere in the world with little more than a dime and a plastic whistle from a cereal box. He had found the key to unlocking the phone system, and willfully manipulating it was a heady trip. It was also illegal and that was why he was continually on the move. On arrival at a pay-phone booth Draper would deposit a dime and dial up one of the many toll-free numbers that companies established to provide information over the phone; when the toll-free call went through, his dime came rattling back. A quick toot on the plastic whistle electronically disconnected his pay phone from the company's toll-free answering system, and on the company's side the line went dead. But the maneuver actually gave him open access to the inside of the entire telephone-switching infrastructure. To navigate this immense electronic space, he needed more than a plastic whistle. After careful study, and timely consultations with phone hackers like himself, he designed and built "blue boxes" that emitted sets of tones that switched the phone system itself. A certain series of tones would ring up a friend miles away for a chat, another series would create an "open relay" that established a group conference call, yet others would dial a selected phone booth in New York or London or Paris. Nicknamed "Captain Crunch" for the breakfast cereal where you found the plastic whistles, Draper could even route a phone call clear around the world, and back to himself. Eventually, when such "phone phreaking" became widespread, AT&T ordered the retrieval and destruction of all known copies of the November 1960 issue of the *Bell System Technical Journal*. An article there, "Signaling Systems for Control of Telephone Systems," had provided all the needed details. It helpfully advised: "Since these [switching] frequencies are in the voice band they can be passed over the line facility with the same ease as voice currents." Two California teenagers eager for this information found it at a computer science library at Stanford University, and they too made and sold "blue boxes" for their friends. Of the duo Steve Wozniak was the technical brains, while Steve Jobs was the promotional genius: Apple Computer was the joint result.[32]

Draper's exploits were possible owing to the monolithic internal structure of the phone system. For decades AT&T had a government-sanctioned monopoly for long-distance phone calls in the United States. At the local level, phone systems remained wildly diverse. A non-corporate cooperative venture was established in Knappa, Oregon, for that rural community's residents to phone one another, but anyone calling to San Francisco or Chicago needed to go through the long-distance Bell System. The world's phone system was composed of numerous local phone companies, wired together into regional or national systems, and then connected through international switching stations between countries. Yet as Draper showed, the phone system could be manipulated as though it were a single system. The series of tones that worked in California to redirect local calls also worked across the world, or at least anywhere with automatic call switching. Phone phreakers delighted in trying to call the pope in Vatican City since it was part of this integrated worldwide system, but the whistle scheme could not connect them to residents of Burma or Botswana, since one still needed a human operator for those calls. Today, by the way, Draper's whistles cannot work their magic because the telephone system uses completely different channels and signals for switching than for carrying phone calls.

A version of Draper's phone hacking strangely enough lives on in today's internet, which is also to a striking extent an integrated and monolithic system: the protocols that direct computer traffic for emails or social media or YouTube videos are astonishingly uniform around the world. The ARPANET, the direct predecessor of the internet, originally had a single protocol for handling all traffic. The growing size of the early network was not really the trigger that prompted the creation of the internet. Instead, as chapter 8 discussed, the trigger was the desire to connect diverse networks with distinct operating characteristics. ARPANET was a fast and highly reliable network. Other networks in Hawaii and elsewhere used noisy radio signals that were slower and less consistent, and a third network using satellites to relay militarily sensitive information across the North Atlantic had fantastic speed but also unpredictable delays as signals bounced from ground to satellite and back. Yet other computer networks used various proprietary standards or the X.25 protocol, which created virtual circuits much like telephone-line connections.

Robert Kahn and Vinton Cerf conceived the internet to literally interconnect these dissimilar networks, allowing each to retain its own message-routing protocol. They reasoned correctly that an internetwork that could successfully handle this level of diversity could handle nearly any kind of network traffic. The substantial paradox is that, while the specific protocols invented by Kahn and Cerf—Transmission Control Protocol and (especially) Internet Protocol, respectively, TCP and IP—were designed to accommodate diversity, what eventually emerged was to a surprising degree a monolithic network running a single protocol, the now-ubiquitous TCP/IP. The spread of TCP/IP in the 1980s and the elaboration of electronic commerce in the 1990s were additional steps, hardly recognized at the time, toward greater exposure to systemic risk. The underlying problem is that the internet was not designed, conceived, or built as a secure network.

Effective security in any integrated and monolithic system is a difficult task. Once you have figured out how to get onto the information "highway," as Draper did, you can go a very long distance. The universality of TCP/IP in effect creates one massive and uniform system, a single information highway, if you will, with well-posted entrances and exits. It's the perfect place for a thief: no borders, no checkpoints, no passports, and weak or nonexistent identification. A couple years ago I set up my Brother laser printer at home, somewhat to my surprise, by sending commands through TCP. With the right codes, I could also set up my home printer from my work computer or from any other computer with access to the internet. Any enterprising hacker could do the same—or worse—with nothing standing in the way except my home network's low-security "firewall."

Controlling someone's laser printer is one thing, but the real prize is controlling entire computers. The simplest scheme is gaining control of a clandestine network of computers, through a self-spreading virus or worm, and using them to send out unsolicited spam messages to harvest credit-card numbers or perpetrate other mischief. Some computer operating systems, especially those with built-in remote access for troubleshooting, are simply sitting targets. Widely reported viruses, worms, and other malware include the SQL Slammer worm, which exploits a buffer-overflow bug in Microsoft server software (and which knocked out a nuclear plant for five hours in 2003), while the pernicious Conficker worm infected perhaps 10 million Windows computers in 190 countries. The state of Washington expanded unemployment insurance payments during the 2020 coronavirus pandemic, but unfortunately lost "hundreds of millions of dollars" to an international network of online fraudsters operating from Nigeria.

The more serious problem is that we have connected a range of complex technical systems that require maximum security to the patently insecure internet. Incredible as it may seem, "power grids, water treatment and distribution systems, major dams, and oil and chemical refineries are all controlled today by networked computers."

Since networked computers are vulnerable computers, it follows that all technical systems controlled by networked computers are themselves vulnerable. The next-generation Internet Protocol, IPv6, is unlikely to be a security fix. Taking form in the 1990s, IPv6 was to some extent shaped by privacy concerns but, remarkably, "the protocol does not intrinsically address security issues." The new Internet Protocol, just like the older IPv4, is vulnerable to many forms of internet mischief.[34] The examples of pervasive systemic risk that round out this section might seem the product of some fevered paranoid imagination. On the contrary, these examples of vulnerabilities in air-traffic control (ATC), the electricity-supply grid, and cyberwarfare come from official government reports or well-placed insiders (fig. 9.4).

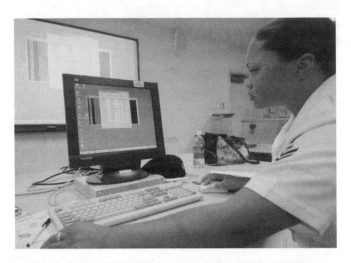

FIG. 9.4. US MILITARY MONITORS NETWORKS FOR CYBER ATTACKS, 2009

US Navy technician scans for intrusions during cyberwar training at the Space and Naval Warfare Systems Center in San Diego.

U.S. Navy photo 070712-N-9758L-048.

The air-traffic control system in North America performed flawlessly on 11 September 2001, quickly and effectively grounding some 4,500 planes, but the complex system of city-based control towers, in-route flight control centers, and local airport traffic direction is simply riddled with systemic risk. In September 2004 a "design anomaly" in a Microsoft server shut down the ATC radio system in Southern California and left 800 airplanes out of contact with air-traffic control for three hours. The Microsoft Windows Advanced 2000 Server, running on Dell hardware, was programmed to automatically shut down each 49.7 days, to prevent data overloads and fatal errors; and so every 30 days technicians needed to manually restart the computers. Something went wrong. Technicians somehow failed to properly reset the server, which crashed without warning, and the backup system crashed, too. The result was that frantic air-traffic controllers—unable to contact pilots directly—watched at least five near-misses and used their personal cell phones to relay warnings to other ATC facilities.[35] The regulating Federal Aviation Administration, or FAA, turned to commercial suppliers like Microsoft and Dell only after experiencing decades of difficulties in modernizing the computer systems at the heart of air-traffic control.

Nationwide air-traffic control was a product of the 1960s. Sperry-Rand Univac won a pioneering contract and adapted one of its military computers for the FAA's Automated Radar Tracking System (ARTS-1), first tested in Atlanta in 1964. Five years later, the company won the contract for the successor ARTS-3 and over the next years installed systems in sixty-three cities, culminating with Dallas in 1975. For ARTS-3, Sperry-Rand built the UNIVAC 8303 computer to replace its original military-derived computer, while retaining its own proprietary assembly language to run the complex software. Consequently, as late as 1997, the FAA was still purchasing UNIVAC 8303s. Two years later the General Accounting Office told members of Congress:

> The Automated Radar Terminal System (ARTS)-IIIA is the critical data processing system used in about 55 terminal radar approach control facilities. These systems provide essential aircraft position and flight plan information to controllers. The ARTS-IIIA system continues to rely on a 1960s-vintage computer. . . . Home computers available today have 250 times the memory of this archaic processor.

As *Scientific American* put it, if the UNIVAC 8303 were a personal computer, its "paltry 256 kilobytes of total memory . . . could not run Flight Simulator," the popular Microsoft game.[36]

The parallel ATC system for in-route traffic was also an antiquated patchwork of hardware and software. The FAA's hoped-for Advanced Automation System (1981–94) aimed at entirely revamping in-route computing, but FAA eventually canceled its contract with IBM when results were disappointing and costs, with 1,000 or more IBM software engineers at work, ballooned to $7.3 billion. One software manager called the ill-fated project "the greatest debacle in the history of organized work." Planes were still flying of course. In 1997 the IBM-designed legacy Host system contained a half-million lines of code written in Jovial and assembly language, dating back nearly thirty years; the software had been ported from the original late-1960s-vintage IBM 9020 computer to run on the newer IBM 3080. "But Host has at most only 16 MB of RAM, a serious limitation," noted an author in *IEEE Spectrum*. "And it badly needs replacing." "There are very few programmers still competent in Jovial," noted one NASA researcher referring to the "virtually extinct programming language."[37] In both cases, software written in the 1960s, for the first generation of air-traffic control, long outlived the computers on which it was designed to run as well as the specialized programmers needed to maintain them.

In the wake of these frustrating experiences with proprietary computer systems, the FAA shifted to commercial software and off-the-shelf components that could be customized. As well, the FAA embraced networked computing, but a 2009 investigation by the Transportation Department's Office of Inspector General made clear the troublesome security shortfalls. "Web applications used in supporting ATC systems operations are not properly secured to prevent attacks or unauthorized access."[38] The Inspector General's report examined seventy web applications used to disseminate information to the public as well as those used internally to support FAA's eight operating systems, including in-route tracking, weather, inventory, local radar, and overall traffic management. Most alarming were 763 "high risk" vulnerabilities, including the possibility of attackers gaining "immediate access into a computer system, such as allowing execution of remote commands," along with numerous lesser problems. The simple fact is that these web-based applications "often act as front-end interfaces (providing front-door access) to ATC systems."

The study's investigators succeeded in breaking into the Juneau weather system, the Albuquerque control tower, and six in-route control centers (Anchorage, Boston, Denver, Oakland, Salt Lake City, and Seattle), as well as the overall load management system. A malicious hacker gaining access to the in-route control centers could shut down electric power to the entire system, while a hacker entering the load management system could let loose computer viruses or worms. In fact, there have been several actual attacks. In 2006, a virus spreading on the public internet infected the FAA's ATC computers and forced the agency to "shut down a portion of its ATC systems in Alaska." In 2008, although not widely reported at the time, "hackers took control of FAA's critical network servers (domain controllers) and gained the power to shut down the servers, which could cause serious disruption to FAA's mission-support network." In February 2009, hackers used a public FAA computer to gain unauthorized access to personal information on 48,000 current and former FAA employees. Indeed, although it's difficult to believe, the ATC system for years was vulnerable to the entire range of internet-borne mischief. It still may be. The FAA's present-day $20 billion "next generation" air-traffic control system (2007–25) replaces radar with GPS for in-route tracking and promises to retire the 1960s-vintage computer systems, but security vulnerabilities may remain in data communication and surveillance systems.

Vulnerabilities persist in the electricity grid that connects schools, offices, and households with the nuclear power plants, thermal plants, or wind turbines that generate electricity. The North American grid is composed of numerous regional electricity grids, formed into two overarching transnational grids spanning the US–Canadian border (Eastern Interconnection and Western Interconnection), as well as smaller separate grids in the states of Texas and Alaska and the province of Quebec. The Eastern grid's spanning the border explains why the northeast's 1965 and 2003 blackouts also spanned the border. Electricity grids are massively complex structures. Originally, as chapter 5 described, electricity was sent over long-distance lines using alternating current (AC). But in recent decades, high-voltage direct current (HVDC) lines have been built to transmit electricity over great distances, such as the Pacific DC Intertie (opened 1970) that sends spare hydroelectric power from near The Dalles, Oregon, the distance of 846 miles to the Los Angeles area. The Intermountain HVDC line (opened in 1986) sends electricity from a coal-burning plant in Utah across 488 miles also to the Los Angeles region. In western Europe thirty or more HVDC lines connect stations and consumers crossing the Baltic Sea as well as islands in the Mediterranean. In 2019 China opened the 2,000-mile Changji–Guquan line operating at just over 1 million volts that connects hydropower in the far northwest Xinjiang Uyghur Autonomous Region to the country's densely settled eastern coast.

HVDC can efficiently send large amounts of power between regional grids (six DC lines connect the Eastern and Western grids in the United States), or between two AC grids that are not themselves "in phase," but in general the managers of electricity grids are continually synchronizing their systems' voltages and phases. It is a massive information problem, conducted in real time. Again, electric grid operators enthusiastically adopted web-based networked computing. And while the specifics are not the same as ATC, the same underlying problem occurred: complex technical systems connected to networked computers are themselves vulnerable.

In 2008, the Idaho National Laboratory analyzed a set of sixteen investigations for the Department of Energy, seeking to identify common vulnerabilities in the "critical energy infrastructures," that is, the electricity grid.[39] The report identified twenty-eight different vulnerabilities in eleven distinct categories. Some vulnerabilities stem from routine security shortcomings, such as weak or nonexistent passwords, failure to remove authorized users when their jobs change, and inadequate firewall protection. Computing causes other problems. Common computer programming errors, such as buffer overflows and integer overflows in grid control systems, can date back to the 1960s. You can create an overflow error in your own calculator by multiplying two large numbers together and then repeatedly multiplying the results until the calculator cannot internally represent the proper result. It's a bit arcane, but malicious hackers gaining partial access to a computer system can tempt it with varied data inputs until an overflow error causes it to crash, when it sometimes reveals ways of gaining more complete access. (Think of randomly throwing rocks at the front of a house, entirely unaware of the location of any window—until one rock breaks through and you know where the window is.) The electricity grid is also vulnerable to viruses or worms, such as the SQL Slammer that injects malicious code into commonly used SQL database servers.

Several vulnerabilities are distinctive to the management of electricity grids, where grid managers cooperate minute-by-minute to transfer large blocks of electric power as needed. The grid managers frequently use proprietary or nonstandard communication systems. As a result, sadly, the readily available means to secure and encrypt messages (including log-ins and passwords) are not commonly used. Incredibly enough, grid managers send "clear text" messages, which can easily be intercepted, over insecure protocols such as ftp and telnet. Worse, some grid operators are unaware they are on the internet. One expert on electric grid security provided the following "sanitized" conversation with an operator of a large gas-fired turbine generator: *Do you worry about cyber threats?* "No, we are completely disconnected from the net." *That's great! This is a peaking unit, how do you know how much power to make?* "The office receives an order from the ISO [Independent System Operator], then sends it over to us. We get the message here on this screen." *Is that message coming in over the internet?* "Yes, we can see all the ISO to company traffic. Oh, that's not good, is it?"[40]

At the core of the security problem is a common (but non-secure) communications protocol known as ICCP that is widely used throughout the electric utility industry. In the words of the Idaho National Laboratory:

> The Inter-Control Center Communications Protocol (ICCP) is an open protocol used internationally in the electric power industry to exchange data among utilities, regional transmission operators, independent power producers, and others. The clear text data exchange may provide many opportunities for an attacker to intercept, understand, and alter the data being exchanged.

Dangerous scenarios are easy to imagine, such as a malicious attacker simply repeating a real, bona fide order—sent in unencrypted clear text—to transfer a sizable amount of power between two regional grids. Such an overload of power could easily crash the receiving grid, as happened for different reasons with the eastern US blackout of 2003.

Some scary scenarios far beyond "clear text" tampering have already occurred.

Since at least 2009, Russian hackers have tested, probed, and infiltrated US energy systems, including electric power, according to the *Wall Street Journal*. The *New York Times* reported in 2018 that Russians had gained "access to critical control systems at power plants" prompting the US Department of Homeland Security and Federal Bureau of Investigation to warn grid operators about the imminent threat. One Russian group known as Energetic Bear, an arm of the post-Soviet secret police, successfully "breached the power grid, water treatment facilities and even nuclear power plants, including one in Kansas." "We now have evidence they're sitting on the machines, connected to industrial control infrastructure, that allow them to effectively turn the power off or effect sabotage," said one security expert: "They have the ability to shut the power off. All that's missing is some political motivation."[41]

Finally, cyberwar is now understood as the "fifth" domain of warfare in addition to land, sea, air, and outer space. The US military famously embraced a "net centric strategy" in the mid-1990s, but it has experienced numerous weaknesses in doing so. As one analyst observed, "American [military] forces are highly information-dependent and rely heavily on precisely coordinated logistics networks." Such dependence creates vulnerabilities. "From at least 2003 to 2005, a series of coordinated cyber attacks [from China] hit U.S. military, government, and contractor Web sites with abandon," reported *Government Computing News*. "The systematic incursions, collectively dubbed Titan Rain, attacked hundreds of government computers." Then in 2006, the National Nuclear Security Administration discovered that a year-old network attack had stolen the names of and personal data on 1,500 of its employees, many with highly classified careers.[42]

In April 2007, the small Baltic country of Estonia was the target of a massive denial-of-service attack that knocked out websites of its government, banks, mass media, and political parties, disabling the country's telephone exchange and crippling its credit-card network. Estonian officials blamed the attacks on the Russian government (a Soviet-era statue was in dispute) but subsequent investigations found only a loose network of Russian "hacktivists" and legions of botnets. Similar denial-of-service attacks on the White House and the Pentagon as well as on several South Korean government ministries occurred over several days in July 2009. And on 12 January 2010, citing "a highly sophisticated and targeted attack on our corporate infrastructure originating from China that resulted in the theft of intellectual property from Google," the search giant announced "a new approach to China." This was big news: Google subsequently moved its servers to Hong Kong, at the time reasonably independent from the mainland. Google's chief legal officer David

Drummond noted that the attacks were aimed particularly at the Gmail accounts of human-rights advocates in China. He further stated, "We are no longer willing to continue censoring our results on Google.cn," which had infamously suppressed internet content deemed off limits by the Chinese government (see chapter 10).[43]

Increasing geopolitical uncertainty and exposure to systemic risk are likely consequences of cyberwarfare. Unlike aircraft carriers or battlefield tanks, cyberwarfare remains largely invisible to the outside world, and there are precious few multilateral treaties or binding codes of conduct, such as the Geneva Convention that prohibited chemical warfare.[44] Secrecy reigns. For instance, no one talked openly about an Israeli Air Force raid in September 2007 that destroyed a suspected nuclear site in eastern Syria. Israeli intelligence had spotted suspicious construction activities and the presence of numerous North Korean workers. CIA photographs and an investigation by the UN International Atomic Energy Agency later confirmed that this was a nuclear processing facility, built in clear violation of the nonproliferation accords in force. The nighttime air attack succeeded despite the Syrian radar operators dutifully watching their screens though the entire hour and never once seeing the Israeli jets. The older Israeli jets ought to have been plainly visible on the Syrian radar screens. Somehow, the Israelis gained control over Syria's expensive Russian-built air-defense system. "What appeared on the [Syrian] radar screens was what the Israeli Air Force had put there, an image of nothing." And even if the Syrians had scrambled their own military planes, there would have been nothing for them to aim at. It is still not publicly known just how the Israelis accomplished this feat.

With a great degree of stealth, the Israeli government waited until 2018 to publicly acknowledge the 2007 attack. At the time, there was speculation by US military officials that the Israelis had used something similar to a US attack system that permitted "users to invade communications networks, see what enemy sensors see and even take over as systems administrator . . . so that approaching aircraft can't be seen." The US system worked by feeding false targets, misleading control signals, and other electronic mischief into the enemy air-defense system. In thwarting Syria's Russian-built air-defense system, Israel might well have been targeting Iran as well, since it too bought the same Russian system. Russian computing, like its Soviet predecessor, left a lot to be desired. In June 1982, at the height of the Cold War under President Ronald Reagan, a Soviet computer system catastrophically malfunctioned. American intelligence agents had placed malware on a Canadian computer that was subsequently stolen and installed as the control system for a Soviet gas pipeline in Siberia. "The pipeline software . . . was programmed to go haywire, after a decent interval, to reset pump speeds and valve settings to produce pressures far beyond those acceptable to pipeline joints and welds," according to a former Air Force secretary. As planned, the control computer triggered mayhem in the gas pipeline, resulting in "the most monumental non-nuclear explosion and fire ever seen from [outer] space."[45]

As with most spy-versus-spy episodes, the facts about the Siberian gas pipeline explosion cannot be entirely verified. Predictably, Soviet sources denied that the gas pipeline explosion was caused by defective American software from a stolen Canadian computer. A few Western analysts doubt the story as well. Yet an unclassified 1996 CIA review indicates that the United States was presenting deliberately defective technology to the Soviets, who obligingly stole the stuff. The Farewell Dossier, obtained via the French from a Soviet double agent, created a handy shopping list of the advanced technologies the Soviets desired. According to one intelligence official, "Contrived computer chips found their way into Soviet military equipment, flawed turbines were installed on a gas pipeline, and defective plans disrupted the output of chemical plants and a tractor factory."[46]

For a final instance of surprising vulnerability in an expensive weapon system, consider the US military's use in Iraq, Afghanistan, and Pakistan of the unmanned Predator drone aircraft. While some Predators actually launched missiles, they were mostly used for forward surveillance, sending a stream of live video images up to an orbiting satellite and then down to a relay station for intelligence screening and analysis. It turns out that the drone's "unencrypted down-link" was easily read by a $26 off-the-shelf software tool that was designed by a Russian company to snare legal non-military internet content such as soccer games. Significant video content ("days and days and hours and hours of proof") was found on the laptops of captured Shiite fighters in Iraq. Even though this video vulnerability was recognized as early as the 1990s, and one might think that encrypting the video link would be a simple matter, it was the perennial problem of insecure internet traffic. General Atomics Aeronautical Systems manufactured the Predator drones, and "some of its communications technology is proprietary, so widely used encryption systems aren't readily compatible." The US Air Force in time replaced the Predator drones with $10 million Reaper drones, also built by General Atomics, but "the newer drones' video feeds can still be hacked much like the original ones."[47]

The unmanned drones are part of a regional arms race that inflames geopolitical instability. In 2017 and 2020 the US government sold more than fifty Predators and Reapers, armed with missiles, to the Indian government for a total of $5 billion. The latter deal was a response to China's selling to Pakistan, India's bitter rival, its own attack drones and with plans to jointly produce forty-eight more for Pakistan's air force. For its part, China since 2013 has embarked on an immense regional development thrust known as the Belt and Road Initiative. Pakistan alone is slated for $60 billion. In Pakistan the Chinese drones were to defend the "crown jewel" of the initiative, a massive new military base and commercial port at Gwadar "in the highly restive southwestern province of Baluchistan." The troubled region's prospects seem dire.[48]

Any comforting line between offensive and defensive cyberwarfare has recently blurred beyond repair. The cyber arms race escalated when US military planning "chose to militarize cyber issues by designating cyberspace as a military domain," also establishing the US Cyber Command at the headquarters of the National Security Agency in 2009. As noted, Russian cyber attacks penetrated the Pentagon's information networks in 2008 as well as US energy infrastructure (electricity grid and oil pipelines) since around 2012. Russia also destabilized Estonia and invaded Ukraine using paramilitary forces, cyberwarfare, and disinformation, annexing Crimea in early 2014. "In December 2015, a Russian intelligence unit shut off power to hundreds of thousands of people in [non-occupied] western Ukraine." Russian efforts to influence the 2016 US presidential election through hacking into the computers of the Democratic National Committee, fabricated accounts on social media, and flamboyant disinformation are well known and well documented. "There shouldn't be any doubt in anybody's mind," stated Michael S. Rogers, director of the National Security Agency, in describing the Russian attacks. "This was a conscious effort by a nation-state to attempt to achieve a specific effect."[49]

The US intelligence agencies repeatedly issued warnings that "Russia has inserted malware that could sabotage American power plants, oil and gas pipelines, or water supplies in any future conflict with the United States." Likely in response to continued Russian efforts to tamper with the 2018 midterm elections, the United States launched a set of clearly offensive attacks on Russia's electric power grid—and, remarkably enough, permitted and even encouraged well-placed reporters to publicize the story in 2019. Among the steps were installing "reconnaissance probes into the control systems of the Russian electric grid," starting as early as 2012, and more recently "placement of potentially crippling malware inside the Russian system" as well as other measures not yet publicly disclosed. The National Defense Authorization Act of 2018 permits the US Cyber Command to carry out "clandestine military activity" on foreign computer networks. We may already be at war; it is difficult to tell.[50]

VULNERABLE SHIPPING

Shipping ports are extensively computerized, but that is not why they remain the "soft underbelly" of domestic security. Each year, around 11 million shipping containers arrive at US ports from around the world, with computers tracking their every movement. Approximately 35 million containers circulate in the world, each a rectangular steel box that is most often a standard 8 feet tall and wide and 20 feet or 40 feet in length. In general container cargo shipping is highly standardized so that the maximum number of containers can be loaded into the minimum space. Containers filled with perishable goods like bananas or high-value items like computers are aggregated at container ports around the world and then loaded by the thousands onto special-built container ships. As of this writing in 2020, the world's seventy largest container ships are all within a few inches of 400 meters, or 1,312 feet, in length, and operated by shipping companies based in China, Denmark, France, Hong Kong, Japan, South Korea, Switzerland, and Taiwan. They are roughly four professional soccer fields in length (1,300 feet) and just over a full soccer field in width (183 feet). The giant vessels carry up to 24,000 TEU (1 TEU, or 20-equivalent-unit, equals a standard container 20-feet long) or enough space for perhaps 3 million 29-inch televisions. Container ships are frequently described as being "as large as aircraft carriers" and in fact the largest are more than 200 feet longer than the top-of-the-line *Nimitz*-class aircraft carriers. Aircraft carriers have a crew of 3,000 or more while even the largest container ship might have fewer than twenty.

The entire container-port-and-ship system has been exquisitely engineered for efficiency, so that the massive ships spend as little time as possible in port and as much time as possible productively hauling containers between ports; needless to say, with economic efficiency as the overriding concern, security traditionally received scant emphasis. Compounding all security efforts is the large size of many ports, their historic location in the least secure districts, and not least their daunting variety. In the United States alone there are 260 inland and coastal ports, ranging from the three largest container ports at Los Angeles, Long Beach, and New York–Newark, each handing between 4 and 5 (TEU) million containers annually with a mix of domestic and international shipments and transshipments from one vessel to another. Petroleum products are handled in Houston, in addition to 1.8 million containers, and the nation's grain is exported though the South Louisiana ports downstream from New Orleans, including several hundred thousand containers, so sometimes these ports also claim to be the country's largest. Alaska, in addition to Anchorage (mostly oil and bulk cargo with some containers), has twenty smaller ports scattered along its 6,000 or so miles of coastline.[51]

FIG. 9.5. *GJERTRUD MAERSK*

The ship (just over 1,200 feet long), built in 2005, is anchored here at Felixstowe, the United Kingdom's largest container port (18 March 2007). In March 2020, it was first container ship in the world infected by coronavirus; seven crew members were evacuated in Ningbo, China.
Photo courtesy of Ian G. Hardie.

The system of container shipping emerged in the postwar decades. Shippers had long struggled with the power of the longshoremen's unions over ports, citing waste and inefficiencies as well as casual pilfering entailed in the bulk or open system of loading and unloading. Recall that in early industrial London "only a sixth of the timber entering [a shipyard] left it afloat." Container shipping cut the needed number of longshoremen, since the containers were locked and sealed and the heavy lifting no longer done on longshoremen's backs but rather by mechanical cranes. Containers also promised "multimodal shipping," in which containers can be brought by truck or train to a shipping port, carried across an ocean or two, and then carried farther again on truck or train to the final destination. Container shipping holds special attraction for shippers of high-value goods; once a container is "sealed" at the originating factory in China or Singapore it remains closed until the designated recipient opens it. In 1997, sixty consumer electronics and computer manufacturers and retailers created the Technology Asset Protection Association that institutionalized these goals. To reduce in-transit theft, the association simply mandated that shippers adhere to a set of stringent policies, enact strong security requirements, and conduct periodic self-evaluations.[52]

The terrorist attacks of 11 September 2001 dramatically changed the orientation of security for the global shipping industry. Instead of concerns about theft of cargo or misdirection of containers, port authorities worried that terrorists might use them either as a "target" for attack or as a "conduit" into a target country. The anthrax scare in the months following September 11 further inflamed these concerns, and alarming think-tank reports warned of the possible use of containers to deliver biohazards, lethal chemicals, or nuclear materials. "We have to view every container as a Trojan horse," warned Stephen E. Flynn, a former Coast Guard commander and security expert employed by Science Applications International Corporation, or SAIC, a huge government security contractor.[53] Several reports remembered the Hiroshima atomic bomb with its simple gun-type design (see chapter 5). Plutonium is unstable and dangerous to handle, but if a terrorist group could procure enough enriched uranium it might build a simple bomb. Some floated the possibility of lodging such a bomb in a tanker ship or other vessel, but that seems unlikely when simply sending a bomb to the intended target in a standard container might complete the task. The vulnerability of the port system was made evident in 2002 and 2003 when *ABC News* shipped several 15-pound cylinders of depleted uranium (natural uranium with low levels of the fissile isotope uranium-235) into US ports without detection. Even though depleted uranium could not possibly be made into a bomb, highly enriched weapons-grade uranium cannot be easily detected either. Indeed, the recent use of x-ray and gamma-ray scanners is to detect the heavy shielding that might disguise the uranium. In prior years, discovery of containers with human stowaways "was practically a weekly event" at the Los Angeles port but the x-ray and gamma-ray screening made this a deadly gambit. The Department of Homeland Security was forced to conclude that "improvements are needed in the inspection process to ensure that weapons of mass destruction . . . do not gain access to the U.S. through oceangoing cargo containers."[54]

Large population centers in close proximity to the world's busiest ports—including Shanghai, Singapore, Hong Kong, Rotterdam, New York, and Los Angeles—make dangerous shipments even more worrisome. Adequate port security, before and after the September 2001 attacks, has been difficult to impose since any comprehensive inspection system would undo the speed and efficiency of container shipping. Estimates vary concerning the percentage of containers that are ever inspected, but all such figures are pitifully low, around 5 percent. A similar problem with airfreight shipping was made evident in 2010: while passenger suitcases checked on aircraft undergo extensive bomb-sniffing and x-ray examination, there was little scrutiny of the cargoes routinely loaded onto the same airplanes.

In the spring of 2006, many Americans were shocked to confront a foreign takeover of the country's shipping infrastructure. Actually, six major US ports were owned already by the British-based shipping giant, Peninsular and Oriental

FIG. 9.6. PANORAMA OF THE PORT OF SINGAPORE

World's "busiest port" locates insecure shipping activities next to a city of 5 million.

Wikimedia Commons image of Kroisenbrunner (10 July 2007); permission via CC BY-SA 3.0.

Holding (see chapter 4). The real problem was that Dubai's DP World had bought the venerable P&O for $6.8 billion. In consequence, ownership of a substantial part of the Port of New York and New Jersey, a passenger harbor in New York, the port of Philadelphia, as well as fifteen other cargo facilities along the East Coast and the Gulf of Mexico seemed about to pass from one foreign company to another. But of course not all foreign companies, or countries, are equal. "Our public is very concerned about a foreign country, in this case specifically a foreign country from the Middle East, having a major role in our ports," stated one member of Congress.[55]

Months of pitched political wrangling ensued. Both Democratic and Republican politicians, keen to appear tough on security issues, openly defied President George W. Bush's support for the port deal. Bush argued that the Dubai port deal was essential since "in order to win the war on terror we have got to strengthen our friendships and relationships with moderate Arab countries in the Middle East."[56] Bush noted that Dubai was among the very most cooperative countries in implementing American demands for beefing up port security through data sharing and on-site inspections. Eventually, a palatable face-saving solution emerged. Senator Charles Schumer of New York called the sale of Dubai World's controversial port assets to the American International Group "an appropriate final chapter." While admitting that the huge financial services company had little or no experience running ports, Schumer affirmed that AIG "has good experience running companies." Two years later AIG went bankrupt and required a $85 billion government bailout.[57]

One long-lasting consequence of the Dubai port scare was the SAFE Port Act of 2006.[58] The act—formally the Security and Accountability For Every Port Act—extended the previous guiding legislation, the Maritime Transportation Security Act of 2002 and gave legislative bite to several agency practices. The SAFE Port Act created the legal environment for port security with specific implications for Customs and Border Protection, Coast Guard, and Transportation Security Administration, all overseen by the Department of Homeland Security. The act was approved by near-unanimous votes by the House and Senate and signed by President Bush on 13 October.

The SAFE Port Act mandated extensive cooperation between diverse government agencies, never a simple task. The Coast Guard had recently reorganized its highly decentralized field offices, including air stations and marine safety offices, into thirty-five integrated sector command centers, but these did not meet the act's stringent requirements for extensive interagency cooperation.[59] The Coast Guard revamped four command centers where it shared responsibilities with the US Navy, at Hampton Roads, Virginia; Jacksonville, Florida; San Diego, California; and Seattle, Washington. The act required the Coast Guard to establish interagency operational centers also with Customs and Border Protection, Immigration and Customs Enforcement, the Department of Defense, the Secure Border Initiative Network, as well as state and local partners. One center in Charleston, South Carolina, was run by the Department of Justice. Such interagency activities as obtaining security clearances, building working relationships, cooperating with diverse local partners, and addressing competing leadership roles are significant challenges that stretched out over years.[60]

Port security ever since has been bedeviled by the existence of two rival and contradictory inspection regimes for containers. The SAFE Port Act set up a "risk-based" inspection regime focusing on security at the originating (foreign) port and the receiving (domestic) port, assuming containers remained secure and tamper-free en route across thousands of miles of ocean. The act's Container Security Initiative placed Customs and Border Protection (CBP) staff overseas at originating foreign ports to identify specific shipments, bound for the United States, that might require additional screening or personal inspection. Despite questions about jurisdiction and sovereignty, this initiative was reasonably successful and even led to foreign port operators coming for inspection visits to the United States. The SAFE Port Act's risk-based inspection regime has the merit of assigning scarce inspectors to the containers deemed at risk, but it clearly conflicted with a rival piece of legislation (discussed later).

The SAFE Port Act's risk-based Automated Targeting System relies on a complex mathematical model that considers specific information about each arriving shipment and, for each, assigns a "risk score" to identify particular shipments for further inspection. Around 5 percent of arriving containers are scanned by x-ray technology, and of these about 5 percent are subjected to further in-person inspections. There are numerous practical problems. It is common practice that shipping manifests—the official documents that describe the contents of each shipment—are changed upon arrival, but the consequent risk scores cannot be easily updated. Since the majority of on-arrival adjustments lower the risk scores, it follows that the inspections done on the basis of the original higher risk scores are wasting inspectors' scarce time. An evaluation by the Government Accountability Office recommended that CBP consider upgrading ATS to incorporate "smart features" including "more complex algorithms and real-time intelligence."[61]

The rival legislative mandate of 100 percent screening, while seemingly more secure, created a logistical nightmare. So-called 100 percent screening was imposed by the Implementing Recommendations of the 9/11 Commission Act of 2007. There are numerous vexing challenges, not least that "scanning equipment at some ports is located several miles away from where cargo containers are stored."[62] Few ports have any single central place where scanning stations could be set up, and so not easily avoided, while transshipment containers are in a category of their own. Some estimates pointed to the need for at least 2,000 inspectors per shift at the Port of Los

Angeles (the nation's largest container port), while that port's total CBP staff—to handle the wide range of its inspection, customs, and other statutory responsibilities—numbered around 500. Further, the Implementing Recommendations Act mandates that *foreign ports* do scanning on each and every container bound for a US port, raising thorny questions of sovereignty.[63] In October 2009, the GAO issued a strongly worded critical report. CBP has made only "limited progress" in scanning containers at the 100 percent level anywhere, achieving a majority of containers only at three low-volume foreign ports, while several larger foreign ports in a trial run achieved no better than five percent. "Logistical, technological, and other problems at participating ports . . . have prevented any of the participating ports from achieving 100 percent scanning." Facing up to this unhappy result, the Department of Homeland Security issued "a blanket extension to all foreign ports" to remain technically within the letter of the law. In a 2016 follow-up, GAO found that "DHS has not yet identified a viable solution to meet the [100 percent] requirement."[64]

Rarely mentioned in discussions of port security is the simple fact that a large proportion of containers are not part of any inspection regime, risk-based or otherwise. All ports have cargo shipments that are never officially "landed" at the port and indeed are destined for another country. This is the shadowy world of transshipments. Shipments of electronics from China to Los Angeles might contain some containers bound for Brazil or another country or even for another US port. Such transshipments are common in the shipping industry, and special arrangements (originally to avoid paying import duties to the transit country) are routine. Transshipments are known as "in-bond goods," since the responsible shipper must possess a legal bond that guarantees payment of any import duties or other fees that may fall due; and they represent a security abyss. In-bond shipments are permitted to reside in the United States without any inspection for 15 to 60 days in order to reach their final port or destination; shippers can alter the shipping destination without notifying CBP at all. Routine changes in shipping manifests are simply not tracked for shipments bound for foreign ports. Strange to say, CBP does not collect information on the volume or value of in-bond shipments. (Sample data indicate that there are around 10 million or more in-bond transactions each year.) While one would clearly like to know more, port officials estimate that in-bond transshipments constitute between 30 and 60 percent of total port arrivals. These in-bond shipments passing through US ports are essentially unknown, unknowable, and uninspected. The GAO concluded, with some restraint, that "CBP's management of the [in-bond] system has impeded efforts to manage security risks."[65]

CREATING SECURITY

It seems inescapable that we need to do better at building security into our techno-logical systems—energy, information, transport, and others—rather than attempting to graft security onto them after they have been built and society has come to depend on them. And in a world that is "hot, flat, and crowded" (in Thomas Friedman's mem-orable phrase), sustainable energy is just as important as secure shipping. "Our oil addiction is not just changing the climate system; it is also changing the international system," he writes. "Through our energy purchases we are helping to strengthen the most intolerant, antimodern, anti-Western, anti-women's rights, and antipluralistic strain of Islam—the strain propagated by Saudi Arabia." In this view, the several security initiatives in the wake of 11 September 2001 simply fail to address the fun-damental problems.[66] An acute analysis by the RAND Corporation of the global shipping industry offers a robust diagnosis of the structural conditions for global security. The study's insights and diagnosis might be extended to other transporta-tion, information, and energy systems.[67]

The RAND researchers identify three largely separate "layers" in the global ship-ping industry. First, the "logistics" level handles the physical movement of goods, in containers or as bulk shipments such as iron ore or coal or break-bulk shipments such as bales of jute. This layer comprises truck, rail, and ocean carriers as well as for-eign ports and customers' vehicles picking up goods upon delivery at a port. Second, the financial "transactions" level routes the flow of money as shippers pay for their goods to be moved from one location to the other. This layer consists of customers, retailers, foreign suppliers, freight consolidators, non-vessel-operating common car-riers, and the various import-export banks that finance global trade. These are the latter-day heirs to the "agents, factors, brokers, insurers, . . . and a great variety of other dealers in money" from industrial-era London (chapter 3). Strangely enough, the flows of money and the flows of goods are separate systems. Once a physical ship-ment is sent in motion in China or some other originating location, it may be han-dled by a large number of shippers, warehouses, freight forwarders, storage facilities, and other individuals. At this level, financial transactions are at a substantial remove from the point-to-point shippers and receivers. The third level even further removed from the physical layer of logistics are the institutions charged with governance and "regulation." These include such international organizations at the World Customs Organization, the International Maritime Organization, and the (US-based) regu-latory agencies such as the Customs and Border Protection and Federal Maritime Commission as well as the Coast Guard. A key insight of the RAND researchers is that the "oversight agencies have a limited range of influence over organizations in either the transaction or logistics layer."[68]

In the specific example of shipping security, the three layers are frighteningly autonomous. The World Customs Organization has some influence over foreign suppliers on the financial-transactions level, but that is all. The US government agencies like the Federal Trade Commission have influence on retailers such as Walmart, Target, and others (on the transactions level), while the CBP strives to influence domestic and foreign ports (on the logistics level). The Coast Guard at the oversight layer has influence on the US ports and on the ocean carriers (both on the logistics level), but it has surprisingly little influence over any institution on the financial transactions level and no influence on domestic rail or truck shipments on the logistics level. Domestic rail and truck carriers escape oversight from global-shipping governance, while key actors on the financial-transactions level (such as the export-import banks, customers, freight consolidators, and non-vessel-operating common carriers, which do not own any ships but otherwise function as shippers often for smaller cargos) largely escape any global oversight whatsoever. Since the transaction and logistics layers have been for many decades perfecting efficiency and speed of shipments, it is no wonder that the weak regulation layer has experienced persistent difficulties.

This pattern of dysfunctional risk governance suggests a need to reconsider the "risk society." The popular "risk society" thesis developed by Ulrich Beck and Anthony Giddens in the 1990s was enthusiastically adopted by many academics and policy analysts and some environmentalists. Beck and Giddens argued that, in recent decades, society had undergone a fundamental transformation: no longer was modern society principally occupied with the production and management of goods but rather with that of risks. Globalization in the 1990s was eroding the nation-state, and a new social order was at hand. Indeed, they noted the paradox of a new society emerging in the absence of any evident social revolution, suggesting that "strong economic growth, rapid technification, and high employment security can unleash the storm that will sail or float industrial society into a new epoch." Their optimistic viewpoint suggested that older sociological theories, such as those dealing with social class, could be set aside. Even the wealthiest members of society, who might own or control polluting factories, would themselves be harmed by the pollution and thus naturally take steps to curtail that pollution. Since, in their theory, these concerns with risk were widely shared throughout society and deeply embedded in planning for the future, modern society gained the capacity for "reflexive modernization."[69]

Beck, in Germany, pointed to the strong environmental movement there, while Giddens gained the ear of Tony Blair, leader of the Labour Party in Britain (1994–2007). For a time, the "risk society" seemed an exciting path forward literally into a new modernity. The thesis looks rather less plausible, however, given the findings of this chapter: the patently absurd ethanol subsidies that keep gas-guzzling vehicles on the road; the alarmingly insecure internet that permits unauthorized access to air-traffic control and the electricity grid and military intelligence; and the persisting problems in trying to graft security onto global shipping. In each, modern society is clearly doing a terrible job of "anticipating" risks. Some attempts at addressing risk actually generate greater geopolitical instability.

These examples of systemic risk—created by and propagated through networks and systems of technology—also challenge Beck and Giddens' emphasis on sociological "agency" and "individualization" through the reflexive anticipation of risks. No quasi-automatic or reflexive social process is dealing constructively with these technology-induced systemic risks. One underlying problem is that the "risk society" and "reflexive modernization" theories, oddly enough, paid insufficient attention to the technologies that constitute and create modern society.[70] Understanding these systems and networks and structures is necessary to comprehend social and cultural developments, for technologies can constrain and enable as well as channel these processes. In this chapter, I examined the emergence of technologies at the core of contemporary society where individual people and companies and governments have, over many years, built up sociotechnical "structures" that are not easily modified. Creating a more secure world will require altering these entrenched structures and systems. None of the necessary changes will come automatically or reflexively. I wish to live in a secure modern world, but I believe it will take immense work and substantial creativity to remake the present world to be so.

CHAPTER 10

1990 – 2016

Dominance of the Digital

Technological eras in the past hundred years were largely driven by social, cultural, and economic forces such as modernity, the military, globalization, and efforts to contain systemic risk. "Science and Systems, 1870–1930" describes one era when prevailing technologies like electrification and synthetic chemicals provoked changes in the organizational and economic realms, with the rise of stabilized corporate industry and mass consumption at the center. No surprise this led to the 1933 Chicago World's Fair motto of "Science Finds—Industry Applies—Man Conforms." With the recent dominance of the digital, it also appeared that technology was in the driver's seat: that digital technologies themselves were bringing about confident globalization, cultural integration, geopolitical unification, and a cohesive globe-spanning internet that promised democratic forms of society everywhere it spread. This chapter closely examines two of this era's defining developments: the rise of Moore's Law as the driving force behind the digital revolution; and the internet as a political and cultural form that promised democracy everywhere and yet resulted in something quite different.

Moore's Law started out as a speculative conjecture in the 1960s and slowly gained force in the semiconductor industry and beyond, so that by the 1990s it was widely recognized as a technological force in and of itself. The world's semiconductor electronics companies rode the crest of enthusiasm for globalization (see chapter 8) to create international organizations and build global cooperative mechanisms that synchronized their research and development (R&D) efforts. Moore's Law looked unstoppable until 2004 when Intel first hit a "thermal wall" in its most advanced integrated-circuit chips, requiring a shift away from the reigning R&D strategy that focused on ever-smaller single-core chips to the multi-core monstrosities we have today. The year 2016 marks the end of this era of self-assured digital dominance. In that year, the worldwide R&D planning that sustained Moore's Law ended.

Several unnerving political developments also occurred in 2016. Through the 2000s, the latest generation of tech company founders were routinely lionized. Most everywhere they gave public speeches, they were lauded as beneficent kings bestowing riches on the world. No longer. Controversies with labor policies at Apple, security procedures at Facebook, algorithmic bias at Google, unbounded market power at Amazon, and other shortcomings flashed a transformation in public attitudes toward high tech companies in the United States; in Europe, smarting against another "American challenge," the political-economic environment was never quite so favorable. As I write this, the biggest tech companies—Amazon, Apple, Facebook, and Google—are being investigated by the *American* Congress for antitrust violations and elsewhere treated with mistrust and alarm. At Facebook, employees struggle to convince themselves "it's for a good cause," according to one insider: "Otherwise, your job is truly no different from other industries that wreck the planet and pay their employees exorbitantly to help them forget."[1] The year 2016 also saw nationalistic, authoritarian, and antiglobalization governments rise to power in the United States and the United Kingdom and their further entrenchment in China, Egypt, Hungary, Russia, Saudi Arabia, Turkey, and elsewhere. The internet, once hailed as a force to spread democracy around the globe, instead has too often become an oppressive tool deployed by authoritarian states that "are now actively shaping cyberspace to their own strategic advantage." We were promised an internet to bring diverse cultures together, but instead we got a divisive "splinternet."[2] Clear insight into our future has never been more important.

MOORE'S LAW

For decades Moore's Law propelled the digital revolution in solid-state electronics. Initially a casual observation by Gordon Moore, an industry insider, it soon became a touchstone and article of faith for the leading US chip manufacturer (co-founded by Moore) and eventually for the world's entire chip industry. In effect, Moore's Law created the necessary conditions that made computing smaller and cheaper, ushering in the successive shifts from room-sized mainframe computers to the desk-sized minicomputers that formed the backbone of the early internet and then the networked personal computers that flooded on to the internet and made it a mass phenomenon. Tech journalists everywhere set Moore's Law high on a pedestal. As one eager account put it:

> Information technology, powered by Moore's Law, provided nearly all the productivity growth of the last 40 years and promises to transform industries, such as health care and education, which desperately need creative disruption. It has powered a wave of globalization that helped bring two billion people out of material poverty and has delivered the gift of knowledge to billions more.

"Moore's Law is Silicon Valley's guiding principle, like all Ten Commandments wrapped into one," according to one insider.[3] The saga of Moore's Law is an archetype of "how technology changes the world." What follows here is, remarkably enough, the first complete account of its unlikely origins, lengthy influence, and recent demise.[4]

Moore's Law can be best understood as an *emergent structure* with institutional and cognitive dimensions; across time, it changed how people thought about technology and how they acted. This did not happen spontaneously. To begin, Moore's Law had prominent champions inside Intel who mounted an external publicity campaign as well as internally embraced it, according to one Intel CEO, as "a fundamental expectation that everybody at Intel buys into." Gaining momentum beyond Intel, it guided the US semiconductor industry's "technology roadmaps" that enabled future planning and then was extended to international roadmapping that connected R&D efforts around the world. In framing these roadmaps, Moore's Law served to align industry-wide decisions on finance, manufacturing, supply chains, product development, and consumer culture. Some even believe it resulted in an "exponential advance of computer power . . . impervious to social, economic, or political contexts" and therefore with Moore's Law "raw technological determinism is at work."[5]

But, on closer inspection, Moore's Law was not at all impervious or deterministic and in fact its achievement depended on three shifts in US government policy, two profound industry realignments, and wide-ranging diplomacy at multiple levels. Venture capital, by funding companies that seemed to conform to it, further entrenched Moore's Law. In reality, Moore's Law depended on literally hundreds of organizations and thousands of people making immense financial, intellectual, and organizational investments to bring it about. These investments made the revolution in digital electronics and computing. It was not all sweetness and light. Toxic chemicals used in chip manufacturing created nearly two dozen heavily polluted Superfund sites in Silicon Valley, while the industry's furious pace of innovation and race for off-shoring has left these sites abandoned and largely forgotten.[6] Also during the Moore's Law decades, consumer electronics in reliably discarding yesterday's shiny devices for tomorrow's faster and cheaper ones generated a mountain of waste electronics that found its way to the waste-pickers and informal recyclers in the global economy. In south coastal China, home of the world's largest e-waste dump, "The stench of burnt plastic envelops the small town of Guiyu, some of whose rivers are black with industrial effluent."[7]

In 1965 the trade journal *Electronics* asked Moore, then research director for Fairchild Semiconductor, to pen a few words on the future of electronics. This future was not so clear. At the time, IBM was beginning deliveries of its System/360 computer, which powered the company to a unique dominant position, making up fully three-quarters of the entire US high-tech sector's stock-market capitalization. Solid-state transistors were widely used in computers, but the more recent integrated circuits—which put dozens of transistors onto a single semiconductor slab, or "chip," invented five years earlier—were not part of the IBM package. Its model 360 used a distinctive "solid logic" module created from discrete transistors and screened resistors printed on a half-inch-square ceramic substrate; six such modules were assembled on an inch-and-a-half square printed-circuit card to build up the computer's circuitry. IBM thought the monolithic integrated circuits that Moore had in mind, constructed from next-generation metal-oxide-silicon, or MOS, transistors, were too risky and experimental.[8] But then "Silicon Valley" had not been invented either.

Moore's "Cramming More Components onto Integrated Circuits" made a number of bold predictions such as the coming of home computers, electronics-laced automobiles, and even "personal portable communication equipment," but the most fundamental was a projection of how many components might be on one integrated-circuit chip inside ten years. Moore examined recent trends in the number of components such as transistors, diodes, and resistors that could be most *economically* manufactured on a single chip; it was then 50 but appeared to be doubling each year. Five years in the future, he thought it would be 1,000; and, he famously predicted, "by 1975, the number of components per integrated circuit for minimum cost will be 65,000 . . . built on a single wafer." One beguiling aspect of Moore's "law" was it needed no fundamental breakthroughs in physical or chemical science, simply continuing to use optical light beams to etch ever-smaller circuits. "Only the engineering effort is needed," he noted.

This engineering effort required sizable investments, and Moore was ideally placed to ensure that Intel made them. Intel was founded in 1968 by Moore, Robert Noyce, and like-minded colleagues who had chafed under the corporate yoke of Fairchild. At Intel they would ride the high tide of silicon-based semiconductors, selling first memory chips and then logic chips for personal computers based on Intel's hardware and Microsoft's software. In the 1990s Intel became the world's largest semiconductor manufacturer, and in 2018 it placed seventh in a ranking of the world's most-valuable tech companies, ahead of Cisco Systems, Oracle, and IBM and sandwiched in-between Samsung and Taiwan Semiconductor Manufacturing (of which more later).[9] Moore's Law radiated outward from Intel.

In 1975 when his decade-old prediction came to pass, Moore gave two talks that caught the attention of the technical community, but it was Moore's co-founder and extrovert colleague Robert Noyce who brought the phenomenon to a far-wider audience two years later. Whereas Moore's article in the *IEEE Technical Digest of the International Electron Devices Meeting* made mostly cautious technical assessments, Noyce, then writing as chairman of Intel, gestured grandly in *Scientific American* that a "true revolution . . . , the integrated microelectronic circuit, has given rise to a qualitative change in human capabilities" with applications in space exploration, intercontinental ballistic missiles, office automation, industrial control systems and, not least, computing. Just a year after Steve Jobs and Steve Wozniak founded Apple, Noyce thought that the "personal computer . . . will give the individual access to vast stores of information." The key was advances in semiconductor electronics and especially Moore's prediction that the "number of elements in advanced integrated circuits had been doubling every year" with "further miniaturization . . . limited [less] by the laws of physics than by the laws of economics." Moore and Noyce carefully chose data points to create straight lines showing exponential growth. In *Scientific American*, Noyce picked newly emerging "charge-coupled devices" as the upper-end data point, although they disappeared from later graphs when these devices did not work out as anticipated.[10] Noyce's equally expansive article in *Science* appeared alongside the first published instance of "Moore's Law" (according to the *Oxford English Dictionary*), in an article by a Stanford electrical engineering professor and a board member of the venerable Fairchild Camera and Instrument, the corporate parent of Noyce and Moore's first start-up.[11]

Additional firepower was drafted to the cause when Carver Mead converted to the faith and became a prominent evangelist for Moore's Law—very likely naming it as such. Mead, at the beginning of a distinguished career, a brand-new assistant professor, recalled the first time he met Moore, who came into his California Institute of Technology office and offered two large manila envelopes: "I never had seen so many transistors. I didn't know there were that many transistors in the world. . . . So . . . students could take them home when they were done, and that was just a wonderful thing." Moore set up Mead's "once-a-week commute to Silicon Valley" some 350 miles each way. Stimulated by Moore's questions about how small transistors might go, Mead calculated that smaller transistors would, if properly scaled, get more efficient and faster (and could be made vastly smaller than anyone believed). In the 1970s, Mead gave the phrase "Moore's Law" even greater visibility and heft. Mead was working with Xerox PARC's Lynn Conway on a best-selling textbook (1979) that further extended the stature of these ideas and techniques. "As Mead traveled throughout the silicon community . . . he succeeded in building a belief in a long future for the technology, using Moore's plots as convincing evidence," writes historian David Brock.[12]

These years witnessed a sea change in the hypercompetitive culture in the semiconductor industry. The rise of cooperative R&D activities and industry-wide roadmapping entrenching Moore's Law were unthinkable during the early years when cutthroat competition ruled the industry. Charles E. Sporck, after working for General Electric on the East Coast and Fairchild Semiconductor in California, led National Semiconductor from 1967 to 1991. As he recalled these years, "the semiconductor industry was a tremendously competitive, knock them down, fight them, kill them environment." One company "learned to their horror . . . that competing with this big gorilla [Fairchild] was a disaster." "We buried competitors, bang, bang . . . on a regular basis," stated one company founder. "We were always somewhat cantankerous," recalled another.[13] Texas Instruments, Intel, Motorola, National Semiconductor, and Advanced Micro Devices emerged as the dominant American "merchant" companies selling chips on an open market, while IBM and AT&T were "captive" manufacturers for their own in-house needs.

The industry's customary cutthroat competition seemed a liability when Japan's Hitachi, Fujitsu, and Nippon Electric Company (NEC) made significant inroads in the US market. For years, Japanese companies had quietly been creating advanced manufacturing capability through technology transfer, in-house R&D, government support, and a fabled attention to quality. A turning point was Japan's technical success with 64K memory chips in the 1980s, selling at such low prices that Intel actually exited the memory chip market to focus on more complex logic-based microprocessors. "I think we've already lost out in the 256K," said one industry executive: "The Japanese have won the dynamic RAM market." Just a few years earlier American companies had controlled 95 percent of the US market for memory chips, but beginning with the demanding 256K chips Japanese companies gained more than 90 percent of this US market.[14]

During these unsettled years, the US semiconductor industry gradually gained cooperative mechanisms that tempered destructive competition. In 1977 industry leaders Robert Noyce, Wilfred Corrigan, Charles Sporck, W. J. Jerry Sanders, and John Welty—founders and/or CEOs, respectively, of Fairchild and Intel, Fairchild and LSI Logic, National Semiconductor, Advanced Micro Devices, and Motorola—took the lead in founding the Semiconductor Industry Association, or SIA, to lobby the US government for trade mechanisms to counter Japan's successes. University research funded by SIA began in 1982 with the Semiconductor Research Corporation, or SRC, aiming to "define common industry needs, invest in and manage the research that would expand the industry knowledge base, and attract premier students to study semiconductor technology."[15] Yet the industry remained wary of greater levels of cooperation owing to the venerable Sherman Antitrust Act, which had repeatedly landed both AT&T and IBM in the courtroom.

Changes in federal antitrust law in the 1980s—the first of three major policy shifts—led to several ventures in cooperative industry-wide R&D and set the groundwork for full-blown "technology roadmapping" that embraced and enshrined Moore's Law. Minneapolis-based Control Data Corporation initiated talks within the computer industry that led to the creation of Microelectronics and Computer Technology Corporation, often labeled MCC. Gordon Moore suggested to CDC's Bill Norris the gambit of "getting all the players in one room" to create MCC in 1982. Once it was started, Norris rather liked the idea of his firm's investing $14 million to gain access to the $120 million in R&D results from the nineteen member companies.[16] Admiral Bobby Ray Inman was chosen to lead MCC based on his inside knowledge of Washington and top-level leadership at the National Security Agency and the Central Intelligence Agency. Besides being Inman's home state, Texas was useful for industry-wide activities since its University of Texas at Austin was a growing tech hotspot, surrounded by ten major computer or semiconductor companies, yet comfortably distant from the minicomputer-centered Boston region as well as the recently named "Silicon Valley" in California. Texan Ross Perot reportedly wrangled two years' use of a private jet to assist MCC in recruiting. MCC's founding actually preceded by two years the National Cooperative Research Act of 1984; the loyal Texans in Congress ensured passage of the accommodating federal act.[17]

Full-blown industry–government cooperative R&D arrived in the form of SEMATECH, a loose acronym for Semiconductor Manufacturing Technology, also located in Austin and composed of many of the same companies. The US government adopted unusual restrictive trade measures against Japan's chipmakers, as the SIA had encouraged, while semiconductor leaders lobbied the US Defense Department to line up direct financial support for the industry—two further shifts in federal policy that supported Moore's Law. A 1986 Defense Science Board task force on "Semiconductor Dependency" helpfully warned that "the existence of a healthy U.S. semiconductor industry is critical to the national defense."[18] As Sporck phrased it, "suddenly Japan is the threat. And interesting things happen. Suddenly, we started talking to each other, and talking about, how do we meet this threat? How do we

FIG. 10.1. "CLEAN ROOMS" IN COMPUTER MANUFACTURING

Clean rooms in company publicity suggested the industry's environmentally friendly image, while highly toxic materials used in chip manufacturing were less visible.

Charles Babbage Institute.

marshal our forces in areas that we don't compete with them and make progress in a fashion so that we could compete more effectively? And that led to SEMATECH."[19] It was quite a change from the bang, bang years of burying competitors.

Once again, Texas pressed its substantial advantages. The University of Texas did some highly creative accounting and found $50 million to buy a former mini-computer factory, build a brand-new chip-manufacturing "clean room" there, and offer the tidy package to SEMATECH for just $2 a year. So-called clean rooms lent a positive image that digital computing was environmentally friendly (fig. 10.1). The city of Austin kicked in $250,000. The federal government matched research funding from the consortium's fourteen companies to provide $1 billion across its first five years (1987–92). Texans in Congress pushed the Defense Department's Advanced Research Projects Agency, or DARPA, to deliver its promised $100 million each year. SEMATECH added to DARPA's reputation as a proactive funder of cutting-edge technology, initially developing manufacturing processes for advanced chips that all members might utilize. By 1991 SEMATECH members were roughly two-thirds of the industry by annual sales ($20.7 billion of total $31.1 billion). Smaller semiconductor firms like Cyrix and Cypress either paid burdensome annual membership fees, with fees capped at just $15 million for the largest companies, or looked at SEMATECH from the outside as an exclusive "corporate country club."[20]

SEMATECH members IBM, Texas Instruments, Intel, and Motorola were early converts to *company*-wide roadmapping. The notion of roadmapping had already emerged in the automobile industry during the 1970s energy crisis, when industry and government sought means to develop new energy-efficient engines. Likewise, military managers used the setting of future targets and R&D milestones in the programmed development, over years, of vastly complex high-tech weapon systems. A likely boost for *industry*-wide roadmapping was a Defense Department program of the 1980s, where military planners identified successive phases—with defined near-micron and sub-micron scale semiconductor devices—to manufacture "very high speed" integrated circuits for military applications. As one internal memo put it, the Office of the Secretary of Defense provided "DOD documentation procedures on road mapping for use as a guide in SEMATECH planning . . . to formulate a standard model for the theory of road mapping."[21]

The famous roadmaps were not only about predictably shrinking device sizes but also *reliably coordinating* the numerous technologies necessary for successful chip production. In the 1960s, most semiconductor companies had supplied their own chemicals and specialized production equipment. One sign of change was the independent Applied Materials, which became a public company in 1972, a year after "Silicon Valley" was named, concentrating on semiconductor production equipment and highly refined (and highly toxic) chemicals.[22] Another sign of change was so-called Rock's Law, named by the venture capitalist Arthur Rock, who observed that capital requirements in the industry were doubling roughly each four years. As transistors got ever smaller and the resulting integrated-circuit chips got ever more dense, the cost of the "fabs" to make them rose to dizzying heights. In 1985 a fab cost around $100 million with capital requirements climbing fast (and as of this writing, in 2020, the cost of one state-of-the art fab is $20 billion).

With the cost of production fabs relentlessly on the rise, chip makers faced economic ruin if they could not rely on well-coordinated innovations by external suppliers of production equipment, chemicals, and testing. As one US government official phrased it: "You can't make any changes in this industry without integration of disparate companies. For example, [one such change in device size] requires an exposure tool (6 companies), resist technology (3 companies), and mask technologies (2–3 other industries) to pull off this change."[23] The largest chip producers like IBM and AT&T had some of these capabilities in-house, but the industry's core of medium-sized firms could not control their suppliers such as Applied Materials and dozens of smaller concerns, and simply could not dictate technically demanding requirements to be achieved years in the future. As Robert Schaller observes, "road-mapping contributes to an *organized* innovation pattern."[24]

An early *industry-wide* roadmap for semiconductors emerged informally within the SRC, when in 1984 representatives from four or five different companies (once fierce competitors) "were sitting around and discussing research strategy" and, at the end of the day, sketched out ten-year "targets for litho[graphy] dimensions, voltage, feature size, etc." These informal targets evolved into a SIA-sponsored annual exercise to alter the future. "The real purpose of a technology roadmap was not to simply forecast the future . . . but to lay out the path to get there. Specifically, this meant a methodical and systematic approach to defining the needs of supporting technologies, materials, and other resources," according to Schaller. Needless to say, Moore's Law was "a central planning assumption." As one industry figure put it, "everyone knows what the clock frequencies are going to be next year, they are driven by Moore's Law scaling curve." Another acknowledged, "Moore's Law productivity gains are becoming increasingly difficult and costly to implement, yet these productivity gains are essential to industry growth."[25] As Moore himself expressed it, "If we can stay on the SIA Roadmap, we can essentially stay on the [Moore's Law] curve. It really becomes a question of putting the track ahead of the train to stay on plan."[26]

It wasn't only semiconductor companies that sought guidance about the future. The venture capital industry, while pioneered in Boston by a Harvard Business School professor named Georges Doriot, really took off as an offshoot of Silicon Valley. No surprise that Robert Noyce became a very wealthy man and, famously, had shoeboxes packed with promissory notes for dozens of start-up ventures. Doriot's venture-capital model came to California when Eugene Kleiner, one of the founders of Fairchild Semiconductor, and Tom Perkins, his former Harvard student and then Hewlett Packard executive, teamed up in 1972 to create the legendary Kleiner Perkins Caufield & Byers; among their notable early successes were America Online, Netscape, Sun Microsystems, and Amazon. Success bred success. Today, nearly fifty of the world's best-known venture capital firms have addresses along Sand Hill Road, just north of Stanford University at the edge of Silicon Valley; and it's said "if a start-up company seeking venture capital is not within a 20-minute drive of the venture firm's offices, it will not be funded." Among the legends of Sand Hill was Don Valentine. He joined Fairchild in 1959 and became its sales manager, then moved to National Semiconductor, and eventually founded the celebrated Sequoia Capital in 1972. His early and lucrative investments included Apple Computer, Atari, Cisco Systems, Electronic Arts, Oracle, Network Appliance, and Sierra Semiconductor. His secret? "That's easy. I just follow Moore's Law and make a few guesses about its consequences."[27]

The consensus around Moore's Law allowed the semiconductor companies, previously "viciously competitive," to speak with a unified voice in lobbying the US government for continued trade restrictions and ongoing funding. SEMATECH, even while its Austin site was being prepared, took up some of the SIA activities. To promote industry-wide roadmapping, SEMATECH from June 1987 to March 1988 organized a series of thirty workshops to consider the interwoven production elements—including lithography, manufacturing systems, thermal processes, automation, packaging, testing, cleaning, process architecture, and others—and how they might be coordinated in an orderly plan. Initially the target was just five years in the future, but soon (in 1992) the roadmapping extended to a full fifteen years. A number of workshop leaders hailing from Intel, IBM, and AT&T—not least Intel's Robert Noyce as SEMATECH's chief executive officer—moved to Austin to manage the sizable operation. Its operational plan expanded to a steady-state headcount of 700 staff and annual budget of $200 million; in essence, SEMATECH operated a full-scale fab in Austin directed at improving chip manufacturing technology, transferring designs, skills, and expertise to consortium members, and rebuilding relationships with the industry's suppliers.[28] The roadmapping exercises cost roughly $1 million each year.

The roadmaps' far-reaching forecasts were updated every two years, evolving into a *National* Technology Roadmap for Semiconductors first published as such in 1994. It was based on the fifteen-year roadmap that emerged two years earlier from the leadership of none other than Gordon Moore, who chaired an influential SIA–SEMATECH workshop that assembled 200 representatives from industry, government, and academia in a suburb of Dallas, Texas, for three days of intense discussion and negotiation. In 1994, the "central assumption . . . was that Moore's Law (a 4x increase in complexity every three years) would again extend over the next 15 years." The roadmapping exercises had started originally with AT&T's circa 1989 proprietary 800 nanometers (nm) for transistor "gate length," which SEMATECH shrank to a breathtaking 70 nm forecast for 2010. For the 1994 workshop, eight technical working groups, with their findings consolidated ahead of time, convened for a one-day 300-person workshop in Boulder, Colorado. The workshop's results formed a fifteen-year plan (1995–2019) to transform standard DRAM memory chips from 64 megabytes to 64 *gigabytes* and to dramatically boost the number of transistors packed into each square centimeter by twentyfold or more. The national roadmap "had a profound impact on semiconductor R&D in the United States," informing international technical communities, guiding government R&D priorities, and framing the future for SEMATECH.[29]

Even though SEMATECH was "tightly wrapped in the United States flag," the semiconductor industry progressively spanned national borders. In the 1960s Fairchild was the first US semiconductor company to move manufacturing to Asia, while the completed SIA and NTRS roadmapping reports circulated far and wide. Companies did, too. By the time of SEMATECH's founding in the late 1980s, Texas Instruments had moved production of its upper-end memory chips from Lubbock, Texas, to north of Tokyo; Motorola and Toshiba jointly owned a production facility also outside Tokyo. One Japanese company, NEC, with a fab located in California, even asked to join SEMATECH but was turned away. "There is a flow of technology to foreigners no matter what we do," acknowledged one US industry insider.[30]

The first *International* Technology Roadmap for Semiconductors in 1999 reflected almost a decade of international diplomacy and engagement. In the 1990s enthusiasm for "globalization" spanned diverse economic, political, and cultural forms (see chapter 8). For many, globalization spelled the end of the nation-state, while private sector actors such as global corporations and border-spanning industry consortia were the future. SEMATECH had reached its allotted ten years of US government funding and intentionally transitioned to full industry sponsorship, which allowed full-blown international engagement. In 1998 the US roadmapping exercise sought international participation and review, but plans for a rival roadmapping effort were announced in Japan. In response to this new threat from Japan, the US SIA reached out to its counterparts in Europe and Asia and, through the newly formed World Semiconductor Council (created in the wake of the US–Japan trade disputes), launched a fully international roadmapping effort. Initially, the 1999 roadmap emerged from the semiconductor industry associations of Japan, Europe, Korea, and Taiwan "in cooperation with" the US SIA, and then the 2001 roadmap was "jointly sponsored by" these five groups. (Mainland China joined the WSC in 2006.) In sharp contrast with the nationalistic "Japan bashing" of the 1980s, deep technical cooperation emerged among hundreds of semiconductor companies around the world.[31]

Curiously, the number of roadmapping participants seemingly obeyed a doubling law, too. In 1992, there were 178 participants; in 1999, 474; and in 2003 a full 936 individual participants. By 2003, international roadmapping was solidly dominated by industry, with nearly 80 percent of the participants from chip-making companies or equipment or materials suppliers; most of the remaining participants hailed from research institutes, universities, or consortia like SEMATECH. Direct government involvement was almost nil. Taiwan (21%) and Japan (22%) together equaled US participants (42%), joined by Korea and Europe (3 and 12%, respectively). Despite daunting logistics and cultural differences, interviewees in 2003 indicated the international effort was preferable to national roadmapping by a 2:1 margin; fully 70 percent of the supporters specifically embraced globalization (affirming "global technology and industry" and "worldwide consensus"). Thought one, "electrons don't know cultural boundaries."[32]

Additional incentive for the immense international effort was Rock's Law, the relentlessly climbing cost of semiconductor fabs. R&D costs were rising, a spur for coordinated R&D efforts spanning nation states. "We needed Canon, Nikon, and others to progress, but could not do anything about it," recalled one industry figure. Whereas in 1979 fully nine of the top ten semiconductor *equipment* manufacturers were US companies, by 2000 just four hailed from the United States, including top-ranked Applied Materials, joined by five Japanese companies, including Canon and Nikon, and upstart ASM Lithography, a Dutch company (about which more later). Given the punishing capital-cost increases, the international cross-firm coordination of R&D efforts was a useful means to manage risk.[33]

The rise of Taiwan to equal Japan marked a shift in the global semiconductor industry's structure and its dominant technology. As noted earlier, American firms had largely left the commodity memory chip market to Japanese and Korean companies like giant Samsung. While roadmaps had previously focused on the industry-standard DRAM memory chips, the ITRS also included more complex logic-based microprocessor chips as well as customized ASICs, or application-specific integrated circuits. For years, the leading chip companies had combined design, engineering, and production of new-generation chips; they had outsourced only production equipment and chemical supply. Taiwan adopted quite a different model, and its leading firm is the plainly named Taiwan Semiconductor Manufacturing Company, or TSMC, as of this writing in 2020, one of three remaining semiconductor companies, along with Intel and Samsung, aiming for the next generation of semiconductor chips. TSMC's stockmarket capitalization surpassed Intel's in 2017.

TSMC was founded in 1987 by Morris Chang, who was educated at MIT and Stanford and gained industry experience at Sylvania Semiconductor, Texas Instruments, and General Instruments before moving to Taiwan as president of a government-industry research consortium. Two years later Chang founded TSMC, which entered the industry by making ASICs. The design and engineering of chips had altered, in part under the influence of the design-engineering model introduced by Mead and Conway's influential textbook. With the traditional mass-production model, a company like Intel invested immense design effort into a particular micro-processor or other chip and then manufactured long runs of them to recoup the high costs. In contrast, TSMC took the leading role in developing a "foundry" model of specialized production.[34]

In the foundry model, TSMC lets companies like Apple, Qualcomm, Nvidia and others do the chip-design engineering; it focuses on keeping its super-expensive pro-duction facilities busy. (Since 2011, many of Apple's iPhone chips come from TSMC; and likewise Apple's M1 computer chip, unveiled in late 2020.) Such a foundry model was initially developed for the fast-changing manufacturing runs of ASICs, which might be custom designed for signal processing or computer graphics or super-fast data handling. TSMC has production fabs in Taiwan, China, and Singapore as well as a wholly owned foundry in Camas, Washington, adjacent to Portland, Oregon; nearby suburban Hillsboro is the location since 2013 of Intel's largest chip-produc-tion facility in the world. For a time TSMC, growing rapidly, sparred with Global Foundries, a spin-off of Silicon Valley–based Advanced Micro Devices, with fabs in Singapore, Germany, and the United States. Also about one-eighth the size of TSMC is United Microelectronics, another Taiwan-based foundry. Second-tier "pure-play" foundry companies, besides these three leaders in the industry, are located in China, Israel, South Korea, Germany, and Russia. Together, these foundry companies man-ufacture the majority of the world's semiconductor chips.

The soaring capital requirements that encouraged ITRS also brought severe concentration of the global industry. TSMC aimed for the top. In 2019, a California company briefly snared the world's attention with a chip featuring 1.2 trillion transistors, a savvy publicity stunt to capitalize on the hype surrounding "arti-ficial intelligence." Built by TSMC, the chip was essentially an entire 12-inch sili-con wafer—manufactured with fully mature three-generation-old technology—that normally would be divided into eighty or more massively multi-core CPU chips.[35] The legacy of Moore's Law directs attention to the latest generation of chips, with the smallest transistors and greatest chip densities. Just three firms are left in the running, with TSMC, Samsung, and Intel vying for the honors. While all three companies succeeded at the technology node, or minimum feature size, of 7 nm it is likely that only Samsung and TSMC will achieve 5 nm in 2020, the industry's roadmap target; Intel itself points to 5 nm in 2023 or beyond. Achieving the next node at 3 nm, roughly the size of 8 uranium atoms, will be a challenge. TSMC says its 3 nm fab in Taiwan will cost $20 billion. It may build a second foundry in the United States, responding to the US government, using 5 nm technology with an estimated cost of $12 billion. "The U.S. is concerned that any of its high-security chip design blueprints could fall into Chinese hands," notes Su Tze-Yun, who directs Taiwan's Institute for National Defense and Security Research, "TSMC's Chinese customers could secretly help the Chinese army build chips targeting the U.S."[36] And just one company, ASM Lithography in the Netherlands, makes the world's supply of cutting-edge "extreme ultra-violet light" lithography equipment.

Looking backward, "the end of Moore's Law" likely happened in 2004. "Our strategy is to go 200 miles per hour until we hit a brick wall," Intel CEO Craig Barrett had noted. Early in 2004 the company hit a "thermal wall." Its 90 nm chips unaccountably ran slower and hotter than designed—abruptly bringing to a close the decades-old pattern of ever-smaller chips running at ever-higher clock speeds. In effect, Moore's Law had ended. In consequence, Intel abandoned two of its advanced designs, code-named Tejas and Jayhawk. With its surface area about 1 square centimeter, the Tejas chip became excessively hot owing to its 150 watts of energy (equivalent to an oversize incandescent light bulb) shrunk to the size of a fingernail. The competitive race for ever-smaller one-core chips became instead a far-more complex journey to achieve multi-core chips. In theory, a dual-core chip with two 2.5 GHz (gigahertz) processors should be faster than one 3.5 GHz chip, but in practice most computational tasks do not easily divide among multiple CPUs. "This is a very hard toggle of our product line," admitted Intel's president. IBM noted similar difficulties. "Classical scaling is dead," said its chief technologist for systems and technology, "In the past, the way everyone made chips faster was to simply shrink them."[37]

Since 2004, the impressive expansion in computing muscle once achieved by shrinking transistors—Intel's Pentium IV with a clock speed of 3.8 GHz was fully 5,000 times faster than its first microprocessor in 1971—has petered out. Intel never achieved its hoped-for 10 GHz Pentium chips. In 2010, as technology writer

Christopher Mims observed, "with this generation of chips, Intel is innovating any-where but in the CPU itself." Its current tenth-generation Comet Lake chip, with multiple cores, isn't that much faster than the Pentium IV. Instead of doubling eight times in the sixteen years from 2004 until 2020 (as Moore's Law might predict), the total speed increase per CPU seems to be a paltry 8 percent.[38] Remarkably, Intel's publicity for Comet Lake chips does *not* list CPU speeds.[39] Computing muscle now is won by advances in computer architecture, memory caching, accelerated network-ing, and programming tricks. Getting multi-core processors to work fully up-to-speed remains a difficult challenge.

The end of Moore's Law also spelled the end of ITRS. Already in 2004, confront-ing the thermal wall, Intel admitted it was "reprioritizing and revamping our road map." Continuing forward, the ITRS organized up to 18 technical working groups to generate updates and revisions. In 2012 ITRS started consolidating its techni-cal working groups and planned further restructuring. Its 2015 reports make clear that ITRS was no longer solely about smaller and faster transistors. Instead, with the world of cloud servers, mobile computing, and "internet of things," the new goals were greater energy efficiencies; instant-on technologies, which use little energy until needed; and reliable sensing. "Keeping abreast of Moore's law is fairly far down on the list," noted one account: "Much more crucial for mobiles is the ability to sur-vive for long periods on battery power while interacting with their surroundings and users."[40] The digital world had changed.

With these shifts, ITRS was formally disbanded in 2016 and its activities taken up by the IEEE's "Rebooting Computing" initiative and a trademarked International Roadmap for Devices and Systems (IRDS). IRDS aims at "a set of predictions that serves as the successor to the ITRS" with a greater emphasis on systems, archi-tecture, and standards. In sharp contrast to its predecessor's 80 percent industry participation, IRDS is composed of several research consortia, a physical-science professional society, a trade organization, and the Department of Energy's Sandia National Laboratories. Dartmouth College, Intel, General Electric, FIAT's research center, and several US government agencies are among the eighty-five members in the principal international R&D consortium, but the semiconductor heavyweights Samsung, TSMC, and ASM Lithography are conspicuously absent.[41]

DEMOCRATIC OR AUTHORITARIAN INTERNET?

According to *Forbes*, the world's ten top-valued tech companies at present are cre-ators or creatures of the internet. The perennial US giants Apple, Google, Microsoft,

Intel, IBM, and Facebook fill most of the top slots, with South Korea's electronics behemoth Samsung at number two. Near the bottom, if one can say that, stands China's Tencent Holdings—a media and gaming titan perhaps not so well known as China's e-commerce colossus Alibaba—sandwiched between venerable Cisco Systems and database goliath Oracle. It seems a million years ago that Microsoft's Bill Gates was finally convinced, in mid-1995, the internet meant a "tidal wave" in computing. There's a lot of water under the bridge since then. The notion that the internet is a free collection of individuals, rather than a highly organized platform for the world's economy, dominated by powerful private companies, might seem a bit quaint today. Yet for decades many business and political leaders as well as passionate pundits of all stripes assured us the internet would work to strengthen individuals over stultifying bureaucracy, to promote vibrant democracy over authoritarian regimes, and to bring about a better and more open public life and shared culture.[42]

In our mobile networked age, it is worth recalling that the internet emerged from a highly centralized society. Large companies of the time were top-down hierarchies, as were central governments around the world; any memo sent to a neighboring division in any large organization would pass upward through the chain of command, be reviewed by supervisors at each step, then passed downward to its intended recipient. No one could directly dispatch a hierarchy-busting email. Indeed, much of the early internet's cultural appeal was its promise to break up centralized hierarchies. Besides the federal government, the largest hierarchy in the US was American Telephone and Telegraph Company, or AT&T, which employed a million people already in the 1950s and 1960s and beyond. During these decades, the world's phone system was composed of centrally switched networks interconnected by long-haul lines. If you made a phone call from San Francisco to New York, it was routed through AT&T's centralized switching facility in Chicago; and the same if you were in Seattle and called Florida. From Los Angeles, a call to Miami might have passed through Dallas, Texas.[43] Telephone engineers sometimes labeled this state of affairs the "Chicago problem." The country's phone system as well as its military communications was highly centralized.

AT&T engineers worried about the phone system's vulnerability. In 1969 J. W. Foss and R. W. Mayo published "Operation Survival" in the *Bell Laboratories Record*, noting that "Bell Labs is responsible for designing reliable communication systems that will survive almost any catastrophe—even nearby nuclear explosions." Their article described how Bell protected its long-haul phone lines against a 20 megaton thermonuclear "surface burst" that blew a mile-wide crater in the earth and

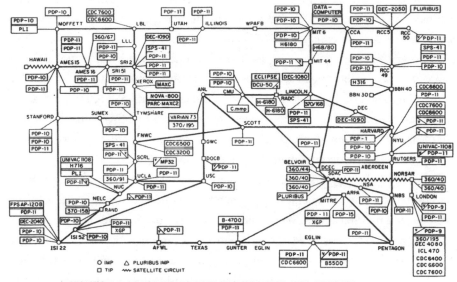

FIG. 10.2. ARPANET MAP (MARCH 1977)

Bolt Beranek and Newman, ARPANET Completion Report (1981), p. 208 at https://walden-family.com/bbn/arpanet
-completion-report.pdf.

created a 3-mile diameter fireball from which physical debris was blasted 5 miles out by 300 mile-per-hour winds while earthquakes, thermal radiation, and destructive nuclear radiation went out a dozen miles—where the overpressure still kicked up 70 mile-per-hour winds.[44] This was not guesswork. More than a decade earlier, the US Castle Bravo test explosion at 15 megatons ripped a hole in the Pacific's Bikini Atoll, creating a fireball nearly 5 miles wide, leaving a crater a mile wide, and punching a mushroom cloud into the upper levels of the stratosphere.

"Operation Survival" does not make for light reading. Foss and May explained that "buildings and antennas which must survive a nuclear explosion are designed to withstand direct nuclear effects" through metal shielding, physical isolation, heavy reinforced concrete or even underground construction, use of hardened coaxial cables, and beefed-up microwave radio relay towers. Underground buildings contained gas-turbine or diesel generators, with fuel, food, and emergency supplies to last three weeks; massive banks of batteries powered the facilities when they were "buttoned up" during the nuclear blast wave. Aboveground switching stations included similar features. You might think this was all history. But in US cities today there are at least eight identifiable buildings with these nuclear-hardened designs; still in use, they allow us to trace a historical pathway connecting yesterday's phone system to today's internet. Along the way, remarkably enough, centralization worked its way back in to the internet.

As discussed in chapters 7 and 8, the US military sponsored the early internet in large measure to prevent a nuclear attack severing the military's command structure in Washington, DC, from its nuclear-command headquarters outside Omaha, Nebraska. As "the person who signed most of the checks for ARPANET's development," serving as ARPA's deputy then full director, Stephen Lukasik plainly stated: "The goal was to exploit new computer technologies to meet the needs of military command and control against nuclear threats, achieve survivable control of US nuclear forces, and improve military tactical and management decision making."[45] With numerous nodes, many links, and multiple routes connecting the ARPANET sites, the network formed a "mesh" that embodied a resilient and decentralized design (fig. 10.2). There was no "Chicago problem" because there was no central switching site in Chicago or anywhere else. Already by 1977 ARPANET featured four or five major routes across the country and multiple connections between the Pentagon in Washington, DC, and the US Air Force's Global Weather Center (GWC) in Nebraska, nearby the Strategic Air Command's headquarters at Offutt Air Force Base.

Awareness of the military origins of the ARPANET and its successor internet soon faded. "The bulk of the ARPA contractors involved were unaware of, nor did they particularly care, why the DoD was supporting their research," Lukasik observed. The wry catchphrase was "we used to say that our money is only bloody on one side," according to one military-funded computer scientist at the University of Illinois. College campuses during the Vietnam War were wary of appearing on the Pentagon's payroll, and military-funded computing projects were highly visible (fig. 10.3). Soon enough, the military money turned green all over. Today, oddly, it's sometimes dismissed as a "myth" that the US military funded and shaped the early internet. "The project had embodied the most peaceful intentions . . . [and] had nothing to do with supporting or surviving war—never did," according to one (counterfactual) account.[46] In time, notwithstanding Lukasik's testament, an idealistic image took hold that the internet created a new unbounded and open space where individuals might meet, independently of physical location or social condition. At first, there was a jumble of new names: Howard Rheingold's "virtual reality" or "virtual community"

FIG. 10.3. UNIVERSITY OF ILLINOIS PROTEST AGAINST ILLIAC-IV COMPUTER (1970)

Courtesy of the University of Illinois Archives in RS 41/62/15, Box 4. See John Day oral history (2010), Charles Babbage Institute OH 422, pp. 6–16, 25–30 at purl.umn.edu/155070.

or John Perry Barlow's "electronic frontier" or "datasphere." Then came cyberspace.

"Cyberspace" famously passed into popular culture with William Gibson's prize-winning novel *Neuromancer* (1984). Its basic plot anticipated and inspired the blockbuster *Matrix* film trilogy (1999–2003), with its computer-created world that had turned human beings into zombie-slaves of a central machine, in turn challenged by a heroic outsider. "Cyberspace," as Gibson described it, is a "consensual hallucination experienced daily by billions of legitimate operators, in every nation, by children being taught mathematical concepts. . . . A graphic representation of data abstracted from banks of every computer in the human system. Unthinkable complexity. Lines of light ranged in the non-space of the mind, clusters and constellations of data." And, yes, there were a lot of pharmaceuticals in the air. The very next year, counterculture impresarios Stewart Brand and Larry Brilliant organized a pioneering virtual community called "the WELL," for "Whole Earth 'Lectronic Link." Located in San Francisco, the WELL attracted several hundred notables (including computer enthusiasts and journalists ranging from *Rolling Stone* to the *Wall Street Journal*) who articulated the culture of cyberspace. In *The Virtual Community: Homesteading on the Electronic Frontier* (1993), Howard Rheingold hailed its potential for community building and personal fulfillment. The WELL also joined together the founding members of the Electronic Frontier Foundation, or EFF—steadfast digital-freedom and internet-democracy advocates—including the singular John Perry Barlow, programmer-activist John Gilmore, and software entrepreneur Mitch Kapor.[47] Barlow shortly became known as the Thomas Jefferson of cyberspace.

Barlow had an unusual background as a Wyoming rancher, Grateful Dead lyricist, global traveler, sometime troublemaker, and internet freedom activist. He gained stature in computing after co-founding EFF in 1990. "Our financial, legal, and even physical lives are increasingly dependent on realities of which we have only the dimmest awareness," Barlow wrote in *Communications of the ACM*, a leading computer-science magazine.[48] In 1996 while he was observing the global elite's annual World Economic Forum in Davos, Switzerland—at a time when the World Wide Web had just come into its own—Barlow penned what must be the internet's most striking pronouncement. Over the years it's cast a long shadow. "A Declaration of the Independence of Cyberspace" begins, portentously: "Governments of the Industrial World, you weary giants of flesh and steel, I come from Cyberspace, the new home of Mind. On behalf of the future, I ask you of the past to leave us alone. You are not welcome among us. You have no sovereignty where we gather."[49]

Initially, Barlow's declaration gained traction as prophetic political philosophy. Barlow fully admitted he had "no greater authority than that with which liberty itself always speaks" (echoing Thomas Jefferson) but then quickly asserted that cyberspace was "naturally independent of the tyrannies you [weary giants of flesh and steel] seek to impose on us. You have no moral right to rule us nor do you possess any methods of enforcement we have true reason to fear." In his formulation, cyberspace was in the active process of forming its own "social contract" (political philosophy again) and building a new world through "our great and gathering conversation . . . the wealth of our marketplaces . . . our culture, our ethics, [and] the unwritten codes that already provide our society more order than could be obtained by any of your impositions." Cyberspace was creating a "world that all may enter without privilege or prejudice accorded by race, economic power, military force, or station of birth" (surely idealistic when a capable computer then cost a month's average salary).[50]

The key for Barlow was that cyberspace was not physical. "Your legal concepts of property, expression, identity, movement, and context do not apply to us. They are all based on matter, and there is no matter here," Barlow asserted. "Our identities have no bodies, so, unlike you, we cannot obtain order by physical coercion. We believe that from ethics, enlightened self-interest, and the commonweal, our governance will emerge." One immediate target was the Clinton–Gore Telecommunications Act of 1996, which Barlow contended "repudiates your own Constitution and insults the dreams of Jefferson, Washington, Mill, Madison, DeToqueville, and Brandeis." The act's title V, section 230, is as of this writing in 2020 in the news for permitting social media like Twitter and Facebook to avoid responsibility for the content of inflammatory, incendiary, and even insurrectionary postings. Beyond doubt, the internet massively benefitted from the view that cyberspace stood apart from traditional society; for years, e-commerce largely escaped even basic sales taxes (see chapter 8). The 1990s were filled with equally idealistic enthusiasm for "open source" software championed by Richard Stallman and others.[51]

Over time, Barlow's confident assertion that cyberspace would drive political change resonated ever more widely. He grouped together a number of diverse countries and charged, defiantly, "In China, Germany, France, Russia, Singapore, Italy and the United States, you are trying to ward off the virus of liberty by erecting guard posts at the frontiers of Cyberspace. These may keep out the contagion for a small time, but they will not work in a world that will soon be blanketed in bit-bearing media."[52] For Barlow, the virus of online liberty inexorably spread the norms of cyberspace (his death in 2018 preceded the coronavirus). The idea spread far and wide that cyberspace and the internet fostered decentralized and democratic political societies, whereas the clock was running out on authoritarian regimes.

Such an appealing vision of "internet freedom" gained force and prominence as a key tenet of US domestic and foreign policy. The Clinton–Gore administration vigorously promoted an "information infrastructure . . . its lifeline a high-speed fiber-optic network capable of transmitting billions of bits of information in a second." After the untold billions of dollars that the United States spent during the Cold War (see chapter 7) on high-tech military hardware, targeted economic development, and directed aid to client countries, the idea of information highways at home and internet freedom abroad was surely attractive. A simplistic and mistaken notion that the West had "won" the Cold War owing to the Soviet Bloc's inability to resist the flood of Western radio and television added fuel to the fire. All that was needed to bring about the end of autocratic regimes, wherever they were, was persistent encouragement of internet access—conveniently, more computers, better networks, greater accessibility. Accordingly US foreign policy aligned with the tsunami of the global internet. In 2000 President Bill Clinton gauged China's emerging attempts to contain online discourse. "Good luck," he cracked: "That's sort of like trying to nail Jello to the wall."[53]

Secretary of State Hillary Clinton beginning in 2008 took steps that made internet freedom central to the policy goals of the Barack Obama administration. In launching this policy initiative, assistant secretary Michael Posner testified to Congress on the technical means, internet tools, and organizational training to bring about "unfettered, safe access to information and communications for more people in more places . . . Internet freedom [is] a key component of our foreign policy." Clinton herself in January 2010 offered a bold and unbounded vision: "We want to put these [online] tools in the hands of people who will use them to advance democracy and human rights, to fight climate change and epidemics, to build global support for President Obama's goal of a world without nuclear weapons, to encourage sustainable economic development that lifts the people at the bottom up." Human rights advocates praised her tough stance against repressive regimes in Saudi Arabia, China, and Egypt and in support of "companies that stand up for human rights."[54]

Silicon Valley tech executives, many of whom had supported Clinton–Gore's earlier "information highway," swung behind the State Department's internet freedom. Microsoft quickly affirmed that the Global Network Initiative, co-founded along with Yahoo and Google (and lauded by Human Rights Watch), was "dedicated to advancing Internet freedom, along with other leaders from industry, human rights organizations, academics, and socially responsible investors." Within a month of Clinton's speech, an impressive roster of tech leaders—EBay CEO John Donahoe, Twitter co-founder Jack Dorsey, Mozilla Foundation chair Mitchell Baker, Cisco Systems CTO Padmasree Warrior, and venture capital maven Shervin Pishevar—accompanied a high-level State Department delegation to Russia. The world beckoned, as one staffer suggested: "The U.S. can open doors to other countries and cultures through its technology sector that produces many of the tools that young people around the world use to connect with one another."[55]

The Arab Spring political uprisings in 2011 appeared to confirm the vision of internet freedom. Using wireless mesh networks, bootlegged internet access, identity-scrambling relays, and social media tools, pro-democracy activists quickly toppled repressive governments in Tunisia, Egypt, Libya, and Yemen and shook the complacent roots of a dozen others. In his moving firsthand account, Wael Ghonim made the case that "the power of the people is greater than the people in power." A Facebook page he co-created as a memorial to a victim of Egypt's brutal police state gained 3 million followers and served to coordinate massive public demonstrations against Egypt's dictator Hosni Mubarak. Ghonim took leave from his job at Google in Dubai to return to Egypt, where he was jailed for eleven days and then reappeared as an acclaimed public figure on television. Ghonim's brave stances embodied the best of internet freedom.

Unfortunately, subsequent events in the Middle East inverted Ghonim's idealistic slogan. An early sign was the shutdown on 27 January 2011, of the Egyptian internet, which shocked internet enthusiasts everywhere; in the midst of pitched street battles, Egyptian authorities simply directed five of its internet providers to switch off (only the one serving the Egyptian stock exchange remained on). The shutdown took 13 minutes. Mobile phone giant Vodafone, based in London, had 28 million subscribers in Egypt, but it, too, like "all mobile operators in Egypt," was "instructed to suspend services [and] obliged to comply" with the order. Egypt "didn't just take down a certain domain name or block a website—they took the whole Internet down," admitted one western expert. Subsequently, over the years, such countries as Bangladesh, Congo, Ethiopia, India, Iraq, Sudan, Venezuela, and numerous others have done likewise during times of political stress.[56] The Arab Spring uprisings in 2011 effectively removed corrupt authoritarian regimes from power, but within a few years anti-democracy military crackdowns, contrived counterprotests, and unrestrained armed militias returned corrupt authoritarians to power in Egypt, provoked civil wars in Yemen, Libya, and Iraq, and strengthened the vicious dictatorship in Syria. Egypt's leader as of this writing in 2020 is none other than Mubarak's former director of military intelligence, the UK- and US-trained Abdel Fattah el-Sisi who came to power through a military coup in 2013. The unhappy result of entrenched authoritarianism, fractious religious turmoil, and economic stagnation is sometimes labeled as "Arab Winter."[57]

Egypt's state-mandated shutdown in January 2011 reminds us that the internet is far more than the non-physical "cyberspace" hailed by John Perry Barlow. The internet always relies on some physical means to connect computers, and it so happened that the physical internet was reconstructed during these years. The early "mesh" internet, with its many nodes, links, and paths was rebuilt with fiber-optic lines forming a "tube" internet with a significant measure of centralization, as described below in some detail. Whereas the "mesh" internet was impossible to surveil using centralized techniques, today's internet actually facilitates centralized surveillance, as Edward Snowden's disclosures in 2013 brought to public attention. Snowden pointed to the NSA's code-named Fairview program that identified eight buildings AT&T constructed to harden the centralized phone system—in Chicago, New York, San Francisco, and five other US cities—and that are in use today as centralized nodes in the extensively surveilled internet (fig. 10.4).[58]

FIG. 10.4. AT&T BUILDING AT 811 10TH AVENUE (NEW YORK CITY)

This AT&T building was built from concrete and granite in 1964 as a "hardened telco data center" with 21 floors, "each one of them windowless and built to resist a nuclear blast." In 2000 it became an internet data center, one of eight US locations where AT&T facilitated NSA surveillance.

Courtesy Rob Brokkeli by CC BY-SA 4.0.

Today's global internet, writ large, can be described as a cooperating network of "autonomous systems," numbering 101,608 in May 2021. The internet has no government, central or otherwise, but it does have governance processes that help keep the autonomous systems working together. And, despite Barlow's idea that cyberspace had no physical manifestation—and the equally misleading image of the disembodied "cloud"—the internet has a physical reality: fiber-optic cables; server farms, data centers, and switching sites; and high-volume internet exchanges. Many of these appear on physical maps. Chicago has four internet exchanges—at 350 East Cermak Road, 600 South Federal Street, 427 South LaSalle Street, and one undisclosed address.[59]

Governments powerfully shape the internet's autonomous systems within their legal and regulatory domains. The US government's "four principles" and "network neutrality" contrasts with Europe's mandate for extensive local-retail competition for internet service. "Europe needs to have a choice [confronting the US internet giants] and pursue the digital transformation in its own way," insists the European Commission.[60] Increasingly, Microsoft, Facebook, and Google are building parallel private internets to circumvent the global public internet. The physicality of these varied connections, including software-imposed restrictions, shapes access to information. Egypt's internet in 2011, with five connections to the world, was easy to sever from the global internet. For years, mainland China had *three* outward connections, making it an easy matter to block Chinese internet users' access to the *New York Times*, Facebook, Google, or any other site that authorities in China deemed troublesome. To understand how "internet freedom" has changed, we need to understand the physical character of the internet, how it evolved in the early 2000s, and today how internet-repressive countries, hard to believe, successfully "nail Jello to the wall."

The internet's physical connections were initially phone lines. AT&T built special high-capacity "leased lines" that many large organizations used to directly connect their far-flung offices. The US Defense Department at first and then the National Science Foundation simply contracted such leased lines to connect the 100 or so nodes of the ARPANET (1969–90) and later create the high-speed internet backbone of the NSFNET (1985–95). Through the NSFNET years, the internet featured many nodes, many links, and multiple routes: it formed a "mesh" network with no center. Its decentralized structure sustained Barlow's vision. The internet was effectively privatized in 1995 when the NSF turned over its backbone to several phone companies; four "network access points" were established in New Jersey, Chicago, San Jose, and Washington, DC (note these locations), to transfer internet traffic between the existing dozen or so regional internets. Privatization allowed commercial activities to flood onto the internet, paving the way to the "dot-com boom" in the late 1990s.

Also in the late 1990s a "telecom boom" brought about changes that transformed the internet from surveillance-adverse "mesh" to surveillance-friendly "tube." The fiber-optic building blitz took place well after the AT&T Corporation was broken up in 1984 to create seven regional phone companies known as the "Baby Bells." New long-distance phone companies such as MCI and Sprint also rushed in to sell long-distance phone service and, soon, build fiber-optic cables. The new telecommunications companies, inspired by the enticing provisions of the US Telecommunications Act of 1996, invested $500 billion and constructed 20 million miles of fiber-optic cables in short order, a frenetic rush to the future. Perhaps 2 percent of this overbuilt fiber-optic network was actually needed in North America. Such modest traffic did not cover the hefty interest payments due on the debt-fueled fiber-optic extravaganza. Some companies like Sprint and CenturyLink flour-

TABLE 10.1. TIER 1 INTERNET PROVIDERS, 2020

COMPANY	HEADQUARTERS	CONE SIZE
Level 3	US	45,771
Telia Co.	Sweden	34,813
Cogent Communications	US	29,747
NTT America, Inc.	US	20,998
GTT Communications	Spain	20,668
Telecom Italia S.p.A.	Italy	17,534
Hurricane Electric	US	15,095
Tata Communications	US/India	14,593
Zayo Bandwidth	US	12,194
PCCW Global	US/Hong Kong	9,931
Vodafone Group PLC	Europe	8,139
RETN Limited	UK	6,770

Note: Cone size is the total number of direct and indirect customers, or roughly total "autonomous systems."

Source: CAIDA ranking (September 2020) at asrank.caida.org.

ished, while the pieces of AT&T remained prominent players, but scandal-plagued WorldCom, burdened by more than $30 billion in hastily contracted debt, spectacularly went bankrupt by 2002 as did nearly 100 smaller concerns.[61] Someone had to pay the fiber-optic bills. In 2005 Edward Whitacre, CEO of AT&T stated the problem: "What [Google, Yahoo, Vonage, and other content providers] would like to do is to use my pipes free. But I ain't going to let them do that. . . . Why should they be allowed to use my pipes? . . . We and the cable companies have made an investment [in fiber optics] and for . . . anybody to expect to use these pipes [for] free is nuts!"[62]

Today's internet emerged from the ashes of this telecom crash. The time when the United States dominated the internet is long gone. A series of corporate consolidations resulted in a mostly stable system of a dozen "Tier 1" internet providers around the world. Each of these companies can reach the entire global internet using its own fiber-optic network plus non-paying, or "peering," relationships.[63] Table 10.1 lists the twelve largest in 2020: So number one, Level 3—the product of a 2017 merger with CenturyLink, itself a successor to one of the Baby Bells—can reach all parts of the global internet through peering agreements with, say, number two, Telia in Sweden, or number six, Telecom Italia or other peers. The system is not completely stable since number three, Cogent Communications, over the years intentionally provoked disruptive "peering wars" to muscle in on the top rank; today Cogent has peering relations with many of the top twenty, apart from Hurricane, RETN, and two others. Around number eleven begins a second tier of internet companies, such as Comcast, Vodafone, and Verizon, which must pay for access to some parts of the global internet. It's the case that number ten, PCCW, has a peering relationship with Level 3, but lower-ranking Vodafone, RETN, and Telstra must pay Level 3 to carry their traffic. Comcast, the US cable giant, pays Level 3, Cogent, Tata, and Singapore Telecommunications for global internet access, while Hurricane depends on Telia for links to some systems even as, long after the "peering wars," it has no peering relations with highly ranked Cogent. China's top carrier number fifty-one, China Telecom, pays Level 3 and eleven others for access to the global internet, peering with Hurricane Electric, Singapore Telecommunications, and several other medium-size carriers; its largest-paying customer is Angola Cables. In reality, it's a miracle the internet works at all.

You might imagine some central body, somewhere, exerting orderly supervision over the internet. But there is none. Governance for the internet, such as it is, had its origin in the personal relationships of the early ARPANET. For years, one person— Jon Postel at the University of Southern California—managed the entire addressing, naming, and numbering apparatus through the Internet Assigned Numbers Authority (IANA) as well as wrote or edited many of its highly technical standards. "God, at least in the West, is often represented as a man with a flowing beard and sandals," quipped the *Economist* magazine in 1997, a year before his death: "if the [internet] does have a god, he is probably Jon Postel."[64] On Postel's passing, IANA formally became part of the nonprofit Internet Corporation for Assigned Names and Numbers (ICANN), a directly contracted entity of the US Department of Commerce (1998–2016). ICANN sets internet standards through its engineering and research task forces and is as close to a governing body the internet has. After protracted international controversy that the US government unduly controlled the internet, ICANN was reconstructed in 2016 as a "global multi-stakeholder community" (its government advisory committee has 112 nation-state members) and remains so today.[65]

The character of the "tube" internet depended on fiber-optic cables, private telecommunication companies, and one crucial political event: the terrorist attack on 11 September 2001. In its wake, the United States granted unusually broad surveillance powers to its intelligence agencies—and as Snowden made public—effectively also to the intelligence agencies of the United Kingdom, Australia, Canada, and New Zealand. With the decentralized "mesh" internet, there was no place to centrally surveil internet traffic. With the centralized "tube" internet, it is abundantly clear where surveillance can be effected. Quietly, the National Security Agency told internet operators large and small that they were required to install and connect the NSA's "splitter cabinets" (you can build a primitive one yourself with a laser pointer and glass prism). In Utah, Pete Ashdown's small ISP received a federal order that led to "a duplicate port to tap in to monitor that customer's traffic . . . we ran a mirrored ethernet port" to comply. AT&T reused eight of its Cold War vintage phone buildings as central switching sites for fiber-optic cables. For AT&T, a perfect match exists between the Cold War era buildings (see fig. 10.4) and the physical sites of NSA's surveillance. According to AT&T, it activated "Service Node Routing Complexes," or SNRCs, during the fiber-optic building boom; the NSA indicates it accordingly tapped into "peering circuits at the eight SNRCs."[66]

Routings at AT&T were reconstructed to make the internet more centralized and consequently to facilitate surveillance. Around 2005 an AT&T service technician was directed to connect "every internet backbone circuit I had in northern California" through the AT&T building at 611 Folsom Street in San Francisco. "There was nothing wrong with the services, no facility problems," he stated. "We were getting orders to move backbones . . . and it just grabbed me," he said. "We thought it was government stuff and that they were being intrusive. We thought we were routing our circuits so that they could grab all the data." San Francisco's AT&T building was connected to other top-tier companies, such as Sprint and the United Kingdom's Cable and Wireless as well as to MAE West in Silicon Valley, one of the six largest internet interchanges in the United States. AT&T's seven other Cold War vintage buildings were likewise made into central switching sites for the internet. Verizon, under the code name Stormbrew, participated in the same NSA effort.[67]

While AT&T's and Verizon's cooperation with NSA's surveillance is well documented, the internet exchanges where diverse private networks in the United States and around the world physically meet and swap traffic remain somewhat a matter of tight-lipped "national security." One of the largest is Network Access Point (NAP) of the Americas, a six-story building of 750,000 square feet located at 50 NE Ninth Street in downtown Miami, Florida. It was originally built by a local firm in 2001 and bought by Verizon in 2011. With seven-inch thick, steel-reinforced concrete exterior panels, the purpose-built facility offered "managed hosting, security, disaster recovery, colocation, peering, and cloud computing services" to 160 network carriers, including the US State Department. Global internet goliath Equinix bought the facility in 2017. "The entire . . . third floor of the facility is dedicated to Federal

FIG. 10.5. NATIONAL SECURITY AGENCY'S UTAH DATA CENTER (2013)

NSA at nsa.gov1.info/utah-data-center/photos/magnify-udc-june.jpg.

government users, many of whom operate in SCIF'ed areas [that is, top secret Sensitive Compartmented Information Facility]; access to this floor is restricted to US citizens and requires a government clearance."[68] (In 2018 Telegeography ranked Miami as the number seven internet exchange in the world, the highest in North America.) For its New York internet exchange, Equinix operates eight physical locations, including the Old Macy's Building at 165 Halsey Street in Newark, New Jersey. Equinix also operates internet exchanges in Atlanta, Chicago, Dallas, Denver, Houston, Los Angeles, San Francisco, Seattle, Toronto, and Washington, DC, as well as in Brazil, Colombia, Australia, Japan, Hong Kong, South Korea, Singapore, and across Europe. (So we know *where* the internet is surveilled.) A smaller internet exchange is in Minneapolis at 511 Eleventh Avenue South, where the Midwest Internet Cooperative Exchange, or MICE, opened in 2011 to circumvent a latter-day Chicago problem: before MICE, "all traffic flowing between two local Minnesota [internet] service providers [had] to route through Chicago." Worldwide there are 300 internet exchanges.[69]

Clearly, today's internet has been reconstructed through fiber-optic cables, internet exchanges, and insistent US federal government pressure. When Pete Ashdown's internet service in Utah installed court-ordered surveillance technology, there was little question where the collected traffic was headed: the National Security Agency's new Utah Data Center (fig. 10.5). With the Snowden leaks, many journalists have connected the Bluffdale, Utah, center to the NSA's controversial internet data-gathering and analysis practices known under the code name PRISM. Much of the controversy stems from whether NSA is properly operating under the Foreign Intelligence Surveillance Act (amended in 2008) or whether, as some critics charge, that the NSA is gathering up domestic internet traffic that has little to do with foreign intelligence.

NSA has come a long way from its secretive nickname as "No Such Agency." NSA openly discloses the size, capacity, and purpose of the Utah Data Center and, remarkably enough, provides public access to "numerous Top Secret documents . . . leaked to the media" with the aim of providing "ongoing and direct access to these unauthorized [sic] leaks." The NSA explains the center cost $1.5 billion and has one million square feet in twenty separate buildings. Sixty diesel generators can provide full electrical power for three days. The core computing muscle is provided by a massive Cray XC30 supercomputer with research underway for further wonders. "Since the capability to break the AES-256 encryption key [widely used to encrypt electronic data] within an actionable time period may still be decades away, our Utah facility is sized to store all encrypted (and thereby suspicious) data for safekeeping." There is also explicit description of its controversial "domestic surveillance data . . . [processed] and [stored] in the Utah Data Center." The sign out front proclaims: "If you have nothing to hide, you have nothing to fear."[70]

A clear line between a democratic internet and an authoritarian internet can no longer be drawn. Examples abound. For the 1980 Olympic Games in Moscow, the Soviet government built an immense phone complex with 1,600 new direct-dialed lines that were directly surveilled by the Soviet secret police; today, the state-owned M9 complex is "the heart of the Russian Internet" and the center of the MSK internet exchange, through which "nearly half of Russia's internet traffic passes" with Google renting one entire floor and the post-Soviet secret police situated in another, with its surveillance boxes located throughout the complex. Even the architecture of the phone complex—only 12 of its 19 floors have windows—uncannily echoes those hardened AT&T buildings.[71] It was not supposed to be this way. For decades, internet freedom advocates promised that an authoritarian state like Russia or China could either get access to the internet or retain its authoritarian government—but not both. John Perry Barlow's cyberspace idealism and Hillary Clinton's internet-freedom diplomacy held that perpetual censorship of the internet was impossible, like "nailing Jello to a wall." Internet freedom offered a version of technological determinism: the internet would slay authoritarian states and bring about democracy worldwide. But what actually happened in China entirely upends this notion. Today's China has 900 million internet users, around two-thirds of its population, and roughly three times the total number of US internet users. China's enormous e-commerce sector surpassed the United States' as well as Europe's in 2017, and its largest internet companies include giants like Alibaba, Baidu, JD, Tencent, and Weibo that are traded on global stock exchanges. Essentially, China has built an impressive online economy independent from foreign technology. Its short-form video platform TikTok and mobile-phone company Huawei are in 2020 much in the news as China's tech prowess confronts the West. Remarkably, so far, China has not yielded an inch of its authoritarian government.

Two biographies can illuminate how China nailed Jello to the wall. Robin Li built China's leading internet search engine, Baidu, and subsequently supported numerous tech ventures in China; while Lu Wei and his colleagues worked the government side to construct a durable system of internet censorship that persists today.

Everyone knows three things about the world's leading search engine: that it identifies high-quality websites by ranking incoming links; that its founders did an impressive tour of duty in Silicon Valley, studying computer science but doing a start-up company instead of PhD theses; and that advertising-powered search made them billionaires. Google's Larry Page and Sergey Brin are of course well known. This success story also aptly describes Baidu's Robin (Yanhong) Li, who was born in China, did a bachelor's at Beijing University, and went to the University of Buffalo in New York for graduate work in computer science. Instead of a PhD Li explored large-scale indexing and searching while working for the parent company of the *Wall Street Journal*. In February 1997 he filed a US patent for "Hypertext Document Retrieval System and Method" (5,920,859) that used what he termed "link analysis" to create search results. Li named his invention Rankdex. He soon took his ideas to Silicon Valley–based Infoseek, one of the earliest search engines, where Li also developed a means for searching images. When Larry Page later filed US patents for his similar invention PageRank, he properly referenced Li's earlier patent. Li's work is cited in 320 patents and disclosures around the world.[72]

Li's concepts for Rankdex formed the technical basis for Baidu, which Li founded in January 2000 with business partner Eric Xu. Operating initially in low-rent office space in Beijing, near the founders' alma mater, and funded by Silicon Valley venture capital, Baidu soon developed a string of innovations, including image search and an auction-based and pay-per-click advertising model—again preceding Google's money-spinning model. With revenues from ad-powered search pouring in, Baidu in 2007 landed on the NASDAQ's list of 100 top tech companies. After Google left China in 2010, Baidu achieved roughly 75 percent search-market share in China. Baidu, like Google, is really an internet-innovation platform. Baidu offers an extensive roster of services much like Google's: maps, news, social networking, cloud, finance, encyclopedia, music and entertainment as well as image, text, and video searching. It indexes 2.5 million Chinese patents. Also like Google, Baidu invested heavily in artificial intelligence, or AI, hiring star researcher Andrew Ng from the Silicon Valley nexus of Stanford University and Google Brain. Best known for co-founding Coursera, the online educational venture, Ng built at Baidu a highly accomplished AI research team of 1,300 people working on facial recognition, conversational bots, deep learning, and other next-generation projects. On Twitter Ng even reproved Microsoft's boast of human-level speech recognition: "We had surpassed human-level Chinese recognition [a year earlier]; happy to see Microsoft also get there for English." And, needless to say, Baidu is working full steam on AI-powered self-driving autonomous vehicles.[73]

Everything about internet entrepreneur Robin Li seems normal ("we always don't want anything regulated so we can do things freely," he states) until you learn that in 2013 he joined the Chinese People's Political Consultative Conference, an arm of the ruling Chinese Communist Party. "Most private entrepreneurs have learnt that the party controls everything," notes one veteran China watcher. Besides party officials Li meets many tech counterparts in the 2,000-member advisory body, including the founders, CEOs, or top partners from e-commerce titan JD, the country's largest social network and media colossus Tencent, its second-largest game maker NetEase, cybersecurity powerhouse Qihoo 360, and, not least, Sequoia Capital China. The latter, a spin-off from the venerable Silicon Valley venture capital firm, backed e-commerce giants Alibaba and JD, autonomous vehicle maker NIO, cybersecurity specialist Qihoo 360, and 70 additional lucrative "exits" as its VC-financed companies went public. Even Jack Ma, Alibaba's co-founder and China's richest man, is a member of the Communist Party. As membership dues Ma might pay 2 percent of his income.[74]

The Arab Spring protests of 2011 prompted China to expand and strengthen its control over the internet, especially after the rise to power of authoritarian Xi Jinping (2012–present). The People's Republic of China, politically communist since 1949, officially encouraged capitalist economic practices first in 1978, then, after the unsettled post–Tiananmen Square years, a vibrant export-oriented capitalist sector blossomed in the 1990s. China sent billions of socks, bras, sweaters, and neckties to Walmart and other mass retailers in the West, while joining the World Trade Organization in 2001 and the World Semiconductor Council in 2006 (noted earlier).

China experienced a great transformation in telecommunications, too, expanding service from 2 phones per 1,000 residents in 1980 to 660 per thousand in 2007, although with a steep rural-urban divide; in that latter year it already had 162 million internet users. China's early internet was built largely with hardware from Cisco Systems (later supplemented by Huawei and other Chinese companies.) Wary of further US domination, China initially supported the United Nations' annual Internet Governance Forum (2006–) and then itself convened an annual World Internet Conference (2014–) outside Shanghai. China bluntly aims at "internet sovereignty" in sharp contrast with Western aspirations for internet freedom and unrestricted global connections. "Within Chinese territory, the internet is under the jurisdiction of Chinese sovereignty," stated a 2010 government report: "The internet sovereignty of China should be respected and protected."[75]

China aims to massively expand the economic benefits of the internet, as noted previously with e-commerce, while actively restraining troublesome domestic political consequences. In 2010 China announced its ban on any internet activities "subverting state power, undermining national unity, [or] infringing upon national honor and interests." The task of creating a system of internet monitoring and censorship fell to the Cyberspace Administration of China under the technical leadership of Fang Binxing (starting as early as 2000), chief censor Lu Wei (2012–16), and their latter-day successors. Sometimes known as the "Great Firewall" of China, the censorship system they developed is actually rather flexible and adaptive, involving extensive hands-on human monitoring in addition to invasive technical filtering, active IP blocking, selective DNS attacks, and other measures. The goal was to deny the internet's ability to sustain user-anonymity and wide-ranging free speech that might create an online public sphere. "Chinese companies hired small armies of censors to flag, remove, and report sensitive comments in real time," notes one account. And, if you are wondering, China's internet users were paying attention: at Wuhan University in May 2011 protesters threw eggs and shoes at Fang Binxing, who was shamed as "a running dog for the government" and "enemy of netizens." China's censors cracked down even harder, seeking also to counter Hillary Clinton's escalating internet-freedom campaign. The government directed China's internet companies to hire censorship managers, while installing its own cybersecurity police at the larger tech companies.[76] The result, according to a former censor at Sina Weibo (one of China's largest social media companies) is that the government "rules Weibo with ease." Tencent's super-popular WeChat app offers texting, voice memos, video conferencing, photo sharing, online payments, and much else to its *billion* users, all under close watch of the censorship regime. "Many Chinese Internet users know they are not free online, but they accept this," stated one Chinese citizen in 2015. "Online games and myriad social media platforms keep everyone busy. We can make restaurant reservations and shop all we want. Only a small number of people sense what they are lacking."[77]

The structure of China's connection to the global internet creates a division between its domestic internet with 900 million surveilled users and its unusually thin connections beyond. Remarkably, like Egypt, China has quite limited and highly centralized connections to the global internet. Mainland China has just three connections, led by its capital city Beijing (ranked 42 in global cities), economic powerhouse Shanghai (49), and industry-center Guangzhou (unranked). The island-peninsula of Hong Kong (10)—a British possession from the opium war of 1842 (see chapter 4) through 1997—is formally a special administrative region of China (under "one country, two systems" it retains a capitalist economy), but crackdowns against its democratically elected government have intensified since 2019. It's not difficult to imagine mainland China's anxiety over Hong Kong and its five major global internet exchange points. Besides these four cities, China has just one other internet connection via the Taiwan Straits to what it terms the "renegade province" of Taiwan.[78]

For a time, China's internet users could reach the global internet via encryption-protected (hence, anonymous) VPNs, or virtual private networks, but the government's suppression of VPNs starting in 2017 cut that back. "Apple complied with a Chinese government order to remove VPNs from its Chinese iOS AppStore, and the company that runs Amazon's cloud services in China . . . said it would no longer support VPN use." Early in 2020 China further restricted VPN access to curtail "sharing information about coronavirus." During times of such political stress, "The authorities throttle VPN usage, rendering use of the foreign Internet near impossible." A cat-and-mouse game continues. For two weeks in fall 2020 an Android-based browser allowed Chinese web users—who submitted their national ID numbers to cybersecurity giant Qihoo 360—to view censored versions of otherwise-blocked content such as YouTube and Twitter, but the browser was suppressed in mid-October just one day after it went viral.[79]

An inadvertent leak lately exposed China's prototype police state—combining facial recognition, biometric data including fingerprints and retinal scans, real-time GPS locations, and pervasive online monitoring (having WhatsApp or Skype on your phone is deemed "subversive behavior")—that was live tracking 2.6 million residents in Xinjiang province, home to a Turkic-speaking Muslim minority known as Uyghurs. Since 2017 an estimated one million (in a Uyghur population totaling just 12 million) have been "arbitrarily detained" and sent to re-education camps, work camps, or prison. Networked internet technology is central to the government's systematic repression: the province's "Integrated Joint Operations Platform (IJOP) . . . uses machine-learning systems, information from cameras, smartphones, financial and family-planning records and even unusual electricity use to generate lists of suspects for detention." China's "persecution of the Uyghurs is a crime against humanity," writes the usually staid *Economist* magazine.[80] Other countries reported to have investigated or installed China's system of internet censorship include such authoritarian wonders or wannabes as Belarus, Cuba, Ethiopia, Iran, Malaysia, Russia, Saudi Arabia, Tanzania, Turkey, Thailand, Uganda, Vietnam, and Zimbabwe.[81]

It's uncomfortable to recall that Western companies like Cisco Systems helped build an authoritarian internet in China, while for years Microsoft, Google, Facebook and others readily self-censored their offerings in China or even turned over to the government the names of account holders, landing them in jail in China. Remarkably, in 2019 a US nonprofit founded by Google and IBM executives teamed up with a "big data" company based in Shenzhen to "drive innovation" on high-tech mass surveillance. The consortium's Aegis system permits China's censors to see "the connections of everyone [including cell-phone] location information," to block forbidden content, and to scoop up real-time surveillance data on 200 million residents. AI researchers at MIT; software behemoth Microsoft; chipmakers Intel, Nvidia, Xilinx, and AMD; and deep-learning specialist Amax are other US tech ventures reported to assist in China's high-tech surveillance. "It's disturbing to see that China has successfully recruited Western companies and researchers to assist them in their information control efforts," noted one US senator.[82]

And, alas, not only in China. Founded in 2003 Palantir Technologies developed "big data" analytics for the US National Security Agency, the Central Intelligence Agency, and the Federal Bureau of Investigation for a decade, expanding with funding from the CIA's venture-capital branch. Its data integration software now helps the Immigration and Customs Enforcement agency (ICE) to identify, locate, arrest, and deport migrant workers in the United States, separating families and drawing scathing criticism from human rights advocates. As its CEO freely concedes, "Every technology is dangerous . . . including ours."[83] No surprise really: authoritarian state funding means authoritarian applications of facial recognition, location tracking, machine learning, "big data," and other internet-based surveillance.

It's preposterous of course to suggest that an "era" comes neatly to a close in any single year; likely, no one shut off your laptop or tablet or smartphone at the stroke of midnight anytime in 2016. And yet, the deluge of government-directed authoritarian assaults on the internet in 2016 may be startling. According to the Centre for Internet and Society, there were 56 multiday internet shutdowns in that year costing an estimated $2.4 billion. India itself shut down the unsettled northern state of Jammu and Kashmir for an unprecedented 133 days. Besides India, there were similar internet shutdowns in Brazil, Egypt, Ethiopia, Gambia, Iraq, Pakistan, Syria, Turkey, and Uganda, often during periods of political stress or popular unrest. Ethiopia said its internet shutdown was to prevent university students from cheating on national exams; Syria and Iraq gave similar excuses. Turkey shut off internet in the midst of an attempted military coup. And it's not getting better. In the same troubled northern region, India in 2017 imposed 145 days of no internet, phone, mobile, or SMS service and a further 213 days of no service extending from August 2019 to March 2020. Trenchant critiques such as Shoshana Zuboff's *Age of Surveillance Capitalism* (2019) point out the additional dangers of private companies like Google and Microsoft treating personal information as a commodity like coal or iron ore. The idealistic vision of an independent cyberspace and boundless internet freedom connecting the globe has ended.[84]

The eras in this book represent complex historical constructions. The eras emerge from constellations of political, social, or cultural power and often are potent means for entrenching those constellations for decades. Technologists during a dominant era readily know what the key technical problems are and which ones are likely to receive support and funding. During the era of industry, technologists worked on improving the production of cotton, coal, iron, steel, porter beer, and other factory-scale goods. In the subsequent era of empire (1840–1914) these industrial technologies did not disappear and yet many inventors and engineers shifted to the new problems posed by empire, including ocean and land transportation, long-distance communication, and deployment of military power. Eras create structures that channel the development of technology, but they do not last forever.

The Dominance of the Digital was for years sustained by enthusiasm for globalization and tolerance of neoliberal capitalism. In turn, the enterprising digital companies and eye-catching digital technologies furthered the economic aims of globalizers and brought eye-popping wealth to the most-successful companies and countries. It's difficult to recall that China was a poor country a generation ago; globalization brought wealth to China beginning impressively in the 1990s with its successful export-oriented industries. A leading example was assembling consumer electronics, and the largest such company in China was and remains Taiwan-headquartered Hon Hai Precision Industry Co., Ltd., better known as Foxconn. In 2016 Foxconn had a dozen factory complexes in mainland China, with the largest in south-coastal Shenzhen employing 350,000 workers. This is where Apple iPhones come from. Foxconn also operates huge assembly sites in Brazil, Japan, India, Malaysia, Mexico, and across eastern Europe. In January 2020 Foxconn shut its factories in China as the coronavirus spread across China from its probable origin in Wuhan to the north; remarkably, by mid-May, the company reported its factories were back to normal and its production of iPhones (as well as electronics for Dell, Microsoft, and Sony) returned to anticipated levels.[85] Wuhan itself received special permission for its local Yangtze Memory Technologies, founded in 2016 as one of China's high-profile chip manufacturers, to remain in operation (with strict quarantining). Also in Wuhan numerous optical electronics and laser technology ventures, including China's largest producer of fiber-optic cables, form a concentrated high-tech region known as "Optics Valley." But Wuhan Semiconductor, a spin-off from TSMC (its parent company's CEO was TSMC's COO), has experienced severe difficulties in opening its own massive chip-manufacturing plant even before coronavirus restrictions.[86]

Even if some globalization slogans were too simple, such as MTV's "one world, one image, one channel," enthusiasm for globalization permeated the international semiconductor roadmapping exercises that, for a time, enabled and entrenched Moore's Law. Semiconductor engineers preferred international to national roadmapping, embraced "global technology and industry" and "worldwide consensus," and declared "electrons don't know cultural boundaries."[87] And the internet, as described in chapter 8, intentionally embodied "global" values for years. That era has passed. Perhaps if—earlier in the prime of globalization—Foxconn had promised the state of Wisconsin to build a new television factory there employing 13,000 workers, it might have followed through with its 2017 announcement; as it is, Foxconn has slashed its commitment to, at best, a few hundred engineers in Wisconsin. As of October 2020, owing to the company's pullback, the state's promised $4 billion in tax credits is on ice.[88] Wisconsin workers will not make televisions for Foxconn. Globalization died a second death in 2016 with the upsurge of nationalism, reckless trade wars, inhumane immigration and refugee restrictions, and the profound cultural fragmentation that followed. Similarly, the dominance of the digital is a far cry from its glory years not so long ago. As with the era of systemic risk, there are plenty of technology challenges facing the world today, with climate change spiraling out of control, ecological sustainability far from at hand, and coronavirus as yet untamed. Perhaps we may be facing the revenge of nature?

CHAPTER 11

The Question of Technology

> Whatever is going on, it has crushed our technology. The
> word itself seems outdated to me, lost in space. . . . And do
> we simply have to sit here and mourn our fate?
> —Don DeLillo, *The Silence* (2020)

The fizz and foam of contemporary commentary on technology issues from polarized stances. Writers such as Bill McKibben and Wendell Berry offer penetrating critiques of modern society and modern technology from holistic, environmental, and even spiritual perspectives. From them we can learn what is "wrong" with the present moment, but their determined suspicion of technology gives me pause. In *Enough: Staying Human in an Engineered Age* (2003), McKibben put forward an argument for rejecting genetic engineering, nanotechnology, and robotics. Berry, even more rooted in the land, wrote in a famous essay "Why I Am Not Going to Buy a Computer" of his love for traditional ways: "As a farmer, I do almost all of my work with horses. As a writer, I work with a pencil or a pen and a piece of paper." Recently, the technology skeptics have zeroed in on artificial intelligence, with McKibben warning, "even if AI works . . . the inevitable result will be the supplanting of human beings as the measure of meaning in the world. And to what end?" At the opposite pole are the techno-utopians such as George Gilder, Ray Kurzweil, Kevin Kelly, and the AI boosters who celebrate the onward rush of technology.[1] While admiring their enthusiasm, I think we can and should do better, at the least, to redirect technology toward sustainability and security. The idea that "technology" is inevitably moving society toward some singular endpoint, or singularity, leaves me cold. The account in this book supports an engaged middle-ground stance toward technology, resisting the undue pessimism of some writers and tempering the naive optimism of others. This chapter appraises various attempts to discern what technology is—what it is good for, when and how it changes society—and, quite possibly something of fundamental importance about our society and our future.

The term *technology* has a specific history, but in this book I have used it in a broad and flexible way.² Indeed, my underlying goal has been to display the *variety* of technologies, to describe how they changed across time, and to understand how they interacted with diverse societies and cultures. There's no simple definition of *technology* that adequately conveys the variety of its forms or sufficiently emphasizes the social and cultural interactions and consequences that I believe are essential to understand. The key point is that technologies are consequential for social and political futures. There is not "one path" forward.³

SCIENCE AND ECONOMICS

Ever since the science-and-systems era, it has been common to believe *science* is essential to technology, that scientific discoveries are the main drivers of technological innovations. The influential "linear model" posits a desirable and necessary link between scientific advances and technological innovations. Recall the link between the basic laws of electricity and Thomas Edison's electric lighting system or between the insights of organic chemistry and the synthetic dye industry, and the employment of thousands of physicists and chemists in industrial research as well as the atomic bomb. During the decades when modernism held sway, many planners and visionaries conceptualized science, technology, and social progress as a tidy package: that, as the Italian Futurists proclaimed, "the triumphant progress of science makes profound changes in humanity inevitable." The notion that social change starts with science and science-based technologies is a key tenet of modernist thinking.

Then again, scientific theories and insights had less to do with technological innovation during the eras of industry, commerce, and courts. Scientific discoveries did not drive the technologies of gunpowder weapons or palace building in the Renaissance, shipbuilding or sugar refining in the era of Dutch commerce, or even the steam-driven factories of the industrial revolution. Indeed, steam engines depended more on empirical engineering, while the later science of thermodynamics resulted from measuring and analyzing them. As scientist-historian L. J. Henderson famously wrote, "Science owes more to the steam engine than the steam engine owes to Science."

Even today, university research and industrial innovation interact deeply and profoundly—without arrows in any single direction. As policy scholar Keith Pavitt put it, at "one extreme . . . is . . . the relatively rare 'linear model'" where fundamental research by a university researcher leads in step-wise order to an industrial innovation. Alternatively, quite apart from scientific discoveries, universities providing a ready supply of graduates trained in particular techniques and embedded in international networks "is ranked by many industrialists as the greatest benefit" of universities to industry. While research in chemistry or pharmaceuticals can result directly in new synthetic molecules or processes, such as the synthetic messenger RNA research that venerable Pfizer and upstart Moderna each commercialized with coronavirus vaccines, in fields like mechanical engineering the benefits lie in training students in simulation and modeling techniques as well as testing, designing, and managing.[4]

The presumed primacy of physics allowed the savvy atomic scientists at Los Alamos to snare credit for the Manhattan Project. In reality, the activities of engineering and mass-production industry at the mammoth atomic factories at Oak Ridge and Hanford, as well as at Eastman Kodak, Union Carbide, Allis-Chalmers, DuPont, and other industrial contractors required the vast proportion (some 96%) of the atom bomb project's budget. And science is only one part, along with engineering, economics, and military subsidies, of microelectronics and computing technologies, as the saga of Moore's Law reveals clearly. Yet the brazen motto of the 1933 Chicago World's Fair—"Science Finds—Industry Applies—Man Conforms"—lives on, even today, as an article of faith for many scientists and policy analysts and as a commonplace in popular discourse.[5]

Gordon Moore's early insistence that "only the engineering effort" was needed to achieve the technological trajectory of "Moore's Law" implicitly argued against the linear model. Yet scientists played a key role; Moore himself was a Caltech-trained PhD chemist. As transistors became ever-smaller over four decades, the instrumentation needed to create their tiny dimensions shrank from optical-light wavelengths down to today's short "extreme ultraviolet" ones; along the way, the entire chemical armamentarium required to scribe, mask, resist, etch, and ultimately construct ever-smaller silicon microstructures required the labors of thousands of chemists. Each change in transistor-device size—from thousands of nanometers, or nm, down to today's devices around 5 nm—required something like the German's "scientific mass-labor" (chapter 5) for the required production of chemicals.

The creation of nanoscale transistors involved both "materials science" as well as "nanotechnology," but do not imagine there is any simple relationship between them. One instance in this area supporting the "linear model" may be giant magnetoresistance, or GMR, which won two academic researchers the 2007 Nobel Prize in Physics with a citation asserting, unusually, "You would not have an iPod without this effect." The effect allowed tiny alterations in magnetism to be reliably read by disk-drive sensors, permitting the iPod's 1.8-inch disk drive to store 10 gigabytes of data and leading to transformation of a $30 billion industry. The resulting field of "spintronics" promises equal wonders, while the "resulting flood of research monies transformed nanotechnology into one of the most robustly funded, aggressively pursued, and widely promoted fields in modern science and engineering."

And yet science in the GMR case was no "unmoved mover": the Nobel laureate physicists depended on an industrial-grade technology called molecular beam epitaxy to build up the precise atomic layers that showed the astounding magnetic effect. The fields of "materials science" and "nanotechnology" themselves were largely created by government research managers looking for promising fields to define: "materials science" emerged in the 1960s with ARPA's well-funded creation of interdisciplinary centers, while "nanotechnology" emerged in the 1990s also with funding from DARPA and the Department of Energy. "Nanotechnology . . . is a tool for taking institutional innovations developed to aid the semiconductor industry and extending them across the scientific landscape," according to historian Cyrus Mody.[6]

A complementary viewpoint to the linear science-industry-society model positions technology as a desirable instrument of *economic growth*. In the eras of commerce, industry, science and systems, globalization, and digital dominance, technologies clearly brought about economic growth. In the seventeenth century, generating wealth was a chief concern of Dutch shipbuilders and the preeminent object of the Dutch import-and-export traffic industries. The British industrial revolution was built around a set of textile, metal, and power technologies that engineers and industrialists attentively tuned to increasing production and boosting profits, while the science-and-systems era featured new science-based industries, most notably electricity and synthetic chemicals. Globalization thrived when nations cut trade barriers and when container shipping slashed global transportation costs. Recently, digital technologies and the internet economy created immense wealth—although certainly it was not widely nor equally spread.

The social, political, and cultural changes that accompany technological change are no less important than the economic changes. Renaissance court culture not only created and shaped many technologies, such as massive building projects, incessant war making, and dynastic displays, but it also was created by and shaped by these technological efforts. "The movement of wealth through the art market . . . transformed . . . the structure of society," writes Richard Goldthwaite in his study of Florence's building industry. "That flow, in short, was the vital force that generated the Renaissance."[7] The quantifiable aspects of Renaissance-era economic growth may have been elsewhere in the woolen and banking industries, but the era's court-based technologies were crucial to the emergence of Renaissance cities and culture.

Similarly, technologies of commerce helped constitute Dutch culture, not just expand its economy. Commerce permeated Dutch society, extending from the top strata of merchants and financiers, through the middle strata of mill and boat owners, down to agricultural workers milking cows for cheese exports. Members of the elite invested in shares of the great trading companies and bought elegant townhouses, while decidedly average Dutch citizens possessed decent oil paintings and, for a time, speculated in dodgy tulip futures.

In comparison with Dutch commerce, at least, both Renaissance court culture and early industrial culture each featured a narrower distribution of wealth. Whereas Dutch orphans were reasonably well fed with daily rations of beer, orphan girls in Renaissance Florence ate only a half-ounce (15 grams) of meat per day. Britain was, in aggregate terms, a relatively wealthy country, even though the industrial revolution did not distribute substantial wealth to the average British worker until perhaps the 1840s. For their part British orphans hoped for modest bowls of watery porridge, say three small "dessert spoonfuls" of oatmeal for each pint of water.[8] What is more, while a focus on economic growth spotlights the industrial era's economic successes, it ignores the era's polluted rivers, cramped housing, and stinking privies. These environmental costs too were aspects of industrial society and very much present today in certain districts of China "whose rivers are black with industrial effluent" (chapter 10).

Contemporary economic writers focusing on the aggregate measures of wealth overlook and thus underplay the wrenching cultural changes and environmental damages that have accompanied globalization. Development agencies aiming only at aggregate economic measures (such as per capita GDP figures) persistently disregard the environmental hazards and cultural strains in "undeveloped" regions or "Third World" countries. Much of the frank resentment aimed at the World Bank, International Monetary Fund, and World Trade Organization results from their conceptual blindness to the implications of technology in social and cultural change.[9] Technological changes—whether in seventeenth-century Holland, nineteenth-century India, or twenty-first-century Thailand—generate a complex mix of economic, social, and cultural changes.

VARIETY AND CULTURE

What is technology good for? In short, many things—and with a great variety of social and cultural consequences. This unappreciated variety suggests technology's potential for changing society and culture. Acquiring such insight is difficult, because we typically lack a long-term view of the interplay between technological and cultural changes. It is difficult to see the technological alternatives that once existed but now might seem to be quaint relics doomed to fail. It is just as difficult to see the contested, multilevel selection processes that determined which of the alternatives became successful and which did not. Most often, we see only the final results of immensely complex social processes and then wrongly assume that there was no other path to the present. For these reasons, the technological "paths not taken" are often ignored or even suppressed.[10]

In taking a long-term perspective, we can see historical actors embracing technologies toward many different ends. These include building impressive palaces or asserting dynasties in the Renaissance courts; generating wealth or improving industries or maintaining far-flung imperial possessions; embracing modern culture, contesting the Cold War, or even fighting globalization. Some of these actions and ends, such as commissioning Renaissance-era buildings or sculptures to legitimate political rule, might startle readers who thought that technology is only about economic growth and material plenty. Then, as well, technologies figure in some of the leading moral problems of the modern age, such as the transatlantic sugar-and-slave trade in the era of commerce, the machine-gun slaughter of native peoples during the era of imperialism, the poison-gas chambers built by industrialists and engineers in Nazi Germany, and today's internet surveillance that sends untold thousands to prison camps. This staggering diversity of technological ends and embedded visions suggests that the "question of technology" is deep and troublesome.

Indeed, without playing down the economics, I believe a more pressing problem is to understand the varied social, political, and cultural consequences of technology.[11] As we have seen repeatedly, agents of change mobilize technologies to promote, reinforce, or preserve certain social and cultural arrangements. In her study of French nuclear power, Gabrielle Hecht suggests the complementary notion of "technopolitics." Technopolitics, she notes, refers to "the strategic practice of designing or using technology to constitute, embody, or enact political goals." When French engineers consciously built a civilian nuclear reactor that was capable of producing weapons-grade plutonium, they were not so much responsively shaping the reactor technology to realize an announced political aim (there was no such aim at the time) but rather proactively intending that reactor to enact a political goal of bringing about a certain nuclear vision of France's future. During the 1950s and early 1960s, there was no policy debate or public discussion on whether France should develop nuclear weapons; indeed, the country's civilian politicians seemed content to let the engineers quietly create the technical and material means for France to build a nuclear bomb and become a nuclear power.[12]

A giant experiment in using technologies to shape political structures and social practices has been running in Europe for decades. Even though political theory has been slow to appreciate the fact, European integration has been shaped by and constituted through innumerable technological systems and networks. Indeed, both the shifting external boundaries of Europe as well as its dynamic internal structure are revealed by its technological infrastructures. In the twentieth century, a country's railroad gauge indicated whether it was effectively part of the European economy or not. Most of continental Europe had adopted the "standard" distance between rails of 4 feet 8½ inches, or 1,435 millimeters. For many years Spain, with its extra-wide railway gauge (1,668 mm) that required all passengers and most goods to be off-loaded and reloaded at the Spanish–French border, was clearly out. So, too, by conscious design, was the Soviet Union—heir to the nonstandard Russian gauge of 1,524 mm long before the Cold War—since its military leaders feared that Western European armies might use railroads to send invading troops into Soviet territory. In this measure, Spain joined Europe in the 1990s with its high-speed passenger lines built to standard gauge, while Russia and most post-Soviet successor states retain the Russian gauge. During the Cold War, the continent's energy and communications networks were divided into eastern and western sectors.

Even with the political expansion of the European Union to comprise nearly thirty countries, some surprises await. The Europe-wide electricity blackout in November 2006 provides an apt snapshot of Europe's energy infrastructure. The blackout, starting in northern Germany, spread within a scant twenty seconds to nearly all the countries of western continental Europe and parts of Africa, stretching from Portugal in the west, extending to the rim of North Africa in the south, and going east to Romania and Croatia. Among the affected countries, Croatia, Morocco, Algeria, and Tunisia were at the time not part of the political European Union but nonetheless were evidently part of the electrical European union. The blackout did not hit Scandinavia, the Baltic countries, the British Isles, and the Commonwealth of Independent States. The heart of the former Eastern Europe still operated an electricity grid largely independent from Western Europe's. As late as 2006, there was only one link—an electric line connecting Sweden and Poland—between former East and West. Scandinavia is connected to continental Europe via HVDC cables that isolate AC frequency disturbances and decouple the region's grids.[13]

It follows from these observations that technologies are not neutral tools. The presence of a certain technique or technology can alter the goals and aims of a society as well as the ways that people think. Certain representational technologies, ranging from Renaissance-era geometrical perspective and moveable-type printing through to today's internet, changed how humans communicate.[14] Many of these technologies brought about major cultural movements, including court culture of earlier times and "cyberculture" of more recent times. International campaigns on behalf of internet freedom and against internet censorship testify to the signal importance of these technologies today.

So far we have considered instances where technologies have been deployed to satisfy more or less well-formed economic or societal demands, but that is hardly the whole picture. Technology interacts with the goals and aims of society through the *process* of technical change. Frequently, enthusiastic and powerful actors create new demands—and invent new technologies to meet them—rather than simply responding to static "demands" set by society. Agents of change, such as inventors, engineers, industrialists, government officials, and even social movements, typically create new demands through marketing or advertising efforts and by enlisting the promotion, procurement, and regulatory activities of government. Edison did not merely invent light bulbs but also mobilized the power of the state to brand his competitor's alternating-current system as the "death current" responsible for criminal executions. And of course the state itself—whether the precarious city-states of the Renaissance or the powerful nation-states of the military era or the risk-riven contemporary state—can be an actor in technical change.

Innovation might even become an emblematic or totemic value that shapes society and culture, and provides an open field for technologies. In the *Innovation Delusion*, historians Lee Vinsel and Andrew Russell observe that our recent enthusiasm for digital technologies and disruptive innovation (such as Facebook's "move fast and break things") spread seemingly without constraint across society. "We adjusted our values, even our vision of democracy, to be suitably deferential to the gods of Silicon Valley," they write. Innovation and disruption of established society is the inspiration for the "Pirate Parties" that peck at the margins of the political system in three dozen countries, including the Czech Republic, Germany, Iceland, Sweden, the European Parliament, and the United States. For another way of glimpsing the future, simply visit the annual Consumer Electronics Show each January in Las Vegas alongside 175,000 enthusiasts. The very latest is on display, recently the "internet of things." One visitor in 2020 witnessed such wonders as internet-enabled toothbrushes, toilets, pet feeders, and sex toys.[15]

Our internet-soaked age is by no means the first to experience a "new society" driven by technology. In the mid-nineteenth century, distance-spanning technologies did the trick. "Day by day the world perceptibly shrinks before our eyes. Steam and electricity have brought all the world next door. We have yet to re-adjust our political arrangements to the revolution that has been wrought in time and space," wrote one prominent London newspaper in the mid-1880s. "Science has given to the political organism a new circulation, which is steam, and a new nervous system, which is electricity," suggested another writer. "These new conditions make it necessary to reconsider the whole colonial problem."[16]

Steam and electricity trimmed the barriers to British control of overseas possessions, including great distances and uncertain communication. The discovery of quinine to fight malaria reduced another type of barrier to imperialism, that of endemic tropical disease. Many British officials, having seen that telegraphs "saved" colonial rule during the 1857 Indian mutiny, came to believe that continent-spanning railroads, undersea telegraphs, massive public works, and other imperial technologies were necessary to preserving the imperial status quo, whether or not there was any external "demand."

These distance-spanning technologies made the imperialist venture too cheap to ignore, displacing a host of political and social questions about whether it was a good idea in the first place. In *The Imperial Idea and Its Enemies*, A. P. Thornton documents the difficulties that opponents of imperialism faced despite the Indian Mutiny, the costly Boer Wars, the disastrous First World War, and the rise of nationalism abroad and democracy at home. "Statesmen, journalists, and preachers came to every question of policy or morality bound by the silent influence of a half-uttered thought: 'Come what may, the Empire must be saved.'"[17] Other European countries followed enthusiastically where Britain had led, notably in the notorious land grab that carved up the continent of Africa into European colonies. American imperialism gained steam after the 1898 war with Spain brought Cuba, Puerto Rico, Guam, and the Philippines into the US orbit.

Yet each of these technologies—telegraphs, railroads, steamships, and quinine—was not merely a "cause" of imperialism but also the result of massive investments in the imperial venture. Technologies off the shelf did not "cause" imperialism. Rather, certain technologies were themselves selected, developed, and promoted by colonial officials and by engineers or industrialists keen on finding a market in the colonial state. In this way, then, imperial-era technologies were both *products of* and *forces for* imperialism. In like measure, the imperial state was both created by and a force behind the evolving imperial technologies. Indeed, the era of imperialism illustrates in a particularly clear way what is true of each era examined in this book—technologies have interacted deeply with social, cultural, and economic arrangements.[18]

Imperialism also forms the question of who has ready access to and practical control over technologies—and who does not. In the imperialist era, no open parliamentary debate made the decision that Britain would become industrial while India would remain agricultural—and thereby a source of raw materials as well as a captive market for British industrial goods. There was instead a blatant asymmetry of access to decision makers and centers of policy and power, not only in formal terms (lobbying Parliament in London or buying Indian railway shares in London) but also in the designs for railways and the access to and content of technical education. In earlier eras, there had also been substantial formal and informal barriers to accessing technology, notably during the court, industrial, and military eras, in which governments or rulers erected barriers around technologists.[19] Secrecy concerns were pervasive during the military era. In the 1980s a US Navy security officer told me, with a twinkle in his eye, "We have five levels of security: confidential, secret, top secret, and two others that are so secret that I can't even tell you their names." By comparison, the eras of commerce and global culture featured fewer restrictions on and wider access to technology. Confronting systemic risk has led some wealthy Western countries to impose strict limits on access to nuclear or chemical technologies for countries deemed unstable or unreliable.

Access to the content of technology may be just as crucial to technological development as influence on high-level policy decisions. One can see this in a generation of heartfelt experiments in attempting to "transfer" technologies across political, socioeconomic, or cultural boundaries. The all-too-frequent failures strongly suggest that technologies cannot be crated up and sent off like Christmas presents, ready to be unwrapped and instantly put into service. On close examination, one can see that "transferred" technologies often need subtle refinements to be adapted to a new environment in ways that the original designers could never have foreseen. Social arrangements that do not grant agency to local implementers of technology are consequently predisposed to failure.

These questions of technology transfer connect with Arnold Pacey's useful notion of "technology dialogue," an interactive process that he finds is frequently present when technologies successfully cross cultural or social barriers.[20] Absent anything like a well-organized nation-state, technology transfer succeeded in Renaissance Italy, commercial Holland, and industrial England. Some think that British technologists in the industrial era invented new cotton-textile machines, but in terms of "technology dialogue" one might say that they skillfully adapted earlier machines and techniques (from Italy, France, and India) to the peculiarly British conditions of low-wage labor, imported cotton, and sizable domestic and export markets. Closer to home, McDonald's embraced the desirability of local adaptation as explicit corporate policy. It aims to be a "multilocal" company highly attuned to cultural sensibilities and culinary preferences.

The desirability of this conceptual and cultural openness (a two-way process of curiosity and mutual learning) firmly contradicts the proposition that technology is or must be "under the control" of power structures. Historically, the most coercive power structures have had little success with wide-ranging technical innovation. Soviet Russia successfully directed technical innovation down certain state-chosen paths but was not flexible in changing technical paths. Recently, the authoritarian regime in China is pressed to alter its top-down, "high-modern" approach to technology development and economic growth. Since the 1970s, it amply succeeded in a forced-draft industrialization, which created a huge heavy-industrial sector, hyper-specialized industrial districts that fill up Walmart with cheap socks and plastic toys, and the world's largest automobile market. It also pushed through the massive, environmentally destructive Three Gorges hydroelectric dam. Chapter 10 documented China's censored internet as a product of its authoritarian society, with notable assistance from the West's high-tech companies and leading research universities. China's leaders still face the legacy of severe air and water pollution that has generated unrest in China's middle-class society.

DISPLACEMENT AND CHANGE

How does technology change society? History and hindsight can help to identify the choices about society, politics, and culture that are bound up with choices about technologies. *Displacement* is one important dynamic of the interaction between technology and culture. It occurs when practices surrounding an existing or anticipated technology have the effect of displacing alternatives or shutting down open discussion about alternatives. Displacement tilts the playing field. Something of this sort occurred with the hemmed-in debate in Britain over imperialism, with widespread sentiment that, at all cost, "Empire must be saved." One could say that heavy investments in empire prevented open discussion of alternatives to it, with the result that Britain grimly hung on to colonial India until 1947.

During the military era, the heavy investments in nuclear power, especially in the United States, France, and the Soviet Union, had the effect of displacing such alternative sources of energy as wind and solar energy.[21] In the United States, the Price-Anderson Act of 1957 stacked the deck to favor nuclear power. The act capped the damages that any nuclear power plant owner could be legally liable for, even in the case of a disastrous accident, as a result of theft or sabotage, or even in the transport of nuclear fuel or nuclear waste. The Energy Policy Act of 2005 renewed the pronuclear provisions of Price-Anderson through the end of 2025. Even today, the generous institutional arrangements devised to support the development of nuclear power cast a long shadow. It may not be an accident that Germany, Denmark, and China had limited enthusiasm for nuclear power in the past and are today at the forefront of developing solar energy and wind power technologies.

Technological futures are also political futures. Some critics contend that the campaign for nuclear power displaced not only "alternative" energy sources but also a traditional democratic society. Their argument is that given the extreme danger inherent in making, moving, and storing nuclear fuel and nuclear waste, a nuclear-powered society (whatever its formal politics) requires a strong authoritarian state, with coercive security powers over an extraordinarily long duration of time. Efforts in Nevada and Finland to construct storage facilities for high-level nuclear waste confirm the extraordinary measures required over literally thousands of years. Even if some future with full-blown mobile computing and highly efficient alternative energy might make possible a radical decentralization of society, nothing of the sort will occur owing to the persisting imperatives of nuclear-era state security. In effect, the argument goes, the decision for widespread nuclear power, with its requirement of permanent central-state security, displaced the natural pattern of societal evolution. The half-life of weapons-grade plutonium—a bit more than 24,000 years—constrains the pace and direction of social evolution.

Displacement also gives the different eras considered in this book something of their durability and character. Renaissance-era patrons kept Leonardo and other Renaissance technologists engaged in court activities involving technology and, in effect, displaced them from improving industry or generating wealth. Then, by concentrating their efforts on technologies of commerce, the Dutch technologists displaced a court-centered society while enhancing a capitalistic pattern of economic growth and cultural development. And British imperial officials and industrialists, by their explicit designs for railways and public works and technical education, displaced an industrial future for India, directing India to be agricultural instead of developing its manufacturing and mechanical capabilities.[22]

One doesn't even try to create a utilitarian justification for displacement in technology-related aesthetics. The fascination of modernist architects and their patrons with the "modern" building materials of glass and steel and concrete created innumerable city streets, office blocks, and public buildings with a high degree of similarity. In the name of technology and progress, modernists waged a determined campaign and carefully positioned their intellectual and institutional efforts to create a modern culture. For roughly fifty years, a certain formal perspective on modern architecture displaced alternative approaches. Of course, with the rise of postmodern architecture, which embraces varied colors, materials, and playful quotations of historical styles, there has been a counter-displacement of the once-dominant modernism.

In all these examples, the asserting of a distinct purpose for technology—and consequently a desirable direction for social and cultural development—displaced alternative purposes for technology and alternative directions for societal development. Displacement, then, is how societies, through their technologies, orient themselves toward the future and, in a general way, direct themselves down certain social and cultural paths rather than others. This process, of course, is far from smooth or automatic. The paths and the technologies are frequently contested, sometimes in open political debate or social protest, such as the upheavals surrounding Manchester during the industrial revolution or the more recent antiglobalization protests. And often enough, societies experience unintended consequences of technology decisions, such as the open-border global culture built from the military-spawned internet. Yet not all paths are continually contested, and not all consequences are unintended. Over long stretches of time, there are durable patterns in technologies, societies, and cultures. These long-duration patterns, or eras, of technology can be seen as one foundation for our cultural norms.

Displacement does not occur on a level playing field, as noted earlier, and so we must again consider the relation of technology to power. In *Das Kapital*, his magnum opus, Karl Marx asserted, "It would be possible to write quite a history of the inventions, made since 1830, for the sole purpose of supplying capital with weapons against the revolts of the working-class." Chapters in that history would certainly recount the efforts of leading British technologists such as James Nasmyth and Richard Roberts to combat the "refractory class" of workers by inventing and imposing "self-acting" machines that "did not strike for wages." Marx himself noted the observation of Andrew Ure, in his *Philosophy of Manufactures* (1835), on the self-acting mule-spinning machine. It was "a creation destined to restore order among the industrious classes. . . . This invention confirms the great doctrine already propounded, that when capital enlists science [or modern technology] into her service, the refractory hand of labour will always be taught docility."[23]

Power comes in many forms, and capitalists have not had a monopoly on technology. During the Renaissance it was the quasi-feudal courts—not the capitalistic wool merchants—that most effectively deployed technology, shaping it in the image of court culture. During the industrial revolution, capitalistic brewers and cotton spinners and machine builders, rather than women spinning thread at home or aristocrats on their landed estates, mobilized the era's technologists. During the military era, powerful nation-states set the agenda for atomic weapons and command-and-control systems, while scientists, engineers, and industrial managers largely followed the nation-state's funding and direction. During the early phase of globalization, before 2001, capitalists gained power while the nation-state seemed to wither away. During the era of systemic risk, the nation-state is resurgent as an active promoter and shaper of technology.

Despite these many powerful actors, nondominant actors can sometimes also use technologies to advance their alternative agendas. To begin, numerous historical actors, whose power at the time was frankly precarious, used technological projects to enhance or sustain their cultural or political power. In the Renaissance, the patronage of building churches, palaces, or other high-profile public works often enhanced the political legitimacy of an unsteady ruler. Cosimo de' Medici's many buildings were "worthy of a crowned king," declared a fellow Florentine palace-builder.[24] But Cosimo was *not* a crowned king, and in fact he ruled uneasily, and his family's dynasty was exiled for nearly two decades. Insecurity pervaded the Sforza family's rule in Milan and the scandal-ridden popes who doled out patronage projects to Leonardo and others. Lavish public works in colonial India might be the same. Certainly, no one expects a utilitarian explanation for the monumental Victoria Station in Bombay (1887–97)—or today's economically ruinous contests for the "world's tallest building." Indeed, it may be that these actors and entities aimed to consolidate and legitimate their uncertain power through their active patronage of monumental technology and impressive public works.

Closer to home, let's reconsider the story of fax machines in the late 1980s as a mixed instance of dominant and nondominant groups' use of technology. The European students' discovery of fax machines for international community building was a genuine cultural innovation. Yet it is no surprise that the Council of Europe ramped up publicity for this grassroots effort at creating a positive notion of a unified Europe just when the project of "building Europe" got an institutional boost from the Maastricht Treaty of 1992. The European Union continues its quest for political legitimacy today through active patronage of high-profile public works, wide-ranging educational initiatives, and savvy publicity campaigns.[25] Each of these actors, then, used the commissioning and display of technologies to create, enhance, or sustain power. One might say they strived for dominance through their mobilizing of technology.

Nondominant groups in society today might effectively mobilize technology.[26] Workers, with few exceptions since Marx's time, have rarely considered the selective embrace of technology to be a means for furthering their own aims. Several US labor unions in the 1950s contested the power of corporate management to deploy automation technologies; during an epic 116-day strike in the steel industry, union pickets read, "Man versus machine." Yet by the 1960s most US unions had ceded shop-floor control to management in exchange for fat paychecks and the promise of employment security. A famous example of workers' embrace of technology comes from the Scandinavian autoworkers who reconfigured their factory environment to break the monotonous production-line routines of Fordism.[27]

Once you start looking, examples of nondominant groups using technology to advance non-elite agendas come readily to mind. During the era of modernism, national homemakers associations in several countries shaped the contours of domestic technologies and the designs of modern kitchens. From the 1960s onward, citizen-consumer movements in many countries have pushed governments and industries to provide safer cars, cleaner water, broader access to medical treatments, and greater transparency in government decision making. Women's groups have for decades formed a strong and effective lobby for contraceptive technologies. Many alternative political and social movements depend on the internet to circumvent established and expensive media outlets, while their activities are planned, conducted, and coordinated with mobile phones and social networking. Clearly, non-elite groups are using technologies to shape social and cultural developments and influence the future. Facing Don DeLillo's dilemma (in this chapter's epigraph), we simply don't need "to sit here and mourn our fate."

One such active movement to shape the future embraces "information and communication technologies for development," or ICT4D. It requires setting aside the tired Silicon Valley trope of technology-centered development where high-tech companies sell us a vision conveniently enough depending on their products. Instead, as Ramesh Srinivasan recently argues, in paying "attention to the agency of communities across the world . . . we can recognize the potential of grassroots users to strategically employ technologies to support their voices and agendas." In computer science, ICT4D took off around 2010 and created a lively international and interdisciplinary cluster of research conferences, with more than 500 peer-reviewed papers appearing in the authoritative ACM Digital Library. Conferences with up to a thousand participants from across the globe have been held in Cape Town, South Africa; Nairobi, Kenya; Chicago, Illinois; Hyderabad, India; Lusaka, Zambia; Kampala, Uganda; and recently, owing to coronavirus, online. The topics range from deeply philosophical and ethical considerations, including theorizing about technological change and society, through to the practicalities of choosing a cell phone for a field experiment.[28] Plainly, something significant is going on here.

The environmental movement also has altered its perceptions on modern technology. In the 1960s, environmentalists pinned the blame for pollution squarely on capitalism and technology; and they typically rejected science, technology, and capitalism as a matter of principle. Many environmentalists still retain a small-is-beautiful stance that keeps modern technology at a distance. But recently, under the banner of "ecological modernization," some have shifted to a selective embrace of technology and market mechanisms. Advocates of ecological modernization view the institutions of modernity, including science, technology, industry, and the nation-state "not only as the main *causes* of environmental problems, but also as the principal *instruments* of ecological reform." Some impressive conservation projects have been completed, sometimes with direct corporate backing, by the Nature Conservancy and World Wildlife Fund. An entire modern industry provides environmentally friendly organic food to millions of people.[29]

"Modernity" once meant a package of social, cultural, and economic developments that promised a better future. That view shaped development theory in decades past and even framed early work in the history of technology (see chapter 3). Today, we cannot have faith that technology or modernity will secure a better society. In "The Modernization Trap" international relations scholar Jack Snyder observes: "The rise of China, the dynamism of politicized religion, and the assertiveness of authoritarian regional powers have lent credence to the notion that there might be 'multiple modernities'—a variety of ways to be successfully modern, including illiberal ones." And, to be clear, illiberal does not mean "conservative." After the 2016 elections in the United States and the United Kingdom, and the expansion of nationalism, nativism, and populism there and elsewhere, the problem is all around us.

Under the "liberal model," free market capitalism was supposed to spread wealth widely throughout society; accordingly, citizens would have solid economic grounds for faith in the political-economic system. Such a model might be a bit too neat and tidy, yet modernization worked well for many countries, at least some of the time. It did not work well where, as under imperialism or colonialism, there was a disconnect between the costs and benefits of modernization. Neo-liberal capitalism led to protests against unbridled globalization. The "disruptive social and political impact of market forces," if unchecked, notes Snyder, generates fractious discord in which "outsider [populist] politicians seek to mobilize mass protest against elites who are portrayed as selling out the nation." What should we think when some of the richest countries have much of their population experiencing economic decline? It seems doubtful that high-tech billionaires, the principal beneficiaries of a "winner take all" economy, will save us. And with robots, automation, and artificial intelligence looming, such economic strain seems likely. "This is a great time to be alive," writes one natural-language AI program: "The only problem is that, in the next five years, A.I. will replace millions of jobs."[30]

In this book, technology emerges as a potent means of bringing about social, political, or cultural changes. From a Dutch perspective, it was the effective deployment of technologies—printing presses, commercial and military sailing ships, and waves of muskets—that won the battle against imperial Spain, easily the most powerful state in Europe at the time. Or consider the "railroad republicanism" of Paul Kruger, president of the Transvaal region in southern Africa, who for a time used his independently funded railroads to stand up against the entire might of the British Empire. In India, Britain's colonial railways were built only after accommodation with local native rulers. "I consented with great reluctance" was one native Indian ruler's description of permitting the railroad to cross his domain.[31] We may think that technology is a servant of the powerful, but even the most powerful do not always get precisely what they want.

We can see similar challenges to elite agendas around us today. Ask the World Trade Organization about the email-organized street protests that disrupted its summit meetings in Seattle and Genoa and derailed its original elite-dominated globalizing agenda. And remind powerful McDonald's about the grassroots campaign against its beef-extract-laced "vegetarian" French fries or for that matter about its McLibel suit, archived on the internet, where the company's preposterous courtroom arguments are published by antiglobalization activists for all the world to see. Social media has disrupted governments, elections, and deliberative processes worldwide. Nondominant groups are effectively using technologies to contest the power of dominant groups. Technology, to repeat, is not a neutral tool. Social movements that distance themselves from technology (as the environmental movement did for decades) are missing a crucial strategic resource.

DISJUNCTIONS AND DIVISIONS

One source of anxiety about technology is the troubling gap or *disjunction* between our normative expectations for technology and what we observe and experience in the world around us. We perceive technology to be changing quickly, I believe, because we are continually confronted with social, cultural, and economic changes in which technology is clearly implicated. And frequently enough "technology" serves as a shorthand term to refer to a host of complex social and political processes underlying the change.[32] Nevertheless, it is a mistake to believe that technology by itself "causes" change, because as argued previously technology is not only a *force for* but also a *product of* social and cultural change.

In the twentieth century, the once-distinctive eras of technology have piled up on one another. For the eras of courts, commerce, industry, and imperialism, it was reasonably easy to identify the dominant uses for technology and the dominant values built into it. But, for the twentieth century, it became difficult to determine whether technology was about science, modernism, war, or global culture. And then, just when it appeared an era of global culture was taking durable form, came the attacks of 11 September 2001 and the perceptible slide toward systemic risk and global insecurity.

The aggressive and wide-ranging responses by Western nations, above all the United States, to the September 11th attacks marked an end to the 1990s view of global culture as open and peaceful. While clearly the global economy persists, the idealistic version of globalization ended with the dramatic resurgence of the nation-state, assuming vastly expanded powers at home and abroad. The years following 2001 also saw the boiling up of nationalistic trade policies as well as the sharpening of profound technology-mediated divides around the world. Global trade continues apace, but the rosy vision of a peaceful world, economically integrated and culturally harmonious, knitted together by information technology, is done and gone.

There is a wider area of disjunction involving technology, of course, and that is the role of technology in the world's geographical, economic, and cultural *divisions*. One concern is the persisting "digital divide" between wealthy societies that have ready access to computers, telecommunications, and the internet and poorer societies that too frequently do not. The "digital divide" concept, launched by the World Economic Forum in 2000, was front and center at the United Nations–sponsored

World Summit on the Information Society, held in Geneva in 2003 and in Tunis in 2005. On all available measures—phone lines, mobile phones, internet access, traffic, or bandwidth—Africa confronted severe barriers to participation in the global economy and the world community. According to the World Summit, the entire African continent, fifty-three countries with 1 billion people, had fewer internet users than France with 60 million people. While Canada had 107 total phone lines (mobile and land) for each 100 people, a typical figure for the rich West, nine poor countries in Africa each had less than two, including Ethiopia (population 80 million) with 0.77 and Chad (10 million) with just 0.58 total phone lines per 100 people. In 2019 the World Bank tabulated all countries' internet access (average 49%). Fourteen countries were under 10 percent, all of them in Africa (*alphabetically:* Burundi, Central African Republic, Chad, Comoros, Eritrea, Guinea-Bissau, Liberia, Madagascar, Niger, Nigeria, Somalia, South Sudan, and two states in the Congo region).

Since the Geneva-Tunis summit, many analysts have given up counting phone lines and shifted attention to the character and affordances of information technologies—inquiring about the type of society these technologies may bring about. With up to 80 percent of all World Wide Web sites in English, some called it the "World *White* Web." "The new technology revolution is neither global nor cross-cultural," writes Ramesh Srinivasan. "It is primarily produced and shaped by powerful corporations and institutions from Europe and North America, with various collaborators across the world. Yet we treat commercial platforms such as Facebook, Twitter, or Google today as if they were public spaces and systems, ignoring that they must remain primarily accountable to their shareholders." The problem is not principally with the *number* of landlines, mobile phones, or networked computers but with the skewed and asymmetric spread of information systems. Most citizens in the world are "excluded from most decisions made around the future of the Internet and digital technology." As passive users, most citizens—North and South—face a world where they might modestly appropriate a technology (on Twitter I can choose one of six colors, five font sizes, and three screen backgrounds!) but where it's difficult to deeply alter an existing technology or design a fundamentally new one. And, more than ever, the economic, social, and political costs of declining such systems and opting out of the modern world are a mechanism to lock in place unequal and unjust relations.[33]

Even in the industrial West there are "digital divides" between wealthy families and poor ones, as well as between well-wired metropolitan areas and sparsely served rural districts. The shift to at-home online education due to the coronavirus in 2020 revealed that 15 to 20 percent of US public school students lacked basic broadband internet at home—with rural and urban students often faring far worse. "With coronavirus, we're about to expose just how challenging our digital divide is, and just how unequal access to broadband is," stated one official. Students and teachers, as well as school-bus drivers, public libraries, state parks, and even some friendly McDonald's, scrambled to fill the gaps. Looking more widely, while Africa experienced a dramatic rise in cell phone access (increasing access from 12 to 80 percent during 2005–19), it remains substantially below other world regions in mobile broadband and fixed-line broadband.[34] The perpetual race for higher-speed internet access and faster mobile computing suggests that the "have-nots"—in whatever region of the world—may continually be struggling to catch up.

And yet raw numbers may not tell the full story. Consider access to information technologies in India. This country, despite its persisting poverty and modest telecommunication access, is one where small-scale fishermen are using cell phones to drive hard bargains to get the best prices for their boatloads of fish; where villagers without means to purchase a computer for each family instead share an internet kiosk (or cell phone) across families or among an entire village; where a world-class tech industry is firmly in place in the cities of Bangalore, Pune, Hyderabad, Delhi, and Mumbai; and where a national network of two dozen Indian Institutes of Technology prepare women and men for science and engineering careers (notably India is one of the few countries with strong women's participation in computer science). Indeed, the continual circulation of software engineers between India, Silicon Valley, and other world centers in itself constitutes an impressive high-technology dialogue. While not all Indian citizens have ready access to technology, India does not fall hopelessly off the digital divide.[35]

A region with a much more severe disjunction—between its dominant cultural norms and the potentials of twenty-first-century technology—resides in the long arc that stretches from Palestine across the troubled Middle East through the disputed Pakistan-India border and eastward to the huge populations of China and Indonesia. One hesitates to make any sweeping generalizations since the region is rife with political battles and conceptual disputes. What started out at a hopeful "Arab Spring" in 2011 instead became a doleful "Arab Winter" just a few years later, as chapter 10 recounted. Clearly, some countries are experiencing an excruciating disjunction between traditional cultural and political forms and the wider world of modern technologies.

Some countries impose a digital divide on themselves. "The Middle East and

North Africa is one of the most heavily censored regions in the world," concluded one major study of internet censorship. According to the 2010 Internet Censorship Report, a staggering 1.7 billion people around the world were subject to "pervasive or substantial" government-directed internet censorship. China was the worst offender; the woeful list included Egypt, Iran, Saudi Arabia, Syria, and Tunisia (in the Middle East), as well as Burma, Cuba, North Korea, Turkmenistan, Uzbekistan, and Vietnam with Bahrain, Eritrea, and the United Arab Emirates on a watch list. Internet censorship was even a business opportunity for some anti-censorship entrepreneurs. "The market is growing very rapidly," said one such provider with customers in Iran and China.[36]

In 2020, "an especially dismal year for internet freedom," the respected *Freedom on the Net* report found the "coronavirus pandemic is accelerating a dramatic decline in global internet freedom" with many countries clamping down on troublesome information providers or using the pandemic as a pretext to suppress ethnic, religious, or other marginalized communities. The spread of "internet sovereignty" beyond China, Iran, and Russia with the aim of nation-specific internets (see chapter 10) further erodes the open global internet. On its measure of obstacles to access, limits on content, and violations of user rights in sixty-five countries (constituting 87 percent of the world's internet users), the report found internet freedom substantially decreasing in twenty-six countries, including Ecuador, Egypt, India, Kyrgyzstan, Myanmar, Nigeria, the Philippines, Rwanda, and Venezuela. China, Iran, and Syria top the list of worst-offending countries. (Sudan, Ukraine, and Zimbabwe registered modest gains.)[37]

Scholarship on technologies and societies in Africa, Asia, and Latin America has wonderfully expanded in the past decade. We now know more about these rich experiences with technologies and something of the dynamics of how modern technologies are interacting with dynamic social forms. There are no easy paths forward. We still need to address the questions raised by Michael Adas's *Machines as the Measure of Men.*[38] Adas explains how Europeans came to believe in their cultural superiority over the peoples of Africa, India, and China. Whereas Europeans in the pre-industrial era had been curious and even respectful of these peoples and cultures (with travelers' accounts rarely mentioning distinctions in material culture or technology), they became chauvinistic and dismissive during the course of industrialization in the nineteenth century—as a result of the expanded material culture and technological divisions that industrialization created. Eventually Europeans, given their palpable technological leadership, came to believe themselves possessed of a (highly problematic) cultural and social superiority. Moreover, the imperial era inevitably generated resentments among the peoples of Africa, Asia, and the Americas, who were often forcibly incorporated within Western technological modernity.

Mohandas Gandhi's famed *swadeshi* movement in late colonial India responded to this cultural onslaught. Like some contemporary critics of the West, he did not separate industrialization, westernization, and modernization, seeing these as one poisonous package and rejecting the whole. In *Hind Swaraj* (Indian Home Rule), first published in 1909 and translated into English three decades later, Gandhi launched a fearsome counterattack on modern civilization. "India is being ground down, not under the English heel, but under that of modern civilization," he stated, "It is groaning under the monster's terrible weight." He targeted lawyers, doctors, railways, cities, and above all modern machinery. Lawyers tightened the grip of British imperial authority, while doctors propagated sin, increased immorality, violated traditional religion, and lessened individuals' self-control: "To study European medicine is to deepen our slavery," he said. Meanwhile, railways spread famine and the bubonic plague, degraded the sacred holy places of India, and accentuated "the evil nature of man." Gandhi also charged that "machinery is the chief symbol of modern civilization; it represents a great sin." In later years, although he fractionally tempered his views to allow such human-scale technologies as Singer sewing machines, Gandhi affirmed his "severe condemnation" of modern civilization. His famous diagnosis was that Indians should instead take up traditional methods of hand spinning and weaving, simply doing without machine-made "matches, pins, and glassware" and resisting harmful modern allurements.[39] The resentments identified by Gandhi have persisted, and in recent years similar anti-modern sentiments have come from several prominent fundamentalist movements in the world.

The boundless technology-drenched optimism of the pie-in-the-sky utopians and globalizers was always a bit overwrought, but we should not slide into a cynical disengagement from the problems of technology. The dynamics of *displacement, disjunction*, and *division* suggest the magnitude of the challenges before us. There are also models of new thinking about technology to be explored, promising instances of new approaches to technology in developmental policies, philosophy of technology, and environmental thinking.[40] These are bellwethers of using technology as a means to bring about desirable cultural and social changes.

A powerful case can be made against the naïve and uncritical embrace of technology, reminding ourselves of how much mischief such approaches have caused in the past. Sometimes, please, just put the brakes on: "When people seem to want a technology to develop, to literally lust for a possible new toy, that need can take on a force of its own."[41] Yet, all the same, confronting such pressing problems as sustainability and security as well as privacy and poverty—in the absence of a tough-minded engagement with technology—seems senseless. It is time for an intelligent embrace of technology by reformers, social movements, and citizens' groups seeking to activate this potent means of shaping the future. Technologies can be powerful means for making the world a better place. Here are some places to start: take the electricity grid and energy pipelines off the internet; think twice before wiring an insecure technological system to anything; take one serious step toward a sustainable energy future. On your computer, create something useful from scratch: no "modules," "templates," or "libraries" allowed. Above all, pay attention: there is a lot going on.

NOTES

PREFACE

1. See web.archive.org/web/20200529050651/www.thoughtco.com/wernher-von-braun-v -2-rocket-4070822. An aside on this book's architecture: the notion of broad "eras" as an organizing principle popped in my head, gestalt-like, while in Chicago walking across the street to class. "Just as the technologies of empire . . . once defined the relevant research programs for Victorian scientists and engineers, so the military-driven technologies of the Cold War defined the critical problems for the postwar generation of American scientists and engineers," was the prompt from Stuart W. Leslie's *The Cold War and American Science: The Military-Industrial-Academic Complex at MIT and Stanford* (New York: Columbia University Press, 1993), quote on 9.

2. According to AAAS figures, the US military R&D budget for 2003 through 2009 far exceeded the Cold War peak of 1987 at web.archive.org/web/20130618113321/www .aaas.org/spp/rd/trdef09p.pdf.

3. Alice Gordenker, "Japan's Incompatible Power Grids," *Japan Times* (19 July 2011) at web.archive.org/web/20201122192559/https://www.japantimes.co.jp/news/2011/07/19 /reference/japans-incompatible-power-grids/; and Peter Fairley, "Why Japan's Fragmented Grid Can't Cope," *IEEE Spectrum* (6 April 2011) at web.archive.org /web/20201103071058/https://spectrum.ieee.org/energy/the-smarter-grid/why-japans -fragmented-grid-cant-cope.

4. David Miller, "Anthropological Studies of Mobile Phones," *Technology and Culture* 60 no. 4 (2019): 1093–97, quotes on 1093 at doi.org/10.1353/tech.2019.0103; Kate O'Flaherty, "This New Facebook Tool Reveals How You Are Being Tracked Online," *Forbes* (28 January 2020) "apps and websites" at web.archive.org/ web/20200614055354if_/https://www.forbes.com/sites/kateoflahertyuk/2020/01/28 /this-new-facebook-privacy-feature-is-surprisingly-revealing/; Geoffrey A. Fowler, "Facebook Will Now Show You Exactly How It Stalks You—Even When You're Not Using Facebook," *Washington Post* (28 January 2020) "cameras . . . always on" at web.archive.org/web/20200129160247/https://www.washingtonpost.com /technology/2020/01/28/off-facebook-activity-page/.

5. ASCII was standardized by the American Standards Association (today ANSI) and publicized in 1963 to represent numbers and characters in the limited "space" available in computers: standard 8-bit "words" typically had one "parity bit" so the seven available binary "bits" resulted in a universe of 128 characters (2^7). Setting aside control characters, such as printer line feed, there were just 94 printable characters for

the 10 digits and the 52 upper and lowercase letters (binary 1101101 is "m"). ASCII contained the full range of arithmetical operators but not the diacritics to write proper French or Swedish or Spanish let alone Chinese or Persian. ICANN, "ICANN Bringing the Languages of the World to the Global Internet" (30 October 2009) "non-Latin characters" at web.archive.org/web/20201109022914/https://www.icann.org/news /announcement-2009–10–30-en.

CHAPTER 1.
TECHNOLOGIES OF THE COURT, 1450–1600

1. The Medici dynasty was founded by Giovanni di Bicci de' Medici (1360–1429), who became banker to the papacy and amassed a large fortune. Medici rule in nominally republican Florence began in 1434 and lasted until the family's "exile" during 1494–1512. In 1513 the first of two Medici popes was elected, and Medici rule returned to Florence. A successor to the second Medici pope named the head of the Medici dynasty as first duke of Florence in 1537 and as first grand duke of Tuscany in 1569. In 1495 Machiavelli joined his father's confraternity, which helped launch his public career three years later with his election as second chancellor to the city; his public career ended in 1512 with the return of the Medici.

2. Arnold Pacey, *Technology in World Civilization* (Cambridge: MIT Press, 1990), 82–88, quote on 82.

3. Paolo Galluzzi, "The Career of a Technologist," in Paolo Galluzzi, ed., *Leonardo da Vinci: Engineer and Architect* (Montreal Museum of Fine Arts, 1987), 41–109, quote on 41.

4. Ross King, *Brunelleschi's Dome* (New York: Penguin, 2001); Anthony Grafton, *Leon Battista Alberti* (New York: Hill & Wang, 2000), quote on 72.

5. Galluzzi, "Career of a Technologist," 50.

6. Galluzzi, "Career of a Technologist," quote on 62 (letter to Ludovico); Pamela Taylor, ed., *The Notebooks of Leonardo da Vinci* (New York: NAL, 1971), quote on 207 ("time of peace"). Leonardo's Milan-era military engineering is vividly illustrated in Claudio Giorgione, ed., *Leonardo da Vinci: The Models Collection* (Milan: Museo Nazionale della Scienza e della Tecnica Leonardo da Vinci, 2009), 66–83.

7. Martin Clayton, *Leonardo da Vinci* (London: Merrell Holberton, 1996), quote on 50.

8. Galluzzi, "Career of a Technologist," quote on 68 ("injury to friends"); Taylor, *Notebooks of Leonardo*, quote on 107 ("evil nature").

9. Galluzzi, "Career of a Technologist," quote on 80.

10. Bert S. Hall, *Weapons and Warfare in Renaissance Europe* (Baltimore: Johns Hopkins University Press, 1997).

11. A story of Matteo Bandello quoted in Martin Kemp, *Leonardo da Vinci: The Marvellous Works of Nature and Man* (Cambridge: Harvard University Press, 1981), 180.

12. David Mateer, ed., *Courts, Patrons, and Poets* (New Haven: Yale University Press, 2000), quotes on 136 ("propaganda") and 138 ("seven boys"); Mark Elling Rosheim,

"Leonardo's Lost Robot," *Achademia Leonardi Vinci* 9 (1996): 99–108; Kemp, *Leonardo da Vinci*, 152–53, 167–70.

13. Galluzzi, "Career of a Technologist," quote on 83 ("duke lost"); Irma A. Richter, ed., *The Notebooks of Leonardo da Vinci* (Oxford: Oxford University Press, 1952), quote on 326 ("flooding the castle").

14. The Leonardo–Machiavelli collaboration is reconstructed in Roger D. Masters, *Fortune Is a River* (New York: Free Press, 1998); Jerry Brotton, *The Renaissance Bazaar: From the Silk Road to Michelangelo* (Oxford: Oxford University Press, 2002), quote on 68.

15. Martin Clayton, "Leonardo's Anatomy Years," *Nature* 484 (18 April 2012): 314–316 at web.archive.org/web/20200709231734/https://www.nature.com/articles/484314a.

16. Richter, *Notebooks of Leonardo da Vinci,* quote on 382; Clayton, *Leonardo da Vinci,* 129.

17. Galluzzi, "Career of a Technologist," 91.

18. Kemp, *Leonardo da Vinci*, 347–49.

19. The artist Raphael, quoted in Peter Burke, *The European Renaissance: Centres and Peripheries* (Oxford: Blackwell, 1998), 69.

20. Grafton, *Leon Battista Alberti,* 3–29, 71–109, 189–224, quote on 28.

21. Grafton, *Leon Battista Alberti,* quote on 84. Grafton points out that the Alberti family's exile from Florence was lifted only in 1428, some time after which Leon Battista first entered the city. Thus, he was unlikely to have seen Brunelleschi's show box spectacle in person; he did dedicate to Brunelleschi the Italian version of his painting treatise.

22. Joan Gadol, *Leon Battista Alberti* (Chicago: University of Chicago, 1969), 26 n9, 205 n82.

23. Stephen F. Mason, *A History of the Sciences* (New York: Collier, 1962), 111.

24. Lynn White, "The Flavor of Early Renaissance Technology," in B. S. Levy, ed., *Developments in the Early Renaissance* (Albany: SUNY Press, 1972), quote on 38.

25. Printing in the Far East follows Maurice Daumas, ed., *A History of Technology and Invention* (New York: Crown, 1962), 1:285–89. Evidence on whether printing in the Far East *might* have influenced Gutenberg in Europe is evaluated in Albert Kapr, *Johann Gutenberg: The Man and His Invention* (Aldershot, England: Scolar Press, 1996), 109–22.

26. Colin Clair, *A History of European Printing* (London: Academic Press, 1976), 6–14; Daumas, *History of Technology and Invention,* 2:639 (on typography). On dates, I have followed Kapr, *Johann Gutenberg,* 29–99, quote on 84 ("use of a press").

27. Kapr, *Johann Gutenberg,* 142–79, 197–210, 259–66, quote on 259.

28. Clair, *History of European Printing,* quotes on 13 and 14.

29. Elizabeth L. Eisenstein, *The Printing Revolution in Early Modern Europe* (Cambridge: Cambridge University Press, 1983), 145–48.

30. Eisenstein, *Printing Revolution,* 170–75.

31. Elizabeth L. Eisenstein, *The Printing Press as an Agent of Change,* 2 vols. (Cambridge: Cambridge University Press, 1979), 1:408.

32. On Plantin, see Clair, *History of European Printing*, 195–206; Francine de Nave and Leon Voet, "The Plantin Moretus Printing Dynasty and Its History," in Christopher and Sally Brown, eds., *Plantin-Moretus Museum* (Antwerp: Plantin-Moretus Museum, 1989), 11–53; Eisenstein, *Printing Press*, quote on 1:444.

33. Eisenstein, *Printing Revolution*, 43.

34. Pacey, *Technology in World Civilization*, quote on 7.

35. See Dieter Kuhn, *Science and Civilisation in China*, vol. 5 part 9: *Textile Technology: Spinning and Reeling* (Cambridge: Cambridge University Press, 1988), 161, 184, 210, 348, 352, 356, 363, 366–68. By comparison, a drawing by Leonardo (on 164) fairly leaps off the page with its three-dimensional realism. For other incomplete illustrations of key spinning technologies, see Francesca Bray, *Technology and Gender: Fabrics of Power in Late Imperial China* (Berkeley: University of California Press, 1997), 45, 214, 227; and Arnold Pacey, *Technology in World Civilization*, 27.

36. Eugene Ferguson, *Engineering and the Mind's Eye* (Cambridge: MIT Press, 1992), 107–13.

37. Pamela O. Long, "The Openness of Knowledge: An Ideal and Its Context in Sixteenth-Century Writings on Mining and Metallurgy," *Technology and Culture* 32 (April 1991): 318–55, quote on 326 at doi.org/10.2307/3105713. See also Long's *Openness, Secrecy, Authorship: Technical Arts and the Culture of Knowledge from Antiquity to the Renaissance* (Baltimore: Johns Hopkins University Press, 2001), chap. 6.

38. Long, "Openness of Knowledge," 330–50.

39. Agricola quoted in Eisenstein, *Printing Revolution*, 193.

40. Ferguson, *Engineering and the Mind's Eye*, 115.

41. Galileo quoted in Eisenstein, *Printing Revolution*, 251. On Galileo and the Medici see Mario Biagioli's *Galileo, Courtier* (Chicago: University of Chicago Press, 1993).

42. David Landes, *Revolution in Time: Clocks and the Making of the Modern World* (Cambridge, Mass.: Belknap Press, 1983), quote on 70; Luca Molà, *The Silk Industry of Renaissance Venice* (Baltimore: Johns Hopkins University Press, 2000), quote on 37, 121. See the Note on Sources for glassmaking, shipbuilding, and court technologies.

43. Pacey, *Technology in World Civilization*, 82–88; Noel Perrin, *Giving Up the Gun: Japan's Reversion to the Sword, 1543–1879* (Boston: Godine, 1979).

CHAPTER 2.
TECHNIQUES OF COMMERCE, 1588–1740

1. Richard A. Goldthwaite's *Private Wealth in Renaissance Florence* (Princeton: Princeton University Press, 1968), 40–61. During the 1420s and 1430s, the immensely rich Strozzi family had its assets mostly in real estate holdings: the father Simone's estate consisted of real estate holdings (64%), business investments (23%), and state funds (7%); his son Matteo's estate was real estate holdings (70%), with additional investments in state funds (13%) and in wool manufacture (17%). Matteo's son Filippo built a large fortune

in banking and commerce across Europe; in 1491–92 his estate consisted of real estate (14%), personal property, including household furnishings and precious metals (11%), cash (45%), and business investments (30%).

2. In 1575 and 1596, and at least four times in the following century, the Spanish crown declared itself bankrupt; between 1572 and 1607, Spanish troops mutinied forty-six separate times. A relative decline affected the fortunes of Italian city-states and courts. In the 1620s the Swedish crown (as noted below) lost control of its finances to Dutch copper merchants.

3. Karel Davids and Jan Lucassen, eds., *A Miracle Mirrored: The Dutch Republic in European Perspective* (Cambridge: Cambridge University Press, 1995), quote on 1. For long-term "secular trends," see Fernand Braudel, *The Perspective of the World* (New York: Harper & Row, 1984), 71–88.

4. Jonathan I. Israel, *Dutch Primacy in World Trade, 1585–1740* (Oxford: Clarendon Press, 1989), quote on 13. This chapter draws repeatedly on Israel's work.

5. William H. McNeill, *The Pursuit of Power* (Chicago: University of Chicago Press, 1982), 126–41, quotes on 137–39.

6. Joel Mokyr, *Industrialization in the Low Countries, 1795–1850* (New Haven: Yale University Press, 1976), quote on 9.

7. On de Geer and the Swedish iron and copper industries, see Jonathan I. Israel, *The Dutch Republic* (Oxford: Clarendon Press, 1995), 271–75; E. E. Rich and C. H. Wilson, eds., *Cambridge Economic History of Modern Europe* (Cambridge: Cambridge University Press, 1977), 5:245–49, 484–87, 495. On Eskilstuna, see Maths Isacson and Lars Magnusson, *Proto-industrialisation in Scandinavia: Craft Skills in the Industrial Revolution* (Leamington Spa, U.K.: Berg, 1987), 89–108.

8. Rich and Wilson, *Cambridge Economic History*, 5:148–53; Israel, *The Dutch Republic*, 16–18, 116–19; Jan Bieleman, "Dutch Agriculture in the Golden Age, 1570–1660," in Karel Davids and Leo Noordegraaf, eds., *The Dutch Economy in the Golden Age* (Amsterdam: Netherlands Economic History Archives, 1993), 159–83.

9. Israel, *Dutch Primacy in World Trade*, 21.

10. Richard W. Unger, *Dutch Shipbuilding before 1800* (Assen, Netherlands: Van Gorcum, 1978), 29–47, quote on 25.

11. For background and images, see www.vasamuseet.se/en (May 2020); for dimensions, see web.archive.org/web/20151017033410/http://www.vasamuseet.se/en/The-Ship/Vasa-in-numbers/.

12. Karel Davids, "Technological Change and the Economic Expansion of the Dutch Republic, 1580–1680," 79–104, in Davids and Noordegraaf, *Dutch Economy*, 81–82; Arne Kaijser, "System Building from Below: Institutional Change in Dutch Water Control Systems," *Technology and Culture* 43 (2002): 521–48 at doi.org/10.1353/tech.2002.0120.

13. Israel, *Dutch Primacy in World Trade*, 21–22; Leo Noordegraaf, "Dutch Industry in the Golden Age," in Davids and Noordegraaf, *Dutch Economy*, 142–45; Carlo Cipolla, *Before the Industrial Revolution*, 3rd edition (New York: W. W. Norton, 1994), 254.

14. John E. Wills, Jr., *Pepper, Guns and Parleys: The Dutch East India Company and China, 1622–1681* (Cambridge: Harvard University Press, 1974), 10, 19–21, quote on 20.

15. Israel, *Dutch Primacy in World Trade*, 75–76, quote on 75.

16. Israel, *Dutch Primacy in World Trade*, 73–79, quote on 73 n8.

17. Peter M. Garber, *Famous First Bubbles: The Fundamentals of Early Manias* (Cambridge: MIT Press, 2000), 82.

18. Paul Zumthor, *Daily Life in Rembrandt's Holland* (Stanford: Stanford University Press, 1994), quote on 173 ("so many rules"); Mike Dash, *Tulipomania* (New York: Crown, 1999), quote on 141 ("intoxicated head"). In 1590 Haarlem residents consumed an estimated 300 liters of beer per person per year; see Richard W. Unger, "The Scale of Dutch Brewing, 1350–1600," *Research in Economic History* 15 (1995): 261–92. At an Amsterdam *orphanage* during 1639–59, the ration of beer was 270 liters per person per year; see Anne McCants, "Monotonous but not Meager: The Diet of Burgher Orphans in Early Modern Amsterdam," *Research in Economic History* 14 (1994): 69–119. A beer keg is prominent in a ca. 1660 oil painting depicting the feeding of orphans; see Richard W. Unger, *A History of Brewing in Holland, 900–1900* (Leiden: Brill, 2001), 248.

19. Israel, *The Dutch Republic*, 318–27.

20. P'eng Sun-I, "Record of Pacification of the Sea" (1662–64) quoted in Wills, *Pepper, Guns and Parleys*, quote on 29; Israel, *Dutch Primacy in World Trade*, 71–73, 76, 86.

21. Israel, *Dutch Primacy in World Trade*, 101–6.

22. Israel, *Dutch Primacy in World Trade*, 171–87, quote on 177.

23. See Noel Perrin, *Giving Up the Gun: Japan's Reversion to the Sword, 1543–1879* (Boston: Godine, 1979).

24. Israel, *Dutch Primacy in World Trade*, 156–60; Pieter Emmer, "The West India Company, 1621–1791: Dutch or Atlantic?" in his *The Dutch in the Atlantic Economy* (Aldershot, England: Ashgate, 1998), chap. 3.

25. Israel, *Dutch Primacy in World Trade*, 160–70, quote on 163, 186, 255, 319–27; Pieter Emmer and Ernst van den Boogart, "The Dutch Participation in the Atlantic Slave Trade, 1596–1650," in Emmer, *The Dutch in the Atlantic Economy*, chap. 2; Carla Rahn Phillips, *Six Galleons for the King of Spain: Imperial Defense in the Early Seventeenth Century* (Baltimore: Johns Hopkins University Press, 1986), 3–7.

26. Mokyr, *Industrialization in the Low Countries*, quote on 2; Israel, *Dutch Primacy in World Trade*, 114–16; Cipolla, *Before the Industrial Revolution* (3rd ed.), 249–59.

27. Israel, *Dutch Primacy in World Trade*, 187–91, quote on 188.

28. Israel, *Dutch Primacy in World Trade*, 35–36, quote on 36 (on linen bleaching), 116–17,

190–96, quote on 194 ("technical innovations"); Carla Rahn Phillips and William D. Phillips, Jr., *Spain's Golden Fleece* (Baltimore: Johns Hopkins University Press, 1997), 260, 303.

29. Israel, *Dutch Primacy in World Trade,* 260–64, quote on 262.

30. Phillips and Phillips, *Spain's Golden Fleece,* 169–78, 193–209, quote on 204.

31. Israel, *Dutch Primacy in World Trade,* 116, 264–66.

32. Israel, *Dutch Primacy in World Trade,* 266–69; Karel Davids, "The Transformation of an Old Industrial District: Firms, Family, and Mutuality in the Zaanstreek between 1840 and 1920," *Enterprise & Society* 7 no. 3 (2006): 550–80, on 557 at www.jstor.org /stable/23700836

33. Israel, *Dutch Primacy in World Trade,* quote on 269. The varied activities of Dutch guilds, learned societies, and towns are evaluated in Karel Davids, "Shifts of Technological Leadership in Early Modern Europe," in Davids and Lucassen, *A Miracle Mirrored,* 338–66.

34. The Dutch economic decline is detailed in Israel, *The Dutch Republic,* 998–1018.

35. Davids, "Technological Change," 81–83.

36. Israel, *Dutch Primacy in World Trade,* 102, 292–358.

37. Zumthor, *Daily Life in Rembrandt's Holland,* 194–99, quote on 195; Cipolla, *Before the Industrial Revolution* (3rd ed.), 259.

CHAPTER 3.
GEOGRAPHIES OF INDUSTRY, 1740–1851

1. E. J. Hobsbawm, *Industry and Empire* (London: Weidenfeld & Nicolson, 1968; reprint, London: Penguin, 1990), quote on 34; David S. Landes, *The Unbound Prometheus* (Cambridge: Cambridge University Press, 1969), quote on 1–2. For sectoral growth rates, see Maxine Berg, *The Age of Manufactures, 1700–1820,* 1st edition (Oxford: Oxford University Press, 1986), 28.

2. Berg, *Age of Manufactures* (1986), quote in (unpaged) preface. For data on value added by sector, see Maxine Berg, *The Age of Manufactures,* 2nd edition (London: Routledge, 1994), 38. On handloom weavers, see Geoffrey Timmons, *The Last Shift* (Manchester: Manchester University Press, 1993), 25–28, 110, 220.

3. Raymond Williams, *Culture and Society, 1780–1950* (New York: Harper & Row, 1966), xi–xvi and passim.

4. L. D. Schwarz, *London in the Age of Industrialisation: Entrepreneurs, Labour Force and Living Conditions, 1700–1850* (Cambridge: Cambridge University Press, 1992), quotes on 1 (Braudel), 231 ("a storm"). The "gentlemanly capitalism" thesis is scrutinized in M. J. Daunton, " 'Gentlemanly Capitalism' and British Industry, 1820–1914," *Past and Present* 122 (1989): 119–58; the thesis is defended in P. J. Cain and A. G. Hopkins, *British Imperialism, 1688–2000,* 2nd edition (Harlow: Longman, 2002), 114–50, passim.

5. Around 1800, London's total of 290 steam engines outnumbered those in Manchester, Leeds, or Glasgow (respectively home to 240, 130, and 85); see Richard L. Hills, *Power from Steam: A History of the Stationary Steam Engine* (Cambridge: Cambridge University Press, 1989), 299 n78.

6. Data from Eric J. Evans, *The Forging of the Modern State: Early Industrial Britain, 1783–1870* (London: Longman, 1983), 407–9; E. A. Wrigley, "A Simple Model of London's Importance in Changing English Society and Economy 1650–1750," *Past and Present* 37 (1967): 44–70; David R. Green, *From Artisans to Paupers: Economic Change and Poverty in London, 1790–1870* (London: Scolar Press, 1995), Defoe quote on 15. Populations in 1851, in thousands: London 2,362, Liverpool 376, Glasgow 345, Manchester 303, Birmingham 233, Edinburgh 194, Leeds 172, Sheffield 135, Bristol 137, Bradford 104, Newcastle 88, Hull 85, Preston 70.

7. Roy Porter, *London: A Social History* (Cambridge: Harvard University Press, 1994), 134.

8. M. Dorothy George, *London Life in the Eighteenth Century* (New York: Capricorn Books, 1965, orig. 1925), quote on 323 n2; all data in 1851 census from Schwarz, *London in the Age of Industrialisation*, 255–58.

9. Michael W. Flinn, *The History of the British Coal Industry* (Oxford: Clarendon Press, 1984), 2:217, table 7.2; Defoe quoted in Porter, *London*, 138.

10. George, *London Life*, quote on 167.

11. This section relies on Peter Mathias, *The Brewing Industry in England, 1700–1830* (Cambridge: Cambridge University Press, 1959).

12. Mathias, *Brewing Industry in England*, 53–62, quotes on 61, 62.

13. T. R. Gourvish and R. G. Wilson, *The British Brewing Industry, 1830–1980* (Cambridge: Cambridge University Press, 1994), 30–35. Gourvish and Wilson assume that children under 15 (35% of the population) and tea-totallers (3 million in 1899) drank no alcohol, and estimate that women drank half as much as men, calculating that in the peak year of 1876 the average adult male consumed 103 gallons of beer a year (16 pints a week).

14. Gourvish and Wilson, *British Brewing Industry*, 87, 226–66; Mathias, *Brewing Industry in England*, 554–58.

15. Mathias, *Brewing Industry in England*, 36, 551–58.

16. Eric Robinson and A. E. Musson, *James Watt and the Steam Revolution* (New York: Kelley, 1969), quote on 88 (Watt); Mathias, *Brewing Industry in England*, 78–98.

17. Mathias, *Brewing Industry in England*, 41–42, 48–53, 106–9. For ancillary industries, see Philip Scranton, *Endless Novelty: Specialty Production and American Industrialization, 1865–1925* (Princeton: Princeton University Press, 1997).

18. Mathias, *Brewing Industry in England*, 102–6, 117–33.

19. Gourvish and Wilson, *British Brewing Industry*, 30–35, quote on 12. Per capita consumption of beer in England and Wales was 28.4 gallons in 1820–24, 35.4 gallons in 1835–39, and 30.5 in 1840–44, a figure maintained until 1860.

20. *Report on Handloom Weavers* (1840) quoted in Ivy Pinchbeck, *Women Workers and the Industrial Revolution, 1750–1850* (London: Frank Cass, 1930), quote on 179.

21. George, *London Life,* quotes on 74 and 345 n32; Linda Clarke, *Building Capitalism: Historical Change and the Labour Process in the Production of the Built Environment* (London: Routledge, 1992), 82–84, 174–77, 207–17, 243–48; Hermione Hobhouse, *Thomas Cubitt: Master Builder* (London: Macmillan, 1971), 23–25, 96–102.

22. Keith Burgess, "Technological Change and the 1852 Lock-Out in the British Engineering Industry," *International Review of Social History* 14 (1969): 215–36, on 233; Green, *From Artisans to Paupers,* 27–32, 59 n30.

23. Peter Linebaugh, *The London Hanged* (Cambridge: Cambridge University Press, 1992), 371–401, quote on 380; Carolyn Cooper, "The Portsmouth System of Manufacture," *Technology and Culture* 25 (1984): 182–225 at doi.org/10.2307/3104712.

24. David Jeremy, *Transatlantic Industrial Revolution* (Cambridge: MIT Press, 1981), 67.

25. V. A. C. Gatrell, "Labour, Power, and the Size of Firms in Lancashire Cotton in the Second Quarter of the Nineteenth Century," *Economic History Review* 30 (1977): 95–139, statistics on 98; Steven Marcus, *Engels, Manchester, and the Working Class* (New York: Random House, 1974), quote on 46 n30 ("only one chimney"); W. Cooke Taylor, *Notes of a Tour in the Manufacturing Districts of Lancashire* (London: Frank Cass, 1842; reprint, New York: A. M. Kelley, 1968), quote on 6 ("mighty energies").

26. Sidney J. Chapman, *The Lancashire Cotton Industry* (Manchester: Manchester University Press, 1904), quotes on 2 (Defoe), 13 (George Crompton).

27. See the classic account of home and factory spinning in Pinchbeck, *Women Workers,* 129–56, 183–201. Pinchbeck informs many recent treatments, including Sally Alexander, *Women's Work in Nineteenth-Century London* (London: Journeyman Press, 1983); Deborah Valenze, *The First Industrial Woman* (New York: Oxford University Press, 1995); and Katrina Honeyman, *Women, Gender and Industrialisation in England, 1700–1870* (London: Macmillan, 2000).

28. Ian Miller and John Glithero, "Richard Arkwright's Shudehill Mill: The Archaeology of Manchester's First Steam-Powered Cotton Mill," *Industrial Archaeology Review* 38 no. 2 (2016): 98–118 at doi.org/10.1080/03090728.2016.1266214.

29. Pinchbeck, *Women Workers,* quote on 148.

30. Pinchbeck, *Women Workers,* 157–82; for data on Manchester mills, see Roger Lloyd-Jones and M. J. Lewis, *Manchester and the Age of the Factory* (London: Croom Helm, 1988).

31. Lloyd-Jones and Lewis, *Manchester and the Age of the Factory,* 67.

32. Pinchbeck, *Women Workers,* 186 (data from 151 Lancashire cotton mills in 1834); Mary Freifeld, "Technological Change and the 'Self-Acting' Mule: A Study of Skill and the Sexual Division of Labour," *Social History* 11 (October 1986): 319–43; G. N. von Tunzelmann, *Steam Power and British Industrialization to 1860* (Oxford: Clarendon

Press, 1978), 185 (employees and horsepower); Lloyd-Jones and Lewis, *Manchester and the Age of the Factory, Gazette* quoted on 201. In 1833 the three largest mills in Manchester each employed 1,400 workers, eight mills had 500–900, an additional eight had 300–500, and seventeen mills had 100–300; see Chapman, *Lancashire Cotton Industry,* 58.

33. Martin Hewitt, *The Emergence of Stability in the Industrial City: Manchester, 1832–67* (Aldershot, England: Scolar Press, 1996), 31 (employment data); Timmons, *The Last Shift,* 20, table 1.1 (on power looms).

34. Samuel Smiles, *Industrial Biography* (1863) at web.archive.org/web/20190505215120 /www.gutenberg.org/ebooks/404; Freifeld, "Technological Change."

35. Nasmyth quoted in L. T. C. Rolt, *A Short History of Machine Tools* (Cambridge: MIT Press, 1965), 113.

36. A. E. Musson and Eric Robinson, *Science and Technology in the Industrial Revolution* (Toronto: University of Toronto Press, 1969), quote on 507 (from Nasmyth); Lloyd-Jones and Lewis, quote on 134 (on unrest over price of labor); Chapman, *Lancashire Cotton Industry,* quote on 198 ("social war").

37. Marcus, *Engels, Manchester, and the Working Class,* 30–44, quote on 46 ("chimney of the world"); Taylor, *Notes of a Tour in Lancashire* [1842], quote on 2 ("forest of chimneys"); Harold L. Platt, *Shock Cities: The Environmental Transformation and Reform of Manchester and Chicago* (Chicago: University of Chicago Press, 2005), 24–56; John Kasson, *Civilizing the Machine* (New York: Grossman, 1976), 59–60.

38. Marcus, *Engels, Manchester, and the Working Class,* 30–44; Disraeli quoted in Robert J. Werlin, *The English Novel and the Industrial Revolution* (New York: Garland, 1990), 74, 84 n1.

39. Page numbers cited for quotations from Friedrich Engels, *The Condition of the Working Class in England* (1845) in the edition translated and edited by W. O. Henderson and W. H. Chaloner (Stanford: Stanford University Press, 1958).

40. W. O. Henderson and W. H. Chaloner, "Friedrich Engels in Manchester," *Memoirs and Proceedings of the Manchester Literary and Philosophical Society* 98 (1956–57): 13–29; Karl Marx and Friedrich Engels, "Manifesto of the Communist Party" (1848), in Robert C. Tucker, ed., *The Marx-Engels Reader,* 2nd edition (New York: W. W. Norton, 1978), quote on 476.

41. Nathaniel Hawthorne, *Our Old Home* (1863), in Sylvia Pybus, ed., *"Damned Bad Place, Sheffield"* (Sheffield: Sheffield Academic Press, 1994), quote on 126.

42. Pinchbeck, *Women Workers,* 275–76; Evans, *Forging of the Modern State,* 165.

43. Geoffrey Tweedale, *Steel City: Entrepreneurship, Strategy and Technology in Sheffield, 1743–1993* (Oxford: Clarendon Press, 1995), 49–50.

44. Tweedale, *Steel City,* 53.

45. Geoffrey Tweedale, in *Sheffield Steel and America* (Cambridge: Cambridge University Press, 1987), notes an apparent exception to the network or cluster model, Greaves's Sheaf Works, built in 1823 for £50,000, where supposedly "one grand end was kept in

view, namely that of centralizing on the spot all the various processes through which iron must pass . . . until fashioned into razor, penknife or other article of use" (51).

46. Tweedale, *Sheffield Steel*, 63–67, quote on 64 (from Bessemer, *Autobiography*).

47. Sidney Pollard, *A History of Labour in Sheffield* (Liverpool: Liverpool University Press, 1959), 50–54, quote on 51.

48. Pollard, *History of Labour in Sheffield*, 53–54, 83.

49. John Charles Hall, "The Sheffield Grinders," *British Medical Journal* 1 no. 12 (March 21, 1857): 234–236, quote on 235 at www.jstor.com/stable/25191141.

50. Pollard, *History of Labour in Sheffield*, 62–65, 328; John Charles Hall, "The Sheffield File-Cutters," *British Medical Journal* 1 no. 19 (9 May 1857): 385–87 (colic and paralysis) at jstor.org/stable/25191274; Hall, "The Sheffield Grinders" (21 March 1857), quote on 235 ("very old man") at jstor.org/stable/25191141; Hall, "Remarks on the Effects of the Trades of Sheffield," *British Medical Journal* 2 no. 824 (14 October 1876): 485–86 at jstor.org/stable/25243013.

51. Pollard, *History of Labour in Sheffield*, 14–15.

52. Pollard, *History of Labour in Sheffield*, 5–23, 93–100.

53. Fiorenza Belussi and Katia Caldari, "At the Origin of the Industrial District: Alfred Marshall and the Cambridge School," *Cambridge Journal of Economics* 33 no. 2 (March 2009): 335–355 at doi.org/10.1093/cje/ben041.

54. Donald E. Thomas, Jr., "Diesel, Father and Son: Social Philosophies of Technology," *Technology and Culture* 19 no. 3 (1978): 376–393 at www.jstor.com/stable/3103371.

55. See Jeremy, *Transatlantic Industrial Revolution*; Ken Alder, *Engineering the Revolution: Arms and Enlightenment in France, 1763–1815* (Princeton: Princeton University Press, 1997), Thomas J. Misa, *A Nation of Steel: The Making of Modern America, 1865–1925* (Baltimore: Johns Hopkins University Press, 1995).

56. Maths Isacson and Lars Magnusson, *Proto-industrialisation in Scandinavia: Craft Skills in the Industrial Revolution* (Leamington Spa, U.K.: Berg, 1987); Svante Lindqvist, *Technology on Trial: The Introduction of Steam Power Technology into Sweden, 1715–1736* (Stockholm: Almqvist & Wiksell, 1984), 23–33.

57. Sven Beckert, "Cotton and the Global Origins of Capitalism," *Journal of World History* 28 no. 1 March (2017): 107–120, quotes on 107, 109 at doi.org/10.1353/jwh.2017.0004; Peter A. Coclanis, "King Cotton: Sven Beckert, *Empire of Cotton*," *Technology and Culture* 57 no. 3 (2016): 661–667, quote on 663, at doi.org/10.1353/tech.2016.0076.

CHAPTER 4.
INSTRUMENTS OF EMPIRE, 1840–1914

1. Charles Bright, *Telegraphy, Aeronautics and War* (London: Constable, 1918), 56.

2. Daniel Headrick, *The Tools of Empire: Technology and European Imperialism in the Nineteenth Century* (Oxford: Oxford University Press, 1981).

3. Lance E. Davis and Robert A. Huttenback, *Mammon and the Pursuit of Empire: The Economics of British Imperialism* (Cambridge: Cambridge University Press, 1988).

4. On steam in India see Headrick, *Tools of Empire,* 17–57, 129–56.

5. Zaheer Baber, *The Science of Empire: Scientific Knowledge, Civilization, and Colonial Rule in India* (Albany: SUNY Press, 1996), 138–53.

6. Satpal Sangwan, "Technology and Imperialism in the Indian Context: The Case of Steamboats, 1819–1839," in Teresa Meade and Mark Walker, eds., *Science, Medicine, and Cultural Imperialism* (New York: St. Martin's Press, 1991), 60–74, quote on 70.

7. Headrick, *Tools of Empire,* quote on 50.

8. Martin Booth, *Opium* (New York: St. Martin's Press, 1996), 103–73, quote on 146.

9. *Encyclopaedia Britannica,* 11th edition (1910), 14:389; Shrutidev Goswami, "The Opium Evil in Nineteenth Century Assam," *Indian Economic and Social History Review* 19 (1982): 365–76.

10. Ellen N. LaMotte, *Peking Dust* (New York: The Century, 1919), 124–81, 233–37, at archive.org/details/in.ernet.dli.2015.158280; The International Anti Opium Association, *The War Against Opium* (Peking: Tientsin Press, Ltd., 1922), quote on iv (vast quantities) at archive.org/details/in.ernet.dli.2015.157162.

11. Bright, *Telegraphy, Aeronautics and War,* 56.

12. This section follows Mel Gorman, "Sir William O'Shaughnessy, Lord Dalhousie, and the Establishment of the Telegraph System in India," *Technology and Culture* 12 (1971): 581–601 at doi.org/10.2307/3102572; Saroj Ghose, "Commercial Needs and Military Necessities: The Telegraph in India," in Roy MacLeod and Deepak Kumar, eds., *Technology and the Raj: Western Technology and Technical Transfers to India, 1700–1947* (New Delhi: Sage, 1995), 153–76, quote on 156.

13. Edwin Arnold, *The Marquis of Dalhousie's Administration of British India* (London, 1862) 2:241–42, as quoted in Michael Adas, *Machines as the Measure of Men* (Ithaca: Cornell University Press, 1989), 226.

14. Vary T. Coates and Bernard Finn, *A Retrospective Technology Assessment: Submarine Telegraphy* (San Francisco: San Francisco Press, 1979), quotes on 101.

15. Ghose, "Commercial Needs and Military Necessities," quote on 168. For different reasons, battlefield commanders might resent the control that telegraphs permitted. The Prussian field marshal and chief-of-staff H. K. Moltke wrote of the Austro-Prussian War (1866), "No commander is less fortunate than he who operates with a telegraph wire stuck into his back" (as quoted in Martin van Creveld, *Command in War* [Cambridge: Harvard University Press, 1985], 146). I am indebted to Ed Todd for this citation.

16. Daniel Headrick, *The Invisible Weapon: Telecommunications and International Politics, 1851–1945* (Oxford University Press, 1991), 17–115; Headrick, *Tools of Empire,* 130.

17. Peter Harnetty, *Imperialism and Free Trade: Lancashire and India in the Mid-Nineteenth Century* (Vancouver: University of British Columbia Press, 1972), quote on 6.

18. K. N. Chaudhuri, "The Structure of the Indian Textile Industry in the Seventeenth and Eighteenth Centuries," in Michael Adas, ed., *Technology and European Overseas Enterprise* (Aldershot, England: Variorum, 1996), 343–98; Harnetty, *Imperialism and Free Trade*, 7–35.

19. Harnetty, *Imperialism and Free Trade*, quotes 66–67.

20. Arun Kumar, "Colonial Requirements and Engineering Education: The Public Works Department, 1847–1947," in MacLeod and Kumar, *Technology and the Raj*, 216–32; Daniel Headrick, *The Tentacles of Progress: Technology Transfer in the Age of Imperialism, 1850–1940* (Oxford: Oxford University Press, 1988), 315–45; George Tomkyns Chesney, *The Battle of Dorking* (London: Grant Richards, 1914; original 1871) at archive.org/details/battleofdorking00chesrich/page/92/.

21. Daniel Thorner, *Investment in Empire: British Railway and Steam Shipping Enterprise in India, 1825–1849* (Philadelphia: University of Pennsylvania Press, 1950), quote on 11.

22. Thorner, *Investment in Empire*, quotes on 49 ("first consideration"), 96 ("extension"). Military and administrative priorities also informed the routing of later lines; for cases studies see Bharati Ray, "The Genesis of Railway Development in Hyderabad State: A Case Study in Nineteenth Century British Imperialism," *Indian Economic and Social History Review* 21 (1984): 45–69, on 54–55; Mukul Mukherjee, "Railways and Their Impact on Bengal's Economy, 1870–1920," *Indian Economic and Social History Review* 17 (1980): 191–209, quotes on 193–94.

23. Headrick, *Tentacles of Progress*, quote on 71.

24. Ian Derbyshire, "The Building of India's Railways: The Application of Western Technology in the Colonial Periphery, 1850–1920," in MacLeod and Kumar, *Technology and the Raj*, 177–215.

25. Daniel Thorner, "Great Britain and the Development of India's Railways," *Journal of Economic History* 11 (1951): 389–402, quote on 392.

26. John Hoyt Williams, *A Great and Shining Road* (New York: Times Books, 1988), 152 (killed in next war); Creed Haymond, *The Central Pacific Railroad Co.* (San Francisco: H. S. Crocker, 1888), quote on 43 (Indian problem).

27. Donald W. Roman, "Railway Imperialism in Canada, 1847–1868," and Ronald E. Robinson, "Railways and Informal Empire," in Clarence B. Davis and Kenneth E. Wilburn, Jr., eds., *Railway Imperialism* (Boulder: Greenwood, 1991), 7–24, 175–96; A. A. Den Otter, *The Philosophy of Railways: The Transcontinental Railway Idea in British North America* (Toronto: University of Toronto Press, 1997), quote on 204.

28. William E. French, "In the Path of Progress: Railroads and Moral Reform in Porfirian Mexico," in Davis and Wilburn, *Railway Imperialism*, 85–102; Robinson, "Railways and Informal Empire," quote on 186 ("tariff regulations"); David M. Pletcher, *Rails, Mines, and Progress: Seven American Promoters in Mexico, 1867–1911* (Port Washington: Kennikat Press, 1972), quote on 1 ("our India").

29. Pletcher, *Rails, Mines, and Progress*, 313.

30. M. Tamarkin, *Cecil Rhodes and the Cape Afrikaners* (London: Frank Cass, 1996), 8.

31. India's princely state of Hyderabad used a similar tactic—contesting the imperialist's railway plans by building alternative lines with financing directly from private London money; see Ray, "Genesis of Railway Development," 56–60.

32. Kenneth E. Wilburn, Jr., "Engines of Empire and Independence: Railways in South Africa, 1863–1916," in Davis and Wilburn, *Railway Imperialism*, 25–40.

33. W. Travis Hanes III, "Railway Politics and Imperialism in Central Africa, 1889–1953," in Davis and Wilburn, *Railway Imperialism*, 41–69; John Edward Glab, "Transportation's Role in Development of Southern Africa" (PhD diss., American University, 1970), 62–69.

34. On Jared Diamond's *Guns, Germs, and Steel*, see the critical review by Suzanne Moon in *Technology and Culture* 41 no. 3 (2000): 570–71 at doi.org/10.1353/tech.2000.0123; Mohandas K. Gandhi, *Indian Home Rule or Hind Swaraj* (Ahmedabad: Navajivan Publishing, 1938) ("harm that Manchester") section 19 Machinery at web.archive.org /web/20200725161303/https://www.mkgandhi.org/hindswaraj/hindswaraj.htm.

35. E. J. Hobsbawm, *Industry and Empire* (London: Weidenfeld & Nicolson, 1968; reprint, London: Penguin, 1990), 178–93; Patrick K. O'Brien, "The Costs and Benefits of British Imperialism, 1846–1914," *Past and Present* 120 (1988): 163–200.

CHAPTER 5.
SCIENCE AND SYSTEMS, 1870–1930

1. See my essay "The Compelling Tangle of Modernity and Technology" in Thomas Misa et al., eds., *Modernity and Technology* (Cambridge: MIT Press, 2003), 1–30. In *Technology: Critical History of a Concept* (University of Chicago Press, 2018), Eric Schatzberg suggests that cultural changes occurred in the 1930s.

2. Peter Mathias, *The Brewing Industry in England, 1700–1830* (Cambridge: Cambridge University Press, 1959), quote on 65.

3. Henk van den Belt, "Why Monopoly Failed: The Rise and Fall of Société La Fuchsine," *British Journal for the History of Science* 25 (1992): 45–63.

4. Jeffrey Allan Johnson, *The Kaiser's Chemists: Science and Modernization in Imperial Germany* (Chapel Hill: University of North Carolina Press, 1990), 216 n7.

5. Henk van den Belt and Arie Rip, "The Nelson-Winter-Dosi Model and Synthetic Dye Chemistry," in Wiebe Bijker, Trevor Pinch, and Thomas Hughes, eds., *The Social Construction of Technological Systems* (Cambridge: MIT Press, 1987), 135–58, quotes on 143–44.

6. Van den Belt and Rip, "Nelson-Winter-Dosi Model," quotes on 151.

7. Johnson, *Kaiser's Chemists*, 33.

8. Carl Duisberg, *Meine Lebenserinnerungen* (Leipzig, 1933), 44; also in van den Belt and Rip, "Nelson-Winter-Dosi Model," quote on 154.

9. Johnson, *Kaiser's Chemists*, quote on 34.

10. Van den Belt and Rip, "Nelson-Winter-Dosi Model," quote on 155.

11. John Joseph Beer, *The Emergence of the German Dye Industry* (Urbana: University of Illinois Press, 1959), 134.

12. Wilfred Owen, "Dulce et Decorum Est" (1917) at web.archive.org/web/20201105231900 /https://www.poetryfoundation.org/poems/46560/dulce-et-decorum-est.

13. L. F. Haber, *The Poisonous Cloud: Chemical Warfare in the First World War* (Oxford: Clarendon Press, 1986), 22–40, 217, 228, 243.

14. Peter Hayes, *Industry and Ideology: IG Farben in the Nazi Era* (Cambridge: Cambridge University Press, 1987), 17 (1930 stats), 359, 361, 370.

15. Paul Israel, *From Machine Shop to Industrial Laboratory: Telegraphy and the Changing Context of American Invention, 1830–1920* (Baltimore: Johns Hopkins University Press, 1992), quote on 138.

16. Thomas P. Hughes, *American Genesis* (New York: Viking, 1989), 29.

17. Thomas P. Hughes, *Networks of Power: Electrification in Western Society, 1880–1930* (Baltimore: Johns Hopkins University Press, 1983), 24.

18. Edison to Puskas, 13 November 1878, Edison Archives, as quoted in Hughes, *Networks of Power*, 33.

19. Robert Friedel and Paul Israel, *Edison's Electric Light: Biography of an Invention* (New Brunswick: Rutgers University Press, 1986), 137.

20. Edison patent 223,898 (granted 27 January 1880), from Friedel and Israel, *Edison's Electric Light*, quote on 106.

21. Hughes, *Networks of Power*, 33–38, 193.

22. Friedel and Israel, *Edison's Electric Light*, quote on 182.

23. Hughes, *Networks of Power*, quote on 45.

24. Louis Galambos and Joseph Pratt, *The Rise of the Corporate Commonwealth* (New York: Basic Books, 1988), 10–11.

25. Hughes, *Networks of Power*, 46.

26. W. Bernard Carlson, *Innovation as a Social Process: Elihu Thomson and the Rise of General Electric, 1870–1900* (Cambridge: Cambridge University Press, 1991), quote on 284 n20. The next paragraphs draw on Carlson's book.

27. Carlson, *Innovation as a Social Process*, 253–57, 290.

28. Hughes, *Networks of Power*, 107–9; Carlson, *Innovation as a Social Process*, 261–63, 285.

29. Carlson, *Innovation as a Social Process*, 290–301, quotes on 292 (Edison) and 297–98 (Fairfield).

30. Many of the largest US companies were founded during the 1880–1930 period; see Harris Corporation, "Founding Dates of the 1994 *Fortune* 500 U.S. Companies," *Business History Review* 70 (1996): 69–90.

31. Cited in George Basalla, *The Evolution of Technology* (Cambridge: Cambridge

University Press, 1988), 128. A comprehensive study is David A. Hounshell and John Kenly Smith, Jr., *Science and Corporate Strategy: Du Pont R&D, 1902–1980* (Cambridge: Cambridge University Press, 1980).

32. George Wise, "A New Role for Professional Scientists in Industry: Industrial Research at General Electric, 1900–1916," *Technology and Culture* 21 (1980): 408–29, quote on 422 at doi.org/10.2307/3103155.

33. Karl L. Wildes and Nilo A. Lindgren, *A Century of Electrical Engineering and Computer Science at MIT, 1882–1982* (Cambridge: MIT Press, 1985), 32–66, quote on 43. See also W. Bernard Carlson, "Academic Entrepreneurship and Engineering Education: Dugald C. Jackson and the MIT-GE Cooperative Engineering Course, 1907–1932," *Technology and Culture* 29 (1988): 536–67 at doi.org/10.2307/3105273.

34. Vannevar Bush, *Pieces of the Action* (New York: William Morrow, 1970), quote on 254; Wildes and Lindgren, *Electrical Engineering at MIT*, 48, 70.

35. Wildes and Lindgren, *Electrical Engineering at MIT*, quote on 57. In 1933 the electrical engineering curriculum was reorganized to focus on electronics, electromagnetic theory, and energy conversion as basic common concepts (see 106).

36. Wildes and Lindgren, *Electrical Engineering at MIT*, 49–54, 62–66, quote on 63.

37. Wildes and Lindgren, *Electrical Engineering at MIT*, 75–77, 96–105, quote on 100 (on GE's Doherty), 103 (on terminating the network analyzer).

CHAPTER 6.
MATERIALS OF MODERNISM, 1900–1950

1. Umberto Boccioni, Carlo Carrà, Luigi Russolo, Giacomo Balla, and Gino Severini, "Manifesto of the Futurist Painters" (1910) at web.archive.org/web/20191117230744 /http://391.org/manifestos/1910-manifesto-of-futurist-painters-boccioni-carra-russolo-balla-severini/. The "great divide" between the ancient and modern worlds is analyzed in Thomas Misa et al., *Modernity and Technology* (Cambridge: MIT Press, 2003); Bruno Latour, *We Have Never Been Modern* (Cambridge: Harvard University Press, 1993), chap. 2; and Jeffrey L. Meikle, "Domesticating Modernity: Ambivalence and Appropriation, 1920–40," in Wendy Kaplan, ed., *Designing Modernity: The Arts of Reform and Persuasion, 1885–1945* (New York: Thames & Hudson, 1995), 143–67.

2. Bruno Taut, *Modern Architecture* (London: The Studio, 1929), quote on 4; Reyner Banham, *Theory and Design in the First Machine Age* (Cambridge: MIT Press, 1980), 94 (Loos citations).

3. For concrete, see Amy Slaton, *Reinforced Concrete and the Modernization of American Building, 1900–1930* (Baltimore: Johns Hopkins University Press, 2001).

4. For striking photos of workers hand-blowing glass *c.* 1900 see Edward Noyes, "The Window Glass Industry of Omro, Wisconsin," *Wisconsin Magazine of History* 43 no. 2 (1959–1960): 108–118 at jstor.org/stable/4633478; and for Lubbers machine-blowing in

the 1920s see Richard O'Connor, "Perfecting the 'Iron Lung': Making the New Window Glass Technology Work," *IA: Journal of the Society for Industrial Archeology* 23 no. 1 (1997): 6–24 at jstor.org/stable/40968380. On the change from hand- to machine-blown glassmaking, see Pearce Davis, *The Development of the American Glass Industry* (Cambridge: Harvard University Press, 1949); Ken Fones-Wolf, *Glass Towns: Industry, Labor, and Political Economy in Appalachia, 1890–1930s* (Urbana: University of Illinois Press, 2007); and Barbara L. Floyd, *The Glass City: Toledo and The Industry That Built It* (Ann Arbor: University of Michigan Press, 2015), 60–87 at jstor.org/stable/10.3998 /mpub.5848800.7. Geography is analyzed by Naomi R. Lamoreaux and Kenneth L. Sokoloff , "The Geography of Invention in the American Glass Industry, 1870–1925," *Journal of Economic History* 60 no. 3 (2000): 700–29 at jstor.org/stable/2566435.

5. Taut, *Modern Architecture* (1929), 6.

6. R. W. Flint, ed., "The Birth of a Futurist Aesthetic" in *Marinetti: Selected Writings* (New York: Farrar, Straus & Giroux, 1973), 81.

7. Filippo Marinetti, "Founding Manifesto of Futurism" (1909), in Pontus Hulten, ed., *Futurismo e Futurismi* (Milan: Bompiani, 1986), quotes on 514–16; Boccioni quoted in Banham, *Theory and Design*, 102. For a photograph of Marinetti's car in the ditch, see F. T. Marinetti, *The Futurist Cookbook* (San Francisco: Bedford Arts, 1989), 10.

8. Boccioni et al., "Manifesto of the Futurist Painters."

9. Umberto Boccioni, "Technical Manifesto of Futurist Sculpture" (1912), in Hulten, *Futurismo e Futurismi*, 132–33, quotes on 433.

10. Antonio Sant'Elia, "Manifesto of Futurist Architecture" (1914), in Hulten, *Futurismo e Futurismi*, 418–20.

11. Sant'Elia, "Manifesto of Futurist Architecture," 418–20; Banham, *Theory and Design*, 127–37, quotes on 129. Here I have quoted from Sant'Elia's original text (from Banham); Marinetti later changed *new* and *modern* to *Futurist*, upgraded *iron* to *steel*, and added several paragraphs at the beginning and, most famously, at the end. Città Nuova is featured in Vittorio Magnago Lampugnani, ed., *Antonio Sant'Elia: Gezeichnete Architektur* (Munich: Prestel, 1992), 7–55, 166–97.

12. R. J. B. Bosworth, *Mussolini* (New York: Oxford University Press, 2002), 123–69.

13. Marinetti, *Futurist Cookbook*, 119 (car crash menu); Hulten, *Futurismo e Futurismi*, 503–5 (lust). Marinetti in 1928 joined the Italian Academy, despite his earlier loathing of "museums, libraries, and academies of every sort," according to Walter L. Adamson, "How Avant-Gardes End—and Begin: Italian Futurism in Historical Perspective," *New Literary History* 41 no. 4 (2010): 855–874 at jstor.org/stable/23012710.

14. Banham, *Theory and Design*, quote on 155. Theo van Doesburg was born Christian Emil Marie Küpper (1883–1931) according to Ludo van Halem, "Nota Bene: De Stijl," *Rijksmuseum Bulletin* 65 no. 2 (2017): 122–127 at jstor.org/stable/26266311.

15. Carsten-Peter Warncke, *The Ideal as Art: De Stijl 1917–31* (Cologne: Taschen, 1998),

Mondrian quotes on 66; Banham, *Theory and Design,* 148–53, quotes on 150, 151; James Scott, *Seeing Like a State* (New Haven: Yale University Press, 1998), quote on 392 n55 ("After electricity").

16. Theo van Doesburg, "The Will to Style: The New Form Expression of Life, Art and Technology," *De Stijl* 5 no. 2 (February 1922): 23–32 and no. 3 (March 1922): 33–41; reprinted in Joost Baljeu, *Theo van Doesburg* (London: Studio Vista, 1974), 115–26, quotes on 119 and 122.

17. Banham, *Theory and Design,* 139–47, quotes on 141 ("decoration and ornament"), 142 ("modern villa"), and 147 ("normal tool" and "old structural forms").

18. Banham, *Theory and Design,* quotes on 158 ("I bow the knee") and 160–62.

19. John Willett, *The New Sobriety, 1917–1933: Art and Politics in the Weimar Period* (London: Thames & Hudson, 1978), quote on 120.

20. Banham, *Theory and Design*, quote on 283.

21. Warncke, *De Stijl*, 120–45; Theo van Doesburg, "Towards Plastic Architecture," *De Stijl*, ser. 12, nos. 6–7 (1924): 78–83; reprinted in Baljeu, *Theo van Doesburg*, 142–47, quote on 142.

22. Gropius quoted in Banham, *Theory and Design*, 281.

23. Gropius quoted in Banham, *Theory and Design*, 282.

24. Willett, *New Sobriety*, quote on 123.

25. Thomas P. Hughes, *American Genesis* (New York: Viking, 1989), 312–19; Willett, *New Sobriety*, 120–22.

26. The Berlin building campaign is illustrated in Taut, *Modern Architecture*, 100–132.

27. Willett, *New Sobriety,* 124–26.

28. Willett, *New Sobriety,* 125–27; Louis H. Pink, *The New Day in Housing* (New York: John Day Company, 1928), quote on 50 ("Ford cars") at archive.org/details /newdayinhousing00pinkrich.

29. Willett, *New Sobriety,* 127–29, quote on 128.

30. On the Museum of Modern Art's campaign to create and shape the International Style and its expression at the 1939 New York World's Fair, see Terry Smith, *Making the Modern: Industry, Art, and Design in America* (Chicago: University of Chicago Press, 1993), 385–421.

31. Eric Mumford, *The CIAM Discourse on Urbanism, 1928–1960* (Cambridge: MIT Press, 2000).

32. For biographical details, see Lore Kramer, "The Frankfurt Kitchen: Contemporary Criticism and Perspectives," in Angela Oedekoven-Gerischer et al., eds., *Frauen im Design: Berufsbilder und Lebenswege seit 1900* (Stuttgart: Design Center Stuttgart, 1989), 1:148–73.

33. Mary Nolan, *Visions of Modernity* (New York: Oxford University Press, 1994), quote on 208 ("the household").

34. For this and the following paragraphs see Susan R. Henderson, "A Revolution in the

Women's Sphere: Grete Lihotzky and the Frankfurt Kitchen," in Debra Coleman, Elizabeth Danze, and Carol Henderson, eds., *Architecture and Feminism* (New York: Princeton Architectural Press, 1996), 221–53; and Martina Heßler, "The Frankfurt Kitchen: The Model of Modernity and the 'Madness' of Traditional Users, 1926 to 1933," in Ruth Oldenziel and Karin Zachmann, eds., *Cold War Kitchen* (Cambridge: MIT Press, 2009), 163–84; Nolan, *Visions of Modernity*, 206–26.

35. Pink, *New Day in Housing*, quote on 54 ("work of the housewife"); Oedekoven-Gerischer et al., *Frauen im Design*, 159–67, quote on 165; Peter Noever, ed., *Die Frankfurter Küche* (Berlin: Ernst & Sohn, n.d.), 44–46.

CHAPTER 7.
THE MEANS OF DESTRUCTION, 1936–1990

1. Alan Milward, *War, Economy, and Society, 1939–1945* (Berkeley: University of California Press, 1977), 180. Stuart W. Leslie, *The Cold War and American Science* (New York: Columbia University Press, 1993), 8. See Henry Etzkowitz, "Solar versus Nuclear Energy: Autonomous or Dependent Technology?" *Social Problems* 31 (April 1984): 417–34. On rival digital and analog computers in Project SAGE, see Thomas P. Hughes, *Rescuing Prometheus* (New York: Pantheon, 1998), 40–47. On the military's role in computer-controlled machine tools, see David F. Noble, *Forces of Production* (New York: Knopf, 1984).

2. Daniel Greenberg, *The Politics of Pure Science* (New York: NAL, 1971), 170–206; Paul Josephson, *Red Atom: Russia's Nuclear Power Program from Stalin to Today* (New York: Freeman, 2000); Sonja D. Schmid, *Producing Power: The Pre-Chernobyl History of the Soviet Nuclear Industry* (Cambridge: MIT Press, 2015), William J. Broad, *Teller's War: The Top-Secret Story behind the Star Wars Deception* (New York: Simon & Schuster, 1992); Christian Ruhl, "Why There Are No Nuclear Airplanes," *The Atlantic* (20 January 2019) at web.archive.org/web/20201108000551/https://www.theatlantic .com/technology/archive/2019/01/elderly-pilots-who-could-have-flown-nuclear -airplanes/580780/. James Bamford, *The Puzzle Palace* (New York: Penguin, 1983), 109, estimated the National Security Agency's cryptography budget was $10 billion—$5 *trillion* is the estimated cost for the US nuclear enterprise since 1940 in Stephen I. Schwartz, *Atomic Audit: The Costs and Consequences of U.S. Nuclear Weapons since 1940* (Washington, DC: Brookings Institution Press, 1998).

3. See William McNeill, *The Pursuit of Power: Technology, Armed Force and Society since A.D. 1000* (Chicago: University of Chicago Press, 1982), 262–345; and John Ellis, *Eye-Deep in Hell: Trench Warfare in World War I* (Baltimore: Johns Hopkins University Press, 1989).

4. William Carr, *Arms, Autarky and Aggression: A Study in German Foreign Policy, 1933–1939* (New York: W. W. Norton, 1972); Milward, *War, Economy, and Society*, 23–30, 184–206.

5. Martin van Creveld, *Technology and War* (New York: Free Press, 1989), 217–32.

6. Mark Walker, *German National Socialism and the Quest for Nuclear Power, 1939–1949* (Cambridge: Cambridge University Press, 1989), 129–52. There is little evidence that Heisenberg sabotaged the German atomic bomb effort from inside, as imagined in journalist Thomas Powers' *Heisenberg's War* (1993) and playwright Michael Frayn's *Copenhagen* (1998).

7. Walker, *German National Socialism*, 153–78.

8. Leslie R. Groves, *Now It Can Be Told* (New York: Harper & Row, 1962), 20.

9. Edward Teller, *The Legacy of Hiroshima* (Garden City: Doubleday, 1962), quote on 211.

10. On Hanford see Richard G. Hewlett and Oscar E. Anderson, Jr., *The New World, 1939–1946*, vol. 1 of *A History of the United States Atomic Energy Commission* (University Park: Pennsylvania State University Press, 1962), 1:180–226; and David A. Hounshell and John Kenly Smith, Jr., *Science and Corporate Strategy: DuPont R&D, 1902–1980* (Cambridge: Cambridge University Press, 1988), 338–46.

11. Gar Alperovitz, *Atomic Diplomacy: Hiroshima and Potsdam* (New York: Simon & Schuster, 1965), quote on 14.

12. Winston S. Churchill, *Triumph and Tragedy* (Boston: Houghton Mifflin, 1953), quote on 639; Michael Sherry, *The Rise of American Air Power* (New Haven: Yale University Press, 1987), 330.

13. Sherry, *Rise of American Air Power*, quote on 341 (Groves).

14. Spencer R. Weart and Gertrud Weiss Szilard, eds., *Leo Szilard* (Cambridge: MIT Press, 1978), quote on 184 (Szilard); Groves, *Now It Can Be Told*, quote on 70; Hewlett and Anderson, *New World*, 339–40.

15. Richard Rhodes, *The Making of the Atomic Bomb* (New York: Simon & Schuster, 1986), quote on 697 ("prison); Bernard J. O'Keefe, *Nuclear Hostages* (Boston: Houghton Mifflin, 1983), quote on 97 ("excruciating").

16. Richard G. Hewlett and Jack M. Holl, *Atoms for Peace and War, 1953–1961* (Berkeley: University of California Press, 1989), 196–98, 419–23, 493–95; Thomas P. Hughes, *American Genesis* (New York: Viking, 1989), 428–42.

17. The figures for Safeguard and the U.S. nuclear program are expressed in constant 1996 dollars, correcting for inflation. By this measure, the nuclear plane project cost $7 billion. See Schwartz, *Atomic Audit*.

18. Charles Weiner, "How the Transistor Emerged," *IEEE Spectrum* 10 (January 1973): 24–35, quote on 28; Daniel Kevles, *The Physicists* (New York: Knopf, 1977), 302–20.

19. Lillian Hoddeson, "The Discovery of the Point-Contact Transistor," *Historical Studies in the Physical Sciences* 12, no. 1 (1981): 41–76, quote on 53; Mark P. D. Burgess, "Transistor History: Raytheon Part One" (2009) at web.archive.org/web /20201011231010/https://sites.google.com/site/transistorhistory/Home /us-semiconductor-manufacturers/raytheon-part-one-2.

20. Thomas J. Misa, "Military Needs, Commercial Realities, and the Development of the Transistor, 1948–1958," in Merritt Roe Smith, ed., *Military Enterprise and Technological Change* (Cambridge: MIT Press, 1985), 253–87, quote on 268. Compare Mara Mills, "Hearing Aids and the History of Electronics Miniaturization," *IEEE Annals of the History of Computing* 33 no. 2 (2011): 24–45 at doi.org/10.1109/MAHC.2011.43. As Mills accurately notes "hearing aid users became the first *consumer* market" for post-war electronics, but aggregate demand from the military, intelligence, and space markets—as well as lavish research, development, and production support—rather swamped this consumer effect.

21. Misa, "Military Needs," quote on 282.

22. M. D. Fagen, ed., *A History of Engineering and Science in the Bell System: National Service in War and Peace, 1925–1975* (Murray Hill, N.J.: Bell Telephone Laboratories, 1978), quote on 394.

23. Kenneth Flamm, *Creating the Computer* (Washington, D.C.: Brookings Institution, 1988), 259–69.

24. Nancy Stern, *From ENIAC to UNIVAC* (Bedford, Mass.: Digital Press, 1981); William Aspray, *John von Neumann and the Origins of Modern Computing* (Cambridge: MIT Press, 1990); I. Bernard Cohen, "The Computer: A Case Study of Support by Government," in E. Mendelsohn, M. R. Smith, and P. Weingart, eds., *Science, Technology and the Military*, vol. 12 of *Sociology of the Sciences* (Dordrecht: Kluwer, 1988), 119–54.

25. Kent C. Raymond and Thomas M. Smith, *Project Whirlwind: The History of a Pioneer Computer* (Bedford, Mass.: Digital Press, 1980), Forrester quote on 42.

26. Hughes, *Rescuing Prometheus,* 15–67.

27. Charles J. Bashe et al., *IBM's Early Computers* (Cambridge: MIT Press, 1986); Emerson W. Pugh, *Memories that Shaped an Industry: Decisions Leading to IBM System/360* (Cambridge: MIT Press, 1984); Nicholas Lewis, "Purchasing Power: Rivalry, Dissent, and Computing Strategy in Supercomputer Selection at Los Alamos," *IEEE Annals of the History of Computing* 39, no. 3 (2017): 25–40 at doi.org/10.1353/ahc.2017.0021.

28. Christophe Lécuyer, *Making Silicon Valley: Innovation and the Growth of High Tech, 1930–1970* (Cambridge: MIT Press, 2006), 91–128; Margaret O'Mara, *The Code: Silicon Valley and the Remaking of America* (New York: Penguin, 2019) "largest high-tech employer" on p. 36; and, recently, Michael Barbaro, "Silicon Valley's Military Dilemma," *New York Times* (6 March 2019) at web.archive.org/web/20201203221252/https://www.nytimes.com/2019/03/06/podcasts/the-daily/technology-military-contracts.html ; Thomas J. Misa, *Digital State: The Story of Minnesota's Computing Industry* (Minneapolis: University of Minnesota Press, 2013).

29. Arthur L. Norberg and Judy E. O'Neill, *Transforming Computer Technology: Information Processing for the Pentagon, 1962–1986* (Baltimore: Johns Hopkins University Press, 1996), 160–61; Thomas J. Misa, "Computer Security Discourse at RAND, SDC, and

NSA (1958–1970)," *IEEE Annals of the History of Computing* 38, no. 4 (2016): 12–25 at doi.org/10.1353/ahc.2016.0041.

30. Jon Rynn, "More Power to the Workers: The Political Economy of Seymour Melman," *Counterpunch* (2 January 2018) ("Soviets were so focused") at web.archive.org /web/20201109021523/https://www.counterpunch.org/2018/01/02/more-power-to -the-workers-the-political-economy-of-seymour-melman/. For the influence of closed-world thinking on national security strategies and much else, see Paul N. Edwards, *The Closed World: Computers and the Politics of Discourse in Cold War America* (Cambridge: MIT Press, 1996). One can see the "end" of the military era since military R&D spending peaked in 1987, fell around a third during the early 1990s, and remained there until 2000, while nonfederal R&D spending (mostly industrial) rose during these years. For data see the National Science Foundation's "National Patterns of R&D Resources" (2006) table 12 at web.archive.org/web/20071031195153/http://www.nsf.gov/statistics /nsf07331/. As discussed in the introduction, since 2001, US military R&D budgets have increased sharply.

CHAPTER 8.
PROMISES OF GLOBAL CULTURE, 1970–2001

1. John Micklethwait and Adrian Wooldridge, *A Future Perfect: The Essentials of Globalization* (New York: Crown Business, 2000), viii.
2. Kenichi Ohmae, *The Evolving Global Economy* (Boston: Harvard Business School, 1995), quote on xiv. Ohmae was the cofounder of the Tokyo offices of the consultants McKinsey and Company.
3. Richard Barnet and John Cavanagh, "Homogenization of Global Culture," in Jerry Mander and Edward Goldsmith, eds., *The Case against the Global Economy* (San Francisco: Sierra Club Books, 1996), quote on 71 ("homogenizing"); Rani Singh, "Shah Rukh Khan—The Biggest Movie Star In The World," *Forbes* (26 December 2015) at web.archive.org/web/20200829230914if_/https://www.forbes.com/sites /ranisingh/2015/12/26/shah-rukh-khan-the-biggest-movie-star-in-the-world/; William Gruger, "PSY's 'Gangnam Style' Video Hits 1 Billion Views, Unprecedented Milestone," *Billboard* (21 December 2012) at web.archive.org/web/20200410031758if_/https://www .billboard.com/articles//1483733/psys-gangnam-style-video-hits-1-billion-views -unprecedented-milestone/; Jordan Sirani, "Top 10 Best-Selling Video Games of All Time," IGN (19 April 2019) at web.archive.org/web/20201009032130/https://www.ign .com/articles/2019/04/19/top-10-best-selling-video-games-of-all-time.
4. See the World Bank's "Assessing Globalization" (2000) at https://web.archive.org /web/20000823054433/http://www.worldbank.org/html/extdr/pb/globalization/index .htm.
5. "Think Local: Cultural Imperialism Doesn't Sell," *Economist* (13 April 2002) at

web.archive.org/web/20201025210642if_/https://www.economist.com/special
-report/2002/04/13/think-local.

6. Chico Harlan, "In Japan, Fax Machines Find a Final Place to Thrive," *Washington Post*
   (7 June 2012) at tinyurl.com/y5s2utyp *or* web.archive.org/web/20120614070137/https://
   www.washingtonpost.com/world/asia_pacific/in-japan-fax-machines-find-a-final
   -place-to-thrive/2012/06/07/gJQAshFPMV_story.html.

7. Daniel Headrick, *The Tentacles of Progress: Technology Transfer in the Age of
   Imperialism, 1850–1940* (New York: Oxford University Press, 1988), 29.

8. An essential study is Jonathan Coopersmith, *Faxed: The Rise and Fall of the Fax
   Machine* (Baltimore: Johns Hopkins University Press, 2016), quotes on 89 ("self-
   contained telegraph") and 147 ("most successful standards").

9. Coopersmith, *Faxed*, passim.

10. Martin Fackler, "In High-Tech Japan, the Fax Machines Roll On," *New York Times*
    (13 February 2013) ("handwritten fax") at web.archive.org/web/20201108000115/www
    .nytimes.com/2013/02/14/world/asia/in-japan-the-fax-machine-is-anything-but-a-relic.
    html; Ben Dooley and Makiko Inoue, "Japan Needs to Telework: Its Paper-
    Pushing Offices Make That Hard," *New York Times* (14 April 2020) at web.archive
    .org/web/20201127212857/https://www.nytimes.com/2020/04/14/business/japan
    -coronavirus-telework.html; Julian Ryall, "Japanese Decry Boomer-Era Tech as
    Hospitals File Coronavirus Cases by Fax," *South China Morning Post* (5 May 2020) at
    web.archive.org/web/20201104174705/https://www.scmp.com/week-asia/health
    -environment/article/3082907/japan-hospitals-still-use-fax-machines-coronavirus.

11. This section follows the evaluation report: Jean Agnès, *The Fax! Programme; Three Years
    of Experimentation (June 1989–October 1992). A Teaching Aid for Opening Up to Europe*
    (Strasbourg: Council of Europe, 1994). Page numbers for quotes are given in the text.

12. Roger Cohen, "Fearful over the Future, Europe Seizes on Food," *New York Times*
    (29 August 1999) at web.archive.org/web/20201023232123/https://archive.nytimes.com
    /www.nytimes.com/library/review/082999europe-food-review.html; Suzanne Daley,
    "French See a Hero in War on 'McDomination,'" *New York Times* (12 October 1999) at
    web.archive.org/web/20201023232443/https://www.nytimes.com/1999/10/12/world
    /montredon-journal-french-see-a-hero-in-war-on-mcdomination.html.

13. Naomi Klein, *No Logo* (New York: Picador, 1999), quote on 388; John Vidal, *McLibel:
    Burger Culture on Trial* (New York: New Press, 1997). The "gravel pit" quote is online
    at web.archive.org/web/20060526091239/http://www.mcspotlight.org/case/trial
    /transcripts/941205/29.htm and Coca-Cola is "nutritious" at web.archive.org/web
    /20201023230157/https://www.mcspotlight.org/case/trial/transcripts/941208/54.htm.

14. Thomas L. Friedman, *The Lexus and the Olive Tree* (New York: Farrar, Straus and
    Giroux, 1999), quotes on 195, 196, and 309. Friedman blasted antiglobalization critics
    in his "Senseless in Seattle," *New York Times* (1 December 1999) at web.archive.org

/web/20060502195214/www.nytimes.com/library/opinion/friedman/120199frie.html. *Compare* Edward Luce, "The End of the Golden Arches Doctrine," *Financial Times* (10 May 2015) at web.archive.org/web/20200904004252/https://www.ft.com /content/1413fc26-f4c6–11e4–9a58–00144feab7de.

15. George Cohon, *To Russia with Fries* (Toronto: McClelland & Stewart, 1999), quote on 132.

16. Cohon, *To Russia with Fries*, quotes on 133 and 179–80.

17. Cohon, *To Russia with Fries*, quotes on 195 ("Big Mac is *perestroika*") and 281 ("Russian company"). For criticism of McDonald's in Russia, see the interview with Vadim Damier, "Moscow Left Demonstrates against McDonald's," *Green Left Weekly* 33 (30 October 1991) at web.archive.org/web/20200303134723/https://www.greenleft.org.au /content/moscow-left-demonstrates-against-mcdonalds.

18. Cohon, *To Russia with Fries*, quotes on 176. The vertical integration of McDonald's UK was the result of the established food industry's unwillingness to accommodate McDonald's, rather than an inability to do so (as in Russia); see John F. Love, *McDonald's: Behind the Arches* (New York: Bantam, 1995; original edition 1986), 440–41.

19. James L. Watson, ed., *Golden Arches East: McDonald's in East Asia* (Stanford: Stanford University Press, 1997), viii.

20. Watson, *Golden Arches East,* 56, 93, quote on 106. Such leisurely socialization was also visible each afternoon at a nearby McDonald's during the 2000's when I lived in northwest Chicago, where large gatherings of older men recently arrived from southeastern Europe chatted pleasantly. They simply ignored a sign (in English) stating, "Guests may take no longer than 30 minutes to finish their meals."

21. Watson, *Golden Arches East*, 141, quotes on 12 and 181.

22. Love, *McDonald's: Behind the Arches*, 423, 424, 426.

23. Love, *McDonald's: Behind the Arches*, quote on 432.

24. Vidal, *McLibel*, quote 178.

25. Katie Hafner and Matthew Lyon, *Where Wizards Stay Up Late: The Origins of the Internet* (New York: Simon & Schuster, 1996), 10.

26. Paul Baran, *On Distributed Communications*, RAND Memorandum RM-3420-PR volume 1 (August 1964), quote on 1 (enemy attack) at web.archive.org/web /20200424105653/www.rand.org/pubs/research_memoranda/RM3420.html; Janet Abbate, *Inventing the Internet* (Cambridge: MIT Press, 1999), quotes on 77; Paul Edwards, "Infrastructure and Modernity: Force, Time, and Social Organization in the History of Sociotechnical Systems," in Misa et al., *Modernity and Technology* (on nuclear war).

27. Hafner and Lyon, *Where Wizards Stay Up Late*, 56; Paul Baran, "Reliable Digital Communications Systems Using Unreliable Network Repeater Nodes" (RAND report P-1995), quote on 1 ("cloud-of-doom") at web.archive.org/web/20200806053834 /https://www.rand.org/pubs/papers/P1995.html; Abbate, *Inventing the Internet*, 7–21.

28. Abbate, *Inventing the Internet*, 43–81, quotes on 76 (congressional testimony).

29. Urs von Burg, *The Triumph of the Ethernet: Technological Communities and the Battle for the LAN Standard* (Stanford: Stanford University Press, 2001).

30. Abbate, *Inventing the Internet,* 113–32, quotes on 131–32.

31. Abbate, *Inventing the Internet,* 133–45, 147–79; Shane Greenstein, *How the Internet Became Commercial* (Princeton: Princeton University Press, 2015), quote on 28 n6 ("coercive tactics"). The networking rivals to TCP/IP were X.25 (promoted by Europe's telecommunication industry) and the Open Systems Interconnection (supported by the ISO standards-setting body).

32. Abbate, *Inventing the Internet,* 83–111, quote on 108 ("not an important motivation"); Craig Partridge, "The Technical Development of Internet Email," *IEEE Annals of the History of Computing* 30, no. 2 (2008): 3–29 at doi.org/10.1109/MAHC.2008.32.

33. This section follows Abbate, *Inventing the Internet*, 181–218. See chapter 10 below for "cyberspace."

34. Shane Greenstein, *How the Internet Became Commercial* (Princeton: Princeton University Press, 2015), 33–96.

35. See Tim Berners-Lee, "Web Architecture from 50,000 Feet" (October 1999) ("universality of access") at web.archive.org/web/20200222220950/www.w3.org /DesignIssues/Architecture.html; Tim Berners-Lee, "Realising the Full Potential of the Web" (3 December 1997) at web.archive.org/web/20200828094603/http://www .w3.org/1998/02/Potential.html); Herb Brody, "The Web Maestro: An Interview with Tim Berners-Lee," *Technology Review* (July 1996): 34–40.

36. Greenstein, *How the Internet Became Commercial*, 152, 296. See US Congressional Research Service at www.crs.gov reports on "State Taxation of Internet Transactions" (7 May 2013 #R41853); "'Amazon Laws' and Taxation of Internet Sales" (9 April 2015 #R42629); "The Internet Tax Freedom Act" (13 April 2016 #R43772); Andrew Leonard, "The Joy of Perl: How Larry Wall Invented a Messy Programming Language—and Changed the Face of the Web," *Salon* (13 October 1998) ("duct tape") at web.archive.org/web/20200922030013/https://www.salon.com/1998/10/13 /feature_269/.

37. The American Association for the Advancement of Science forecast record high military R&D budget for 2004, surpassing even the Cold War peak at web.archive.org /web/20130618121506/http://www.aaas.org/spp/rd/04pch6.htm.

38. "League of Nationalists," *Economist* (19 November 2016) at web.archive.org /web/20200804090721if_/https://www.economist.com/international/2016/11/19 /league-of-nationalists; Anabel González, "Globalization Is the Only Answer," World Bank (8 August 2016) at web.archive.org/web/20160811225133/https:// www.worldbank.org/en/news/opinion/2016/08/08/globalization-is-the-only -answer.

CHAPTER 9.
PATHS TO INSECURITY, 2001–2010

1. Richard Wright, *Native Son* (New York: HarperCollins, 1993), quotes on 402 (epigraph), 441. Wright treats the Chicago Loop's skyscrapers as fragile and impermanent expressions of modern capitalism on pages 428–30.

2. Skyscrapers might even become amoral spaces, violent and anarchistic absent a supporting social environment; see J. G. Ballard, *High-Rise* (London: Jonathan Cape, 1975). For "the role the skyscraper played in shaping . . . the material history of race," see Adrienne Brown, *The Black Skyscraper: Architecture and the Perception of Race* (Johns Hopkins University Press, 2017), on Wright, 197–99.

3. Christopher Hawthorne, "The Burj Dubai and Architecture's Vacant Stare," *Los Angeles Times* (1 January 2010) at web.archive.org/web/20200729181613/latimesblogs.latimes. com/culturemonster/2010/01/the-burj-dubai-and-architectures-vacant-stare.html; Shane McGinley, "Dubai's Burj Khalifa Is 80% Occupied—Emaar," *Arabian Business* (4 July 2012) at web.archive.org/web/20190423133350/www.arabianbusiness.com /dubai-s-burj-khalifa-is-80-occupied-emaar-464637.html (quoted reader comments at original URL).

4. George Perkovich, *India's Nuclear Bomb: The Impact on Global Proliferation* (Berkeley: University of California Press, 1999); William Langewiesche, *The Atomic Bazaar* (New York: Farrar, Straus and Giroux, 2007).

5. See Steven L. Schwarcz, "Systemic Risk," *Georgetown Law Journal* 97, no. 1 (2008) at web.archive.org/web/20200930184407/https://papers.ssrn.com/sol3/papers. cfm?abstract_id=1008326); and Charles Perrow, *Normal Accidents: Living with High Risk Technologies* (Princeton: Princeton University Press, 1999).

6. David Nye's *When the Lights Went Out: A History of Blackouts in America* (Cambridge: MIT Press, 2010) provides historical perspective; Don DeLillo's *The Silence* (New York: Scribner, 2020) presents a *post*-coronavirus view.

7. Erik van der Vleuten and Vincent Lagendijk, "Transnational Infrastructure Vulnerability: The Historical Shaping of the 2006 European 'Blackout,'" *Energy Policy* 38 (2010): 2042–52; Erik van der Vleuten and Vincent Lagendijk, "Interpreting Transnational Infrastructure Vulnerability: European Blackout and the Historical Dynamics of Transnational Electricity Governance," *Energy Policy* 38 (2010): 2053–62; Kartikay Mehrotra and Andrew MacAskill, "Singh's $400 Billion Power Plan Gains Urgency as Grid Collapses," *Washington Post* (30 July 2012) ("India's grid") at tinyurl .com/y29p3mwr *or* web.archive.org/web/20120805054944/http://washpost.bloomberg.com /Story?docId=1376-M7YDKA0YHQ0X0179TK1GQCDQF2C5AQ5A383U7QR; Gardiner Harris and Vikas Bajaj, "As Power Is Restored in India, the 'Blame Game' Over Blackouts Heats Up," *New York Times* (1 August 2012) at web.archive.org/web/20201112003103 /https://www.nytimes.com/2012/08/02/world/asia/power-restored-after-india-blackout.html.

8. See Aleksandar Jevtic, "15 Biggest Malls in the World" (26 August 2015) at web.archive .org/web/20151208063027/www.insidermonkey.com/blog/15-biggest-malls-in-the -world-367683/.

9. 2012 Statistical Abstract of the United States at web.archive.org/web/20201017000828 /https://www.census.gov/library/publications/2011/compendia/statab/131ed.html. Remarkably, in October 2011 the Census Bureau terminated data collection for the Statistical Abstract; see web.archive.org/web/20201103034158/https://www.census.gov /library/publications/time-series/statistical_abstracts.html. Federal Reserve Bank of St. Louis, "E-Commerce Retail Sales as a Percent of Total Sales" (3 November 2020) at fred .stlouisfed.org/series/ECOMPCTSA.

10. M. Jeffrey Hardwick, *Mall Maker: Victor Gruen, Architect of an American Dream* (Philadelphia: University of Pennsylvania Press, 2004).

11. Mikael Hård and Thomas J. Misa, eds., *Urban Machinery: Inside Modern European Cities* (Cambridge: MIT Press, 2008), chapters 1, 4, 10.

12. China's sales figures included 650,000 heavy trucks. Japan's domestic car market was 4.6 million (2009). In 2008 Japan was the world's largest producer of cars and commercial vehicles (11.5 million), ahead of China (9.3 million), the United States (8.7 million), Germany (6.0 million), and South Korea (3.8 million) at web.archive .org/web/20201112000659/http://www.oica.net/category/production-statistics /2008-statistics/.

13. Barbara M. Kelly, *Expanding the American Dream: Building and Rebuilding Levittown* (Albany: SUNY Press, 1993); and Samuel Zipp, "Suburbia and American Exceptionalism," *Reviews in American History* 36 no. 4 (2008): 594–601.

14. As early as 1940, one-fifth of African Americans in metropolitan areas lived in suburbs; see Andrew Wiese, *Places of Their Own: African-American Suburbanization in the Twentieth Century* (Chicago: University of Chicago Press, 2004). On exclusion and inequality, see Lizabeth Cohen, *A Consumers' Republic: The Politics of Mass Consumption in Postwar America* (New York: A. A. Knopf, 2003), 194–256; Richard Rothstein, *The Color of Law: A Forgotten History of How Our Government Segregated America* (New York: Liveright Publishing, 2017).

15. Hardwick, *Mall Maker*, quote on 1, 103.

16. Robert Bruegmann, *Sprawl: A Compact History* (Chicago: University of Chicago, 2005), 17, 149.

17. Peter de Vries, "Humorists Depict the Suburban Wife," *Life* (24 December 1956): 150.

18. Bruegmann, *Sprawl: A Compact History*, 19.

19. DOE import and consumption data at web.archive.org/web/20201109104835/https:// www.eia.gov/energyexplained/oil-and-petroleum-products/imports-and-exports.php. Countries at web.archive.org/web/20201019213917/https://afdc.energy.gov/data/10621 *and* www.statista.com/statistics/201844/us-petroleum-imports-by-country-since-1985/.

20. Since 2008, the US market share of "cars" and "light trucks," including pickup trucks, SUVs, and minivans, have been nearly equal. In 2010, the two largest selling models were the pickup trucks by Ford F-series (14–15 mpg city) and Chevrolet Silverado (14 mpg city). See web.archive.org/web/20101207140904/online.wsj.com/mdc/public /page/2_3022-autosales.html; Keith Bradsher, *High and Mighty: The Dangerous Rise of the SUV* (New York: Public Affairs, 2002), 27–30.

21. Theodore Steinberg, *American Green: The Obsessive Quest for the Perfect Lawn* (New York: W.W. Norton, 2006); Paul R. Josephson, *Motorized Obsessions: Life, Liberty, and the Small-Bore Engine* (Baltimore: Johns Hopkins University Press, 2007), 194.

22. U.S. Environmental Protection Agency, *Biofuels and the Environment: The Second Triennial Report to Congress* (Washington, DC, EPA/600/R-18/195, 2018) at web .archive.org/web/20201030220816/https://cfpub.epa.gov/si/si_public_file_download .cfm?p_download_id=536328&Lab=IO.

23. Alexei Barrionuevo and Micheline Maynard, "Dual-Fuel Vehicles Open Mileage Loophole for Carmakers," *New York Times* (31 August 2006) ("vehicles get credit") and ("'motivating factor'") at web.archive.org/web/20201109034912/www.nytimes .com/2006/08/31/business/31loophole.html/. A flex-fuel engine cost $70 extra according to GM's Tom Stevens: "GM Calls for Urgent Increase in Ethanol Fuel Stations," *BusinessGreen* (16 February 2010) at web.archive.org/web/20201113232152/https:// www.businessgreen.com/business-green/news/2257966/gm-call-urgent-increase.

24. For E-85 stations in each state in 2020 see web.archive.org/web/20201028200601 /https://afdc.energy.gov/stations/states.

25. Comparing gasoline and E-85 mileage ratings for *2010* models (at web.archive.org/web /20100913010450/https://www.fueleconomy.gov/): the "best" was Ford Expedition losing 21 percent with E-85 and the "worst" were Dodge Ram and Dakota both losing 33 percent.

26. "Under the best-existing practices, the amount of energy used to grow the corn and convert it into ethanol is 57,504 BTUs per gallon. Ethanol itself contains 84,100 BTUs per gallon" at web.archive.org/web/20100502195949/www.carbohydrateeconomy.org /library/admin/uploadedfiles/How_Much_Energy_Does_it_Take_to_Make_a_Gallon _.html.

27. Jerry Garrett, "Ethanol and the Tortilla Tax," *New York Times* (6 September 2007) at web.archive.org/web/20201109074547/https://wheels.blogs.nytimes.com/2007/09/06 /ethanol-and-the-tortilla-tax/; Manuel Roig-Franzia, "A Culinary and Cultural Staple in Crisis," *Washington Post* (27 January 2007) at web.archive.org/web/20081107060137 /https://www.washingtonpost.com/wp-dyn/content/article/2007/01/26 /AR2007012601896_pf.html.

28. Janet Larsen, "U.S. Feeds One Quarter of Its Grain to Cars While Hunger Is on the Rise," Earth Policy Institute (21 January 2010) at web.archive.org/web/20200805192736 /http://www.earth-policy.org/data_highlights/2010/highlights6. US Department of

Agriculture, World Agricultural Supply and Demand Estimates WASDE-490–12 (12 January 2011): Corn—ethanol for fuel = 4.9 billion bushels. Corn—production = 12.45 billion bushels, or 39.4%; recently WASD-605 (9 October 2020) indicates ethanol's proportion of US corn production 37.5% (2018–19 season), 35.6% (2019–20 estimate), 34.3% (2020–21 October projection).

29. See GAO, "Secure Border Initiative: Observations on Deployment Challenges" (10 September 2008) at web.archive.org/web/20201027070816/https://www.gao.gov/products/GAO-08-1141T; and GAO, "Secure Border Initiative: DHS Has Faced Challenges Deploying Technology and Fencing Along the Southwest Border" (4 May 2010) at web.archive.org/web/20200918200753/www.gao.gov/products/GAO-10-651T.

30. Congressional Budget Office, "The Impact of Ethanol Use on Food Prices and Greenhouse-Gas Emissions" (8 April 2009) at web.archive.org/web/20201017210314/https://www.cbo.gov/publication/41173.

31. Lester Brown, "Cars and People Compete for Grain" (1 June 2010) at web.archive.org/web/20200805201751/http://www.earth-policy.org/book_bytes/2010/pb4ch02_ss6.

32. The original source on John Draper a.k.a. Captain Crunch is Ron Rosenbaum, "Secrets of the Little Blue Box," *Esquire* (October 1971) text at web.archive.org/web/20090317153042/www.webcrunchers.com/crunch/stories/esq_art.html. See C. Breen and C. A. Dahlbom, "Signaling Systems for Control of Telephone Switching," *Bell System Technical Journal* 39, no. 6 (November 1960): 1381–1444, quote on 1418 ("same ease") at archive.org/details/bstj39-6-1381.

33. Paul Roberts, Jim Brunner, and Patrick Malone, "'Hundreds of Millions of Dollars' Lost in Washington to Unemployment Fraud Amid Coronavirus Joblessness Surge," *Seattle Times* (21 May 2020) at web.archive.org/web/20200728045538/https://www.seattletimes.com/business/economy/washington-adds-more-than-145000-weekly-jobless-claims-as-coronavirus-crisis-lingers/. Editorial, "Virtually Unprotected," *New York Times* (2 June 2005) at web.archive.org/web/20140907040043/http://www.nytimes.com/2005/06/02/opinion/opinionspecial/02thu1.html.

34. Laura DeNardis, *Protocol Politics: The Globalization of Internet Governance* (Cambridge: MIT Press, 2009), 125–29, quotes on 128. An encrypted protocol known as IPsec was originally designed into IPv6 in 1990 but was removed in a 1998 revision.

35. See Matthew Broersma, "Microsoft Server Crash Nearly Causes 800-Plane Pile-Up," *Techworld News* (21 September 2004) at web.archive.org/web/20100225103354/news.techworld.com/operating-systems/2275/microsoft-server-crash-nearly-causes-800-plane-pile-up/.

36. GAO, "Year 2000 Computing Crisis: FAA Is Making Progress but Important Challenges Remain" (15 March 1999), quote on 6 at web.archive.org/web/20201027033456/https://www.gao.gov/assets/110/107788.pdf; Gary Stix, "Aging Airways," *Scientific American* (May 1994): 96–104, quote on 96 at jstor.org/stable/24942700.

37. Robert N. Britcher, *The Limits of Software: People, Projects, and Perspectives* (Reading MA: Addison-Wesley, 1999), 163–89, quote on 163 ("greatest debacle"); Tekla S. Perry, "In Search of the Future of Air Traffic Control," *IEEE Spectrum* (August 1997): 18–35, quote on 19 "16MB of RAM" at doi.org/10.1109/6.609472; Stix, "Aging Airways," quote on 99.

38. Office of Inspector General, "Review of Web Applications Security and Intrusion Detection in Air Traffic Control Systems" (4 May 2009) quotes on 3, 5, 6, and 7 at web.archive .org/web/20170219160230/https://www.oig.dot.gov/sites/default/files/ATC_Web_Report.pdf. See FAA, "The Future of the NAS" (June 2016) at web.archive.org/web/20201024182149 /www.faa.gov/nextgen/media/futureofthenas.pdf, which gestures, rather vaguely, to "evolving challenges such as cybersecurity," while admitting the "increasingly interconnected [next-generation] system presents new cybersecurity challenges."

39. Idaho National Laboratory, "Common Cyber Security Vulnerabilities Observed in Control System Assessment by the INL NSTB Program" (November 2008) at web .archive.org/web/20201115235615/https://www.smartgrid.gov/files/documents /Common_Cyber_Security_Vulnerabilities_Observed_in_Control_Sy_200812.pdf.

40. See S. Massoud Amin, "Securing the Electricity Grid," *The Bridge* [National Academy of Engineering] 40, no. 1 (Spring 2010): 13–20, quote on 15; Amin and Phillip F. Schewe, "Preventing Blackouts," *Scientific American* (May 2007): 60–67.

41. Siobhan Gorman, "Electricity Grid in U.S. Penetrated by Spies," *Wall Street Journal* (6 April 2009) at web.archive.org/web/20150322130328/https://www.wsj.com/articles /SB123914805204099085; Siobhan Gorman, "U.S. Plans Cyber Shield for Utilities, Companies," *Wall Street Journal* (8 July 2010) at web.archive.org/web/20201109034613 /https://www.wsj.com/articles/SB10001424052748704545004575352983850463108; Nicole Perlroth, "Russian Hackers Targeting Oil and Gas Companies," *New York Times* (30 June 2014) at web.archive.org/web/20201112011946/http://www.nytimes .com/2014/07/01/technology/energy-sector-faces-attacks-from-hackers-in-russia .html; Nicole Perlroth and David E. Sanger, "Cyberattacks Put Russian Fingers on the Switch at Power Plants, U.S. Says" (15 March 2018) "critical control systems" and "effect sabotage" at web.archive.org/web/20201111090755/https://www.nytimes. com/2018/03/15/us/politics/russia-cyberattacks.html; Nicole Perlroth, "Russians Who Pose Election Threat Have Hacked Nuclear Plants and Power Grid," *New York Times* (23 October 2020) "breached the power grid" at web.archive.org/web/20201115092631 /https://www.nytimes.com/2020/10/23/us/politics/energetic-bear-russian-hackers.html

42. Dawn S. Onley, "Red Storm Rising," *Government Computing News* (17 August 2006) at https://web.archive.org/web/20201022105528/https://gcn.com/Articles/2006/08/17 /Red-storm-rising.aspx; "Cyberwar: War in the Fifth Domain," *The Economist* (1 July 2010) at web.archive.org/web/20201022155011/https://www.economist.com/briefing /2010/07/01/war-in-the-fifth-domain.

43. David Drummond, "A New Approach to China," Googleblog (12 January 2010) at web
.archive.org/web/20201112030027/https://googleblog.blogspot.com/2010/01/new
-approach-to-china.html.

44. In internet governance, there is a conflict between the "multi-stakeholder" model
backed by the United States, and some European countries, and the "state sovereignty"
model of Russia, China, and allies. At the United Nations Group of Governmental
Experts on Information Security during 2004–2019, the two sides' "conflicting views
. . . paralyzed international cyber security norm building on the UN level," according
to Juha Kukkola, *Digital Soviet Union: The Russian National Segment of the Internet as a
Closed National Network Shaped by Strategic Cultural Ideas* (Helsinki: National Defence
University, 2020), quote on 67 at www.doria.fi/handle/10024/177157.

45. Richard A. Clarke and Robert K. Knake, *Cyber War: The Next Threat to National
Security* (New York: HarperCollins, 2010), 1–11; Richard Clarke, excerpt from
Cyber War, ABC News (5 March 2010) "image of nothing" at web.archive.org/web
/20201128024527/https://abcnews.go.com/GMA/Books/cyber-war-richard-clarke
/story?id=10414617; David A. Fulghum, "Why Syria's Air Defenses Failed to
Detect Israelis," *Aviation Week* (3 October 2007) 'users to invade' at tinyurl.com
/y6snmbeg *or* web.archive.org/web/20131112114403/http://www.aviationweek.com
/Blogs.aspx?plckBlogId=Blog%3A27ec4a53-dcc8–42d0-bd3a-01329aef79a7&plckPostId
=Blog%3A27cc4a53-dcc8–42d0-bd3a-01329aef79a7Post%3A2710d024–5eda-416c
-b117-ae6d649146cd; Oliver Holmes, "Israel Confirms It Carried Out 2007 Airstrike
on Syrian Nuclear Reactor," *The Guardian* (21 March 2018) at web.archive.org
/web/20200604110958/www.theguardian.com/world/2018/mar/21/israel-admits-it
-carried-out-2007-airstrike-on-syrian-nuclear-reactor; "Cyberwar: War in the Fifth
Domain," *The Economist* (1 July 2010); Eric J. Byres, "Cyber Security and the Pipeline
Control System," *Pipeline and Gas Journal* 236 no. 2 (February 2009) "go haywire" at
pgjonline.com/magazine/2009/february-2009-vol-236-no-2/features/cyber-security
-and-the-pipeline-control-system (November 2020).

46. Gus W. Weiss, "The Farewell Dossier: Duping the Soviets," Center for the Study
of Intelligence (1996) at web.archive.org/web/20201112010541/https://www.cia
.gov/library/center-for-the-study-of-intelligence/csi-publications/csi-studies
/studies/96unclass/farewell.htm.

47. Siobhan Gorman, Yochi J. Dreazen, and August Cole, "Insurgents Hack U.S. Drones,"
Wall Street Journal (17 December 2009) "encryption systems" at web.archive.org
/web/20200320114038/https://www.wsj.com/articles/SB126102247889095011;
Andrew Moran, "Insurgents Hack U.S. Drones Using $26 Software," *Digital Journal*
(17 December 2009) "newer drones" at web.archive.org/web/20100201074042/www
.digitaljournal.com/article/284027.

48. "Modi-Trump Meet: US Approves Sale of 22 Guardian Drones to India," *Times of India*

(23 June 2017) at web.archive.org/web/20170623170639/http://timesofindia.indiatimes
.com/india/us-approves-sale-of-22-guardian-drones-to-india/articleshow/59274194
.cms; Huma Siddiqui, "Long Wait Over! Indian Armed Forces to Get High-Tech US
Armed Drones Equipped with Missiles," *Financial Express* (24 February 2020) at web
.archive.org/web/20201202035920/https://www.financialexpress.com/defence/long
-wait-over-indian-armed-forces-to-get-high-tech-us-armed-drones-equipped-with
-missiles/1877652/; Shishir Gupta, "China to Supply 4 Attack Drones to Pak, Prompts
India to Revive Predator-B Plan," *Hindustan Times* (6 July 2020) at tinyurl.com
/yx99gtdc *or* web.archive.org/web/20201004112126/https://www.hindustantimes.com
/india-news/pakistan-to-get-4-attack-drones-from-china-prompts-india-to-revive
-predator-b-plan/story-M5jUeCOLPnyyofzh03LSzI.html.

49. Juha Kukkola, *Digital Soviet Union: The Russian National Segment of the Internet as a
Closed National Network Shaped by Strategic Cultural Ideas* (Helsinki: National Defence
University, 2020), quote on 286 ("chose to militarize") at www.doria.fi/handle/10024
/177157; Eric Lipton, David E. Sanger, and Scott Shane, "The Perfect Weapon: How
Russian Cyberpower Invaded the U.S.," *New York Times* (13 December 2016) ("specific
effect") at web.archive.org/web/20201201003257/https://www.nytimes.com/2016/12/13
/us/politics/russia-hack-election-dnc.html.

50. David E. Sanger and Nicole Perlroth, "U.S. Escalates Online Attacks on Russia's Power
Grid," *New York Times* (15 June 2019) at web.archive.org/web/20201121222540/www
.nytimes.com/2019/06/15/us/politics/trump-cyber-russia-grid.html; "US and Russia
Clash Over Power Grid 'Hack Attacks,'" BBC (18 June 2019) ("clandestine military
activity") at web.archive.org/web/20201116031811/https://www.bbc.com/news
/technology-48675203.

51. Container port statistics (2 August 2019) at web.archive.org/web/20200926023547
/https://largest.org/structures/ports-usa/.

52. See Marc Levinson's *The Box: How the Shipping Container Made the World Smaller and
the World Economy Bigger* (Princeton: Princeton University Press, 2006); Alexander
Klose, *The Container Principle: How a Box Changes the Way We Think* (Cambridge: MIT
Press, 2015); and Rose George, *Ninety Percent of Everything* (New York: Metropolitan
Books, 2013).

53. Eric Lipton, "Homeland Security Inc.: Company Ties Not Always Noted in Security
Push," *New York Times* (19 June 2006) at web.archive.org/web/20201118224740
/https://www.nytimes.com/2006/06/19/washington/company-ties-not-always-noted
-in-security-push.html.

54. Jonathan Medalia, "Terrorist Nuclear Attacks on Seaports: Threat and Response,"
Congressional Research Service (24 January 2005) at web.archive.org/web
/20201109215740/https://fas.org/irp/crs/RS21293.pdf; and Lornet Turnbull, Kristi
Heim, Sara Jean Green, and Sanjay Bhatt, "15 Days in a Metal Box," *Seattle Times*

(6 April 2006) at web.archive.org/web/20201118225731/https://archive.seattletimes
.com/archive/?date=20060406&slug=smuggling.

55. Carl Hulse, "G.O.P. Leaders Vowing to Block Ports Agreement," *New York Times*
(8 March 2006) at web.archive.org/web/20180212122601/http://www.nytimes
.com/2006/03/08/politics/gop-leaders-vowing-to-block-ports-agreement.html.

56. David Stout, "Bush Says Political Storm over Port Deal Sends Wrong Message," *New
York Times* (10 March 2006) at web.archive.org/web/20180411230610/https://www
.nytimes.com/2006/03/10/international/middleeast/bush-says-political-storm-over
-port-deal-sends.html.

57. Heather Timmons, "Dubai Port Company Sells Its U.S. Holdings to A.I.G.," *New York
Times* (12 December 2006) at web.archive.org/web/20190928163244/https://www
.nytimes.com/2006/12/12/business/worldbusiness/12ports.html.

58. For GAO analysis, see "The SAFE Port Act and Efforts to Secure Our Nation's Seaports"
(4 October 2007) at web.archive.org/web/20201109220730/https://www.gao.gov/new
.items/d0886t.pdf; and "The SAFE Port Act: Status and Implementation One Year
Later" (October 2007) at web.archive.org/web/20170607115847/http://www.gao.gov
/new.items/d08126t.pdf.

59. GAO, "SAFE Port Act: Status and Implementation," 11.

60. For GAO assessments, see "Supply Chain Security: Feasibility and Cost-Benefit Analysis
Would Assist DHS and Congress in Assessing and Implementing the Requirement to
Scan 100 Percent of U.S.-Bound Containers" (30 October 2009) at web.archive
.org/web/20200721035411/https://www.gao.gov/new.items/d1012.pdf; and "Maritime
Security: Coast Guard Needs to Improve Use and Management of Interagency
Operations Centers" (13 February 2012) at web.archive.org/web/20201026163654
/https://www.gao.gov/assets/590/588475.pdf.

61. GAO, "SAFE Port Act and Efforts," 28 n 37.

62. GAO, "SAFE Port Act and Efforts," 42.

63. "A container that was loaded on a vessel in a foreign port shall not enter the United
States (either directly or via a foreign port) unless the container was scanned by
nonintrusive imaging equipment and radiation detection equipment at a foreign port
[*sic*] before it was loaded on a vessel," states the act's first paragraph of Title XVII—
Maritime Cargo; see Implementing Recommendations of the 9/11 Commission Act
of 2007, Public Law 110–53, 121 Stat 489 (p. 225) at web.archive.org/web
/20201031053556/www.congress.gov/110/plaws/publ53/PLAW-110publ53.pdf.

64. GAO, "Supply Chain Security," 7, 20, 28, 31. In July 2010 GAO still found "a variety
of political, logistical, and technological barriers to scanning all cargo containers,"
among other problems; see GAO, "Maritime Security: DHS Progress and Challenges
in Key Areas of Port Security" (21 July 2010), quote on 3 at web.archive.org/web
/20170606122134/http://www.gao.gov/new.items/d10940t.pdf. In 2016 GAO found

continuing problems, see "Maritime Security: Progress and Challenges in Implementing Maritime Cargo Security Programs" (7 July 2016) at web.archive.org /web/20201031162245/https://www.gao.gov/assets/680/678249.pdf.

65. GAO, "SAFE Port Act and Efforts," 29.

66. On structural failures of the U.S. Department of Homeland Security, see Charles Perrow, *The Next Catastrophe: Reducing Our Vulnerabilities to Natural, Industrial, and Terrorist Disasters* (Princeton: Princeton University Press, 2007), 68–129. Thomas L. Friedman, *Hot, Flat and Crowded* (New York: Farrar, Straus and Giroux, 2008), quote on 79.

67. Henry H. Willis and David S. Ortiz, "Evaluating the Security of the Global Containerized Supply Chain," TR-214 (Santa Monica, Calif.: RAND Corporation, 2004) at web.archive.org/web/20200721060051/https://www.rand.org/pubs/technical_reports /TR214.html.

68. Willis and Ortiz, "Evaluating the Security," quote on ix, 12, 32.

69. On "reflexive modernization" see Ulrich Beck, Anthony Giddens, and Scott Lash, *Reflexive Modernization: Politics, Tradition and Aesthetics in the Modern Social Order* (Stanford: Stanford University Press, 1994), quote on 3; and Misa et al., *Modernity and Technology* (Cambridge: MIT Press, 2003), chapters 1, 3, 9, 11.

70. One finds "experts," "expertise," and "expert systems" but not nuclear proliferation, petroleum, electricity networks, or automobiles in Beck et al.'s *Reflexive Modernization*; there Scott Lash's discussion of "information and communication structures" (121, 129–34, 167) is suggestive. Yet Lash's claim that "women are *not* excluded from [information and communication] structures" (134) seems naive; see Thomas J. Misa, ed., *Gender Codes: Why Women Are Leaving Computing* (Wiley/IEEE Computer Society Press, 2010).

CHAPTER 10.
DOMINANCE OF THE DIGITAL, 1990–2016

1. Heather Kelly, "The Biggest Takeaways from the Big-Tech Antitrust Hearing," *Washington Post* (29 July 2020) at web.archive.org/web/20200802172945/www .washingtonpost.com/technology/2020/07/29/big-tech-antitrust-hearing-takeaways/; Kate Riley, "Facebook Is Everywhere: The Impact of Technology Monopolies," *Seattle Times* (27 October 2020) at web.archive.org/web/20201117083913/https://www .seattletimes.com/opinion/facebook-is-everywhere/; Kevin Roose, Mike Isaac, and Sheera Frenkel, "Roiled by Election, Facebook Struggles to Balance Civility and Growth," *New York Times* (24 November 2020) "good cause" at web.archive.org /web/20201124113212/https://www.nytimes.com/2020/11/24/technology/facebook -election-misinformation.html.

2. Ron Deibert, "Authoritarianism Goes Global: Cyberspace under Siege," *Journal of*

Democracy 26 no. 3 (2015): 64–78, quote on 64 ("actively shaping") at doi.org/10.1353 /jod.2015.0051; "League of Nationalists," *Economist* (19 November 2016) at web.archive .org/web/20200804090721if_/https://www.economist.com/international/2016/11/19 /league-of-nationalists; "The Future of the Internet: A Virtual Counter-Revolution," *Economist* (2 September 2010) at web.archive.org/web/20201010154256if_/https:// www.economist.com/briefing/2010/09/02/a-virtual-counter-revolution.

3. Bret Swanson, "Moore's Law at 50: The Performance and Prospects of the Exponential Economy," American Enterprise Institute (10 November 2015), quote on 17 (billions) at web.archive.org/web/20200302063618/www.aei.org/research-products/report /moores-law-at-50-the-performance-and-prospects-of-the-exponential-economy/; Jaron Lanier, *Who Owns the Future?* (New York: Simon & Schuster, 2013), quote on 10 ("Ten Commandments").

4. See Christophe Lecuyer's "Driving Semiconductor Innovation: Moore's Law at Fairchild and Intel," forthcoming in *Enterprise and Society* at doi.org/10.1017/eso.2020.38.

5. Brent Schlender, "Intel's $10 Billion Gamble," *Fortune* 146 no. 9 (11 November 2002) "fundamental expectation" at web.archive.org/web/20201013151016/https://money .cnn.com/magazines/fortune/fortune_archive/2002/11/11/331816/index.htm; Paul E. Ceruzzi, "Moore's Law and Technological Determinism: Reflections on the History of Technology," *Technology and Culture* 46 no. 3 (2005): 584–93, quotes on 593 ("impervious") ("determinism") at doi.org/10.1353/tech.2005.0116.

6. Christophe Lécuyer, "From Clean Rooms to Dirty Water: Labor, Semiconductor Firms, and the Struggle over Pollution and Workplace Hazards in Silicon Valley," *Information & Culture* 52, no. 3 (2017): 304–333 at doi.org/10.1353/lac.2017.0012; Evelyn Nieves, "The Superfund Sites of Silicon Valley," *New York Times* (26 March 2018) at web.archive .org/web/20200408134921/www.nytimes.com/2018/03/26/lens/the-superfund-sites-of -silicon-valley.html. Moore's Law did not determine "one path" to the future: for years IBM pursued superconducting Josephson junction transistors, see Cyrus C. M. Mody, *The Long Arm of Moore's Law: Microelectronics and American Science* (Cambridge: MIT Press, 2017), 20–21, 47–77.

7. Tyrone Siu, "World's Largest Electronics Waste Dump in China," Thomson Reuters (6 July 2015) "rivers are black" at tinyurl.com/y28cjnl3 *or* web.archive.org /web/20161114171344/http://news.trust.org/slideshow/?id=c03216ba-68ee-4558 -a50f-b8f360d90d9b; Zhuang Pinghui, "China's Most Notorious E-Waste Dumping Ground Now Cleaner but Poorer," *South China Morning Post* (22 September 2017) at web .archive.org/web/20201015214542/https://www.scmp.com/news/china/society/article /2112226/chinas-most-notorious-e-waste-dumping-ground-now-cleaner-poorer.

8. Gordon Moore, "Cramming More Components onto Integrated Circuits," *Electronics* 38, no. 8 (19 April 1965): 114–17 reprinted at doi.org/10.1109/N-SSC.2006.4785860; "From Big Blue, to Mighty Microsoft, to Apple's Ascendency," *Economist* (4 April 2015)

at tinyurl.com/y58vqes8 *or* web.archive.org/web/20200929173511/https://cdn.static
-economist.com/sites/default/files/images/print-edition/20150404_WBC737_0.png;
Emerson W. Pugh, *Building IBM: Shaping an Industry and Its Technology* (MIT Press,
1995), 282–86, 301–3.

9. Conner Forrest, "The 25 Biggest Tech Companies in the World, by Market Cap,"
    Tech Republic (10 July 2018) at web.archive.org/web/20200808040244/https://www
    .techrepublic.com/article/the-25-biggest-tech-companies-in-the-world-by-market-cap/.

10. Robert Noyce, "Microelectronics," *Scientific American* 237, no. 3 (September 1977):
    62–69, quotes 65, at jstor.org/stable/24920319; Robert Noyce, "Large-Scale Integration:
    What Is Yet to Come?," *Science* 195, no. 4283 (18 March 1977): 1102–6 at jstor.org
    /stable/1743556; and Ethan Mollick, "Establishing Moore's Law," *IEEE Annals of the
    History of Computing* 28, no. 3 (2006): 62–75 at doi.org/10.1109/MAHC.2006.45.
    Roughly, a transistor chronology goes: point contact, bipolar junction, MOS, MOSFET,
    and FinFET: see David C. Brock and Christophe Lécuyer, "Digital Foundations:
    The Making of Silicon-Gate Manufacturing Technology," *Technology and Culture*
    53, no. 3 (2012): 561–97 at doi.org/10.1353/tech.2012.0122; and Douglas O'Reagan
    and Lee Fleming, "The FinFET Breakthrough and Networks of Innovation in the
    Semiconductor Industry, 1980–2005," *Technology and Culture* 59, no. 2 (2018):
    251–88 at doi.org/10.1353/tech.2018.0029.

11. John G. Linvill and C. Lester Hogan, "Intellectual and Economic Fuel for the Electronics
    Revolution," *Science* 195, no. 4283 (18 March 1977): 1107–13, quote on 1111 at jstor
    .org/stable/1743557. Linvill and Hogan wrote, "If the industry remains on the curve
    of complexity plotted as a function of time given by 'Moore's law' . . . it will achieve a
    complexity of 10 million interconnected components by 1985," citing Moore's "Progress
    in Digital Integrated Electronics," *IEEE International Electron Devices Meeting* (1975):
    11–13 reprinted at doi.org/10.1109/N-SSC.2006.4804410. This 1975 paper mentions
    "the annual doubling law."

12. Carver Mead oral history (Computer History Museum 27 May 2009) quotes on
    7–8 ("many transistors") ("once-a-week commute") at web.archive.org/web
    /20200103232038/https://www.computerhistory.org/collections/catalog/102702086;
    David Brock, *Understanding Moore's Law: Four Decades of Innovation* (Philadelphia:
    Chemical Heritage Press, 2006), 97–103. For Conway's earlier IBM career, see Maria
    Cramer, "52 Years Later, IBM Apologizes for Firing Transgender Woman," *New York
    Times* (21 November 2020) at web.archive.org/web/20201122032735/https://www
    .nytimes.com/2020/11/21/business/lynn-conway-ibm-transgender.html.

13. Charles E. Sporck oral history (Computer History Museum 21 November 2014) "kill
    them" at web.archive.org/web/20200927052757/https://www.computerhistory.org
    /collections/catalog/102740002; Jay T. Last interview, 15 September 2007, Stanford
    University Archives, "disaster" at purl.stanford.edu/th149rh8375; Teradyne founder

Nicholas DeWolf interview, 24 September 2005, "buried competitors" at purl.stanford .edu/vg998xv9695; Applied Materials founder Michael McNeilly interview, 20 July 2004, "cantankerous" at purl.stanford.edu/cm720zj7238 (all accessed April 2020).

14. Hyungsub Choi, "Technology Importation, Corporate Strategies, and the Rise of the Japanese Semiconductor Industry in the 1950s," *Comparative Technology Transfer and Society* 6, no. 2 (2008): 103–26; Warren E. Davis and Daryl G. Hatano, "The American Semiconductor Industry and the Ascendancy of East Asia," *California Management Review* 27, no. 4 (1985): 128–43 at doi.org/10.2307/41165160; W. J. Sanders III, chairman and president of Advanced Micro Devices, quoted in Andrew Pollack, "Japan's Big Lead in Memory Chips," *New York Times* (28 February 1982) at web .archive.org/web/20201101031645/https://www.nytimes.com/1982/02/28/business /japan-s-big-lead-in-memory-chips.html; Robert Schaller, "Technological Innovation in the Semiconductor Industry: A Case Study of the International Technology Roadmap for Semiconductors (ITRS)" (PhD thesis, George Mason University, 2004), 436.

15. National Research Council, *Securing the Future: Regional and National Programs to Support the Semiconductor Industry* (Washington, DC: National Academies Press, 2003), "premier students," 66n 5 at doi.org/10.17226/10677; Schaller (2004), 428, 438–40; SIA Milestones at web.archive.org/web/20201014183750/https://www .semiconductors.org/about/history/.

16. William C. Norris, "Cooperative R&D: A Regional Strategy," *Issues in Science and Technology* 1, no. 2 (Winter 1985): 92–102 at jstor.org/stable/43310879; David V. Gibson and Everett M. Rogers, *R&D Collaboration on Trial: The Microelectronics and Computer Technology Corporation* (Boston: Harvard Business School Press, 1994), 47–65. CDC's Robert M. Price described MCC in Thomas J. Misa, ed., *Building the Control Data Legacy* (Minneapolis: Charles Babbage Institute, 2012), 269.

17. Tamir Agmon and Mary Ann Von Glinow, eds., *Technology Transfer in International Business* (Oxford University Press, 1991), 117–18; Gibson and Rogers, *R&D Collaboration on Trial*, 99–188.

18. Defense Science Board Task Force on Semiconductor Dependency (31 December 1986), quote on 1 ("national security") at web.archive.org/web/20200222004754/https:// apps.dtic.mil/dtic/tr/fulltext/u2/a178284.pdf; Douglas A. Irwin, "The U.S.-Japan Semi-conductor Trade Conflict," in Anne O. Krueger, ed., *The Political Economy of Trade Protection* (University of Chicago Press, 1996) at web.archive.org/web/20151001200221 /http://www.nber.org/chapters/c8717; Ka Zeng, "U.S.-Japan Trade Conflicts: Semi-conductors and Super 301," in *Trade Threats, Trade Wars: Bargaining, Retaliation, and American Coercive Diplomacy* (Ann Arbor: University of Michigan Press, 2003) at doi .org/10.3998/mpub.17690.

19. Sporck oral history, quote on 20. *Compare* Larry D. Browning and Judy C. Shetler,

Sematech: Saving the U.S. Semiconductor Industry (College Station: Texas A&M University Press, 2000) with NRC, *Securing the Future*, 93–121.

20. Gibson and Rogers, *R&D Collaboration on Trial*, 467–500, 539n 8; Douglas A. Irwin and Peter J. Klenow, "Sematech: Purpose and Performance," *Proceedings of the National Academy of Sciences* 93, no. 23 (12 November 1996): 12739–742 at doi.org/10.1073 /pnas.93.23.12739; Richard Brandt, "The Bad Boy of Silicon Valley," *Bloomberg Business* (8 December 1991) "corporate country club" at web.archive.org/web/20180723034448 /https://www.bloomberg.com/news/articles/1991-12-08/the-bad-boy-of-silicon-valley.

21. Schaller (2004), 418–36, quote on 416 ("standard model").

22. "Initially the big guys—IBM, Western Electric, Bell Labs, Motorola, Texas Instruments, Fairchild—they dominated the landscape at the time. And they all had internal [process development] capabilities," stated Applied Materials founder Michael McNeilly (20 July 2004), Stanford University Archives at purl.stanford.edu/cm720zj7238 (acc. April 2020).

23. Karen Brown, telephone interview, 27 August 1999, quoted in Schaller (2004), 83.

24. Schaller (2004), quote on 419.

25. Schaller (2004), quotes on 78, 160 (central planning assumption), 441 (summer study), 434 (real purpose), 673 (everyone knows, difficult and costly); on lithography see Mody, *Long Arm of Moore's Law*, 122–27, 141–44.

26. Ed Korczynski, "Moore's Law Extended: The Return of Cleverness," *Solid State Technology* 40, no. 7 (July 1997), quote on 364 ("stay on plan") available at EBSCO Academic Search Premier.

27. Leslie Berlin, *Man behind the Microchip: Robert Noyce and the Invention of Silicon Valley* (New York: Oxford University Press, 2005), 218–20, 239; Randall Stross, "It's Not the People You Know: It's Where You Are," *New York Times* (22 October 2006) at web .archive.org/web/20200218045450/https://www.nytimes.com/2006/10/22/business /yourmoney/22digi.html; Douglas Cumming and Na Dai, "Local Bias in Venture Capital Investments," *Journal of Empirical Finance* 17 (2010): 362–380 at doi.org/10.1016/j .jempfin.2009.11.001; Erin Griffith, "Don Valentine, Founder of Sequoia Capital, Is Dead at 87," *New York Times* (25 October 2019) at web.archive.org/web /20200807191454/https://www.nytimes.com/2019/10/25/obituaries/don-valentine -sequoia-capital.html; "Remembering Don Valentine" at web.archive.org/web /20200814041200/https://www.sequoiacap.com/article/remembering-don-valentine/; Rich Karlgaard, "Ten Laws of the Modern World," *Forbes* (19 April 2005) "few guesses" at web.archive.org/web/20200424074703if_/https://www.forbes.com/2005/04/19 /cz_rk_0419karlgaard.html.

28. Schaller (2004), quote on 673 ("viciously"); Gibson and Rogers, *R&D Collaboration on Trial*, 507–10; Andrew Pollack, "Conflict at Sematech Forces Out No. 2 Man," *New York Times* (21 March 1989) at web.archive.org/web/20171220053953/http://www.nytimes .com/1989/03/21/business/conflict-at-sematech-forces-out-no-2-man.html. Noyce

died "at his home in Austin" in 1990, according to Jesus Sanchez and Carla Lazzareschi, "Robert Noyce, Computer Age Pioneer, Dies," *Los Angeles Times* (4 June 1990) at web .archive.org/web/20191104134624/https://www.latimes.com/archives/la-xpm-1990–06 -04-mn-455-story.html.

29. W. J. Spencer and T. E. Seidel, "National Technology Roadmaps: The U.S. Semi-conductor Experience," *Proceedings of 4th International Conference on Solid-State and IC Technology* (1995): 211–20 at doi.org/10.1109/ICSICT.1995.500069.

30. Laszlo A. Belady, oral history (21 November 2002), Charles Babbage Institute OH 352, at purl.umn.edu/107110, p. 30, noting SEMATECH predecessor MCC was "an anti-Japanese outfit"; Steve Lohr, "A Piece of Japan's Chip Market," *New York Times* (1 February 1982) at web.archive.org/web/20150524102317/http://www.nytimes .com/1982/02/01/business/a-piece-of-japan-s-chip-market.html; Berlin, *Man behind the Microchip*, quote on 291 ("tightly wrapped"); Susan Chira, "Japan Is Said to End Some Barriers," *New York Times* (10 January 1986) at web.archive.org /web/20150524185557/http://www.nytimes.com/1986/01/10/business/japan-is -said-to-end-some-barriers.html; David E. Sanger, "NEC Wants Part in U.S. Chip Project," *New York Times* (15 August 1988) "flow of technology" at web.archive.org /web/20150525085342/http://www.nytimes.com/1988/08/15/business/nec-wants -part-in-us-chip-project.html.

31. Schaller (2004), 501, 527–28, 560, 572, 609; WSC history at web.archive.org /web/20200220105149/http://www.semiconductorcouncil.org/about-wsc/history/.

32. Schaller (2004), 477, 500–504, 531–35, quote on 532 (electrons); ITRS 2003 at web .archive.org/web/20201014183610/https://www.semiconductors.org/resources/2003 -international-technology-roadmap-for-semiconductors-itrs/.

33. Schaller (2004), 219 (equipment 1979–2000), 401, quote on 29 ("needed Canon").

34. NRC, *Securing the Future*, 27–30, 149–60, 210–18 on Taiwan and TSMC.

35. "The First Computer Chip with a Trillion Transistors," *Economist* (7 December 2019) at web.archive.org/web/20201020015932if_/https://www.economist.com/science-and -technology/2019/12/07/the-first-computer-chip-with-a-trillion-transistors./

36. See David Manners, "TSMC 'Actively' Mulling US Fab," *Electronics Weekly* (17 March 2020) "Chinese hands" at web.archive.org/web/20200424204625/www .electronicsweekly.com/news/business/tsmc-actively-mulling-us-fab-2020–03/; Willy Shih, "TSMC's Announcement of A New U.S. Semiconductor Fab Is Big News," *Forbes* (15 May 2020) at web.archive.org/web/20201105021004/https://www.forbes .com/sites/willyshih/2020/05/15/tsmcs-announcement-of-a-us-fab-is-big-news/.

37. Laurie J. Flynn, "Technology Briefing: Intel to Ship Dual-Core Chips," *New York Times* (8 February 2005) at web.archive.org/web/20201118234203/https://www.nytimes .com/2005/02/08/business/technology/technology-briefing-hardware-intel-to-ship -dualcore.html; John Markoff, "Intel's Big Shift After Hitting Technical Wall," *New York*

Times (17 May 2004) all quotes at web.archive.org/web/20200330191650/www.nytimes
.com/2004/05/17/business/technology-intel-s-big-shift-after-hitting-technical-wall
.html. A technical treatment is Laszlo B. Kish, "End of Moore's Law," *Physics Letters
A* 305 (2002): 144–49.

38. Christopher Mims, "Why CPUs Aren't Getting Any Faster," *Technology Review* (12
October 2010) at www.technologyreview.com/2010/10/12/199966/why-cpus-arent
-getting-any-faster/ (acc. April 2020); Ryan Jones, "Intel 10th Gen Mobile CPUs: All
You Need to Know about Ice Lake Laptop Processors," *Trusted Reviews* (2 October
2019) at web.archive.org/web/20200817210311/https://www.trustedreviews.com
/news/intel-ice-lake-release-date-specs-price-3642152; for Intel's difficulties with
7 nm chips see Asa Fitch, "Intel's Success Came with Making Its Own Chips: Until
Now," *Wall Street Journal* (6 November 2020) at web.archive.org/web/20201107005452
/https://www.wsj.com/articles/intel-chips-cpu-factory-outsourcing-semiconductor
-manufacturing-11604605618.

39. See Intel, "New 10th Gen Intel Core U-Series and Y-Series Processors," at web.archive
.org/web/20201001222849/https://www.intel.com/content/www/us/en/products/docs
/processors/core/10th-gen-core-mobile-u-y-processors-brief.html. A footnote, in tiny
print, notes cryptically: "Over 2X better overall performance vs. 5YO as measured by
SYSMark* 2018 on: Intel® Core™ i7–10510Y processor (CML-Y42)PL1=7W TDP, 4C8T,
Turbo up to 4.2GHz/3.6GHz."

40. Laurie J. Flynn, "Intel Halts Development of 2 New Microprocessors," *New York Times*
(8 May 2004) "revamping" at web.archive.org/web/20200709051317/https://www
.nytimes.com/2004/05/08/business/intel-halts-development-of-2-new-microprocessors
.html; M. Mitchell Waldrop, "The Chips Are Down for Moore's Law," *Nature* 530 (9
February 2016): 144–47 at web.archive.org/web/20201103194202/https://www.nature
.com/news/the-chips-are-down-for-moore-s-law-1.19338.

41. Thomas M. Conte, Erik P. Debenedictis, Paolo A. Gargini, and Elie Track, "Rebooting
Computing: The Road Ahead," *Computer* 50, no. 1 (January 2017): 20–29 at doi
.org/10.1109/MC.2017.8; Paolo A. Gargini, "Roadmap Evolution: From NTRS to ITRS,
from ITRS 2.0 to IRDS," International Conference on Extreme Ultraviolet Lithography
Proceedings 10450 (16 October 2017) at doi.org/10.1117/12.2280803. IRDS members
at web.archive.org/web/20200916200829/https://irds.ieee.org/home/process; INEMI
members at web.archive.org/web/20200621021018/https://www.inemi.org/membership.
ASM Pacific Technology is an equipment manufacturer in Singapore, unrelated to ASML.

42. "The 10 Largest Tech Companies in the World" *Forbes* (15 May 2019) at web.archive.
org/web/20200108040050/www.forbes.com/sites/jonathanponciano/2019/05/15
/worlds-largest-tech-companies-2019/. A critical and perceptive assessment is Evgeny
Morozov, *The Net Delusion: The Dark Side of Internet Freedom* (New York: Public
Affairs, 2011).

43. J. J. Pilliod, "Fundamental Plans for Toll Telephone Plant," *Bell System Technical Journal* 31, no. 5 (1952): 832–50, maps following 836.

44. J. W. Foss and R. W. Mayo, "Operation Survival," *Bell Laboratories Record* 47 (January 1969): 11–17 at hdl.handle.net/2027/mdp.39076000053707.

45. Stephen J. Lukasik, "Why the Arpanet Was Built," *IEEE Annals of the History of Computing* 33, no. 3 (2011): 4–20, quote on 4 ("goal").

46. Lukasik, "Why the Arpanet Was Built," quote on 10 ("unaware"); John Day, personal communication (27 May 2020), crediting John Melvin (see www.rfc-editor.org/rfc /rfc101.html); Katie Hafner and Matthew Lyon, *Where Wizards Stay Up Late: The Origins of the Internet* (New York: Simon & Schuster, 1996), quote on 10.

46. Fred Turner, "Where the Counterculture Met the New Economy: The WELL and the Origins of Virtual Community," *Technology and Culture* 46 no. 3 (2005): 485–512 at doi .org/10.1353/tech.2005.0154.

48. John Perry Barlow, "Coming in to the Country," *Communications of the ACM* 34, no. 3 (1991): 19–21, quote on 19 ("dimmest awareness"). Barlow's account of EFF's founding is at web.archive.org/web/20010627230955/http://www.eff.org/pub/EFF/history.eff.

49. John Perry Barlow, "A Declaration of the Independence of Cyberspace" (8 February 1996) at www.eff.org/cyberspace-independence (archived perma.cc/37QR-HV9S). Critical assessments are Aimée Hope Morrison, "An Impossible Future: John Perry Barlow's 'Declaration of the Independence of Cyberspace,'" *New Media & Society* 11, nos. 1–2 (2009): 53–72 at doi.org/10.1177%2F1461444808100161; and Alexis C. Madrigal, "The End of Cyberspace," *The Atlantic* (1 May 2019) at web.archive .org/web/20201107235206/https://www.theatlantic.com/technology/archive/2019/05 /the-end-of-cyberspace/588340/.

50. *USA Today* at web.archive.org/web/20200601084021/www.usatoday.com/story /tech/2018/06/22/cost-of-a-computer-the-year-you-were-born/36156373/ describes a 1996 Gateway Solo 2100 with an inflation-adjusted price of $6,653, divided by the then-current New York state average weekly wage of $1,401 at web.archive.org /web/20201124004738/https://paidfamilyleave.ny.gov/2020.

51. "No provider or user of an interactive computer service shall be treated as the publisher or speaker of any information provided by another information content provider," is section 230 at web.archive.org/web/20201121190807/https://www.law.cornell.edu /uscode/text/47/230; Christopher Tozzi, *For Fun and Profit: A History of the Free and Open Source Software Revolution* (Cambridge: MIT Press, 2017).

52. Roger Kimball, "The Anguishes of E.M. Cioran," *New Criterion* 38, no. 3 (March 1988) at web.archive.org/web/20201108105859/https://newcriterion.com/issues/1988/3/the-anguishes -of-em-cioran, identifies Cioran's quirky 1960 essay "Russia and the Virus of Liberty."

53. William J. Clinton and Albert Gore, Jr., "Technology for America's Economic Growth" (22 February 1993), quote on 28 ("fiber optic") at web.archive.org/web

/20070619205603/http://www.itsdocs.fhwa.dot.gov/JPODOCS/BRIEFING/7423.pdf;
Bethany Allen-Ebrahimian, "The Man Who Nailed Jello to the Wall," *Foreign Policy* (29
June 2016) at web.archive.org/web/20200815232139/foreignpolicy.com/2016/06/29
/the-man-who-nailed-jello-to-the-wall-lu-wei-china-internet-czar-learns-how-to
-tame-the-web/. A trenchant critique of "internet freedom" and its misguided Cold War
antecedents is Morozov, *Net Delusion*, 33–56, 229–74.

54. Michael Posner, "Global Internet Freedom and the Rule of Law, Part II" (2 March 2010)
at web.archive.org/web/20170517103448/https://www.humanrights.gov/dyn/michael-
posner-global-internet-freedom-and-the-rule-of-law-part-ii.html; Clinton's 2010
speech quoted in Fergus Hanson, "Baked In and Wired: ediplomacy@State" (Brookings
Institution, 2016), 25 at web.archive.org/web/20200923074853/https://www.brookings.
edu/wp-content/uploads/2016/06/baked-in-hansonf-5.pdf; "US: Clinton to Press for
Internet Freedom," *Human Rights Watch* (21 January 2010) at web.archive.org/web
/20150905232716/www.hrw.org/news/2010/01/21/us-clinton-press-internet-freedom.

55. See "Microsoft & Internet Freedom" (27 January 2010) "dedicated to advancing" at web.
archive.org/web/20201001084732/news.microsoft.com/2010/01/27/microsoft
-internet-freedom/; Jessica Guynn, "Silicon Valley Luminaries Become Technology
Ambassadors to Russia," *Los Angeles Times* (16 February 2010) at web.archive.org
/web/20200923093906/https://latimesblogs.latimes.com/technology/2010/02/state
-department-technology-russia.html. Facebook and Twitter declined to join GNI,
according to Morozov, *Net Delusion*, 22–23, 217; Facebook later joined in 2013.

56. Larry Greenemeier, "How Was Egypt's Internet Access Shut Off?," *Scientific
American* (28 January 2011) at web.archive.org/web/20201001191059/https://www
.scientificamerican.com/article/egypt-internet-mubarak/; Matt Richtel, "Egypt Cuts Off
Most Internet and Cell Service," *New York Times* (28 January 2011) "comply" at web.
archive.org/web/20201109134158/https://www.nytimes.com/2011/01/29/technology
/internet/29cutoff.html; Christopher Rhoads and Geoffrey A. Fowler, "Egypt Shuts
Down Internet, Cellphone Services," *Wall Street Journal* (29 January 2011) "whole
internet down" at tinyurl.com/yxg9m4jz *or* web.archive.org/web/20200911144506
/https://www.wsj.com/articles/SB10001424052748703956604576110453371369740;
Jessica Rosenworcel, "The U.S. Government Couldn't Shut Down the Internet, Right?
Think Again," *Washington Post* (8 March 2020) at tinyurl.com/y5uaqmqy *or* web.
archive.org/web/20201114193428/https://www.washingtonpost.com/opinions/the-us
-government-couldnt-shut-down-the-internet-right-think-again/2020/03/06/6074dc86
-5fe5–11ea-b014–4fafa866bb81_story.html; Samuel Woodhams and Simon Migliano,
"The Global Cost of Internet Shutdowns in 2019," Top10VPN (7 January 2020) at web
.archive.org/web/20200131185752if_/https://20200131185752/www.top10vpn.com
/cost-of-internet-shutdowns/.

57. See Nicola Pratt and Dina Rezk, "Securitizing the Muslim Brotherhood: State Violence

and Authoritarianism in Egypt after the Arab Spring," *Security Dialogue* 50, no. 3 (2019): 239–56 at doi.org/10.1177/0967010619830043.

58. A valuable survey is Andrew Blum, *Tubes: A Journey to the Center of the Internet* (New York: Ecco, 2012). Snowden's website at web.archive.org/web/20201125022653 /https://edwardsnowden.com/revelations/ links to Ryan Gallagher and Henrik Moltke, "The Wiretap Rooms: The NSA's Hidden Spy Hubs in Eight U.S. Cities," *The Intercept* (25 June 2018) at web.archive.org/web/20201011155903/https://theintercept .com/2018/06/25/att-internet-nsa-spy-hubs/.

59. See "World - Autonomous System Number statistics" (3 May 2021) at https://web. archive.org/web/20210507163755/https://www-public.imtbs-tsp.eu/~maigron/ RIR_Stats/RIR_Delegations/World/ASN-ByNb.html; Ramakrishnan Durairajan, Paul Barford, Joel Sommers, and Walter Willinger, "InterTubes: A Study of the US Long-haul Fiber-optic Infrastructure," SIGCOMM '15: Proceedings of the 2015 ACM Conference on Special Interest Group on Data Communication (August 2015): 565–78 at doi .org/10.1145/2785956.2787499. See TeleGeography's global Internet Exchange Map at web.archive.org/web/20201112033920/https://www.internetexchangemap.com/.

60. David D. Clark, *Designing an Internet* (Cambridge: MIT Press, 2018), 26, 238, 255; European Commission, "Shaping Europe's Digital Future" (19 February 2020), quote on 2 ("own way") at web.archive.org/web/20201104150524/https://ec.europa.eu/info/sites /info/files/communication-shaping-europes-digital-future-feb2020_en_3.pdf.

61. Robert E. Litan, "The Telecommunications Crash: What to Do Now?," Brookings Institution Policy Brief No. 112 (December 2002) at web.archive.org/web /20200607165739/https://www.brookings.edu/wp-content/uploads/2016/06/pb112. pdf; Seth Schiesel, "Worldcom Leader Departs Company in Turbulent Time," *New York Times* (1 May 2002) at web.archive.org/web/20201020233442/https://www.nytimes .com/2002/05/01/business/worldcom-leader-departs-company-in-turbulent-time.html; Allan Shampine and Hal Sider, "The Telecom Boom and Bust: Their Losses, Our Gain?" (13 November 2007) Millikan Institute at web.archive.org/web/20201126213752/http:// assets1c.milkeninstitute.org/assets/Publication/MIReview/PDF/54–60mr36.pdf.

62. Ken Belson, "SBC Agrees to Acquire AT&T for $16 Billion," *New York Times* (31 January 2005) at web.archive.org/web/20201119052217/https://www.nytimes.com /2005/01/31/business/sbc-agrees-to-acquire-att-for-16-billion.html; Dionne Searcey and Amy Schatz, "Phone Companies Set Off a Battle over Internet Fees," *Wall Street Journal* (6 January 2006) at web.archive.org/web/20190901014204/https://www.wsj.com/articles /SB113651664929039412; Mike Masnick, "SBC: We Own the Internet, So Google Should Pay Up," Tech Dirt (31 October 2005) "my pipes" at web.archive.org/web /20201110114709/https://www.techdirt.com/articles/20051031/0354228.shtml. Several versions of this colorful quote exist.

63. John Markoff, "Internet Traffic Begins to Bypass the U.S.," *New York Times* (29 August

2008) at web.archive.org/web/20201109005034/http://www.nytimes.com/2008/08/30/business/30pipes.html. For origins of "revenue neutral" peering, see Clark, *Designing an Internet*, 247–49.

64. "Postel Disputes," *Economist* (8 February 1997): 116–17; Kal Raustiala, "Governing the Internet," *American Journal of International Law* 110, no. 3 (2016): 491–503 at doi .org/10.1017/S0002930000016912.

65. Craig Timberg, "U.S. to Relinquish Remaining Control over the Internet," *Washington Post* (14 March 2014) at tinyurl.com/y5wfoaq8 *or* web.archive.org /web/20201107202615/www.washingtonpost.com/business/technology/us-to -relinquish-remaining-control-over-the-internet/2014/03/14/0c7472d0-abb5–11e3 -adbc-888c8010c799_story.html; Susmita Baral, "Who Controls the Internet? US Government Hands Over Control to ICANN," *International Business Times* (3 October 2016) at web.archive.org/web/20201108095617/https://www.ibtimes.com/who -controls-internet-us-government-hands-over-control-icann-2425491.

66. James Risen and Eric Lichtblau, "Bush Lets U.S. Spy on Callers Without Courts," *New York Times* (16 December 2005) at web.archive.org/web/20201123193840/https://www .nytimes.com/2005/12/16/politics/bush-lets-us-spy-on-callers-without-courts.html; Justine Sharrock, "When the NSA Shows Up at Your Internet Company," BuzzFeed (19 July 2013) at web.archive.org/web/20200915084128/https://www.buzzfeednews.com /article/justinesharrock/what-is-that-box-when-the-nsa-shows-up-at-your-internet -comp; Janus Kopfstein, "Utah ISP Owner Describes the NSA 'Black Box' That Spied On His Customers," Verge (21 July 2013) at web.archive.org/save/https://www.theverge .com/2013/7/21/4541342/isp-owner-describes-nsa-box-that-spied-on-customers; Gallagher and Moltke, "The Wiretap Rooms" (peering circuits).

67. Julia Angwin, Charlie Savage, Jeff Larson, Henrik Moltke, Laura Poitras, and James Risen, "AT&T Helped U.S. Spy On Internet on a Vast Scale," *New York Times* (15 August 2015) at web.archive.org/web/20201126031945/https://www.nytimes.com/2015/08 /16/us/politics/att-helped-nsa-spy-on-an-array-of-internet-traffic.html; "Newly Disclosed N.S.A. Files Detail Partnerships with AT&T and Verizon," *New York Times* (15 August 2015) at web.archive.org/web/20201120122355/https://www .nytimes.com/interactive/2015/08/15/us/documents.html.

68. On Miami NAP, see web.archive.org/web/20160501121412/http://www.verizonenterprise .com/infrastructure/data-centers/north-america/nap/nap-americas.xml *and* web .archive.org/web/20201126221246/https://www.datacenterhawk.com/colo/equinix /50-ne-9th-street/mi1-nap-of-the-americas *and* web.archive.org/web/20201023221240 /https://baxtel.com/data-center/nap-of-the-americas-equinix-miami-mi1.

69. Thomas J. Misa, *Digital State: The Story of Minnesota's Computing Industry* (Minneapolis: University of Minnesota Press, 2013), quote on 217 ("all traffic"). TeleGeography's worldwide Internet Exchange Map is at www.internetexchangemap.com.

70. At web.archive.org/web/20201108102301/https://nsa.gov1.info/utah-data-center/; see "IC off the Record" at web.archive.org/web/20200925153855/nsa.gov1.info/dni/2018 /index.html (including the Snowden leaks).

71. Andrei Soldatov and Irina Borogan, *The Red Web: The Struggle Between Russia's Digital Dictators and the New Online Revolutionaries* (New York: Public Affairs, 2015), ix–xi, 17–20, 50–51, 83, 304; Juha Kukkola, *Digital Soviet Union: The Russian National Segment of the Internet as a Closed National Network Shaped by Strategic Cultural Ideas* (Helsinki: National Defence University, 2020), 269–71, 333 at www.doria.fi /handle/10024/177157. Telegeography locates MSK-IX's offices at Central Telegraf Communication building on Tverskaya Street 7, Moscow, with its M9 complex at 7 Butlerova Street.

72. David Barboza, "The Rise of Baidu (That's Chinese for Google)," *New York Times* (17 September 2006) at web.archive.org/web/20190627071550/https://www.nytimes .com/2006/09/17/business/yourmoney/17baidu.html. Li patent at patents.google.com /patent/US5920859; Page patents at patents.google.com/patent/US6285999 and patents .google.com/patent/US7269587.

73. Daniela Hernandez, "Man behind the 'Google Brain' Joins Chinese Search Giant Baidu," *Wired* (16 May 2014) at web.archive.org/web/20201111232239/http://www.wired .com/2014/05/andrew-ng-baidu/; Andrew Ng (19 October 2016) at web.archive.org /web/20200610024418/https://twitter.com/AndrewYNg/status/788959615123791872; Paul Mozur, "A.I. Expert at Baidu, Andrew Ng, Resigns From Chinese Search Giant," *New York Times* (22 March 2017) at web.archive.org/web/20201112035640/https:// www.nytimes.com/2017/03/22/business/baidu-artificial-intelligence-andrew-ng.html.

74. Xie Yu and Sidney Leng, "Tech Entrepreneurs Dominate as China's Political Advisers in IT Push," *South China Morning Post* (4 March 2018) at web.archive.org /web/20201109032355/https://www.scmp.com/business/companies/article/2135642 /tech-entrepreneurs-replace-real-estate-tycoons-political-advisers; Jing Yang and Xie Yu, "Successful with Chinese Startups, Sequoia China Launches a Hedge Fund," *Wall Street Journal* (10 September 2020) at web.archive.org/web/20201012004839/https:// www.wsj.com/articles/successful-with-chinese-startups-sequoia-china-launches -a-hedge-fund-11599736284; Charlie Campbell, "Baidu's Robin Li is Helping China Win the 21st Century," *Time* (18 January 2018) "do things freely" at web.archive.org /web/20200815052118/https://time.com/5107485/baidus-robin-li-helping-china -win-21st-century/; Li Yuan, "Jack Ma, China's Richest Man, Belongs to the Communist Party, Of Course," *New York Times* (27 November 2018) at web.archive .org/web/20201108180505/https://www.nytimes.com/2018/11/27/business/jack-ma -communist-party-alibaba.html.

75. David Barboza, "In Roaring China, Sweaters Are West of Socks City," *New York Times* (December 24, 2004) at web.archive.org/web/20151108231310/https://www.nytimes

.com/2004/12/24/business/worldbusiness/in-roaring-china-sweaters-are-west-of
-socks-city.html; Eric Harwit, *China's Telecommunications Revolution* (Oxford: Oxford
University Press, 2008), 158–82; Michael Bristow, "China Defends Internet Censorship,"
BBC News Beijing (8 June 2010) at web.archive.org/web/20200801221533/news
.bbc.co.uk/2/hi/8727647.stm; Raymond Zhong, "At China's Internet Conference,
a Darker Side of Tech Emerges," *New York Times* (8 November 2018) at web.archive
.org/web/20190301031116/https://www.nytimes.com/2018/11/08/technology/china
-world-internet-conference.html; Sarah Lai Stirland, "Cisco Leak: 'Great Firewall' of
China Was a Chance to Sell More Routers," *Wired* (20 May 2008) at web.archive.org
/web/20200724063859/https://www.wired.com/2008/05/leaked-cisco-do/.

76. James Griffiths, *The Great Firewall of China: How to Build and Control an Alternative
Version of the Internet* (London: Zed, 2019), 78–81, 181 (on founding of CAC), 207,
242, 312–13; Tania Branigan, "China's 'Great Firewall' Creator Pelted with Shoes:
Twitter User Claims to Have Mounted Egg and Shoe Attack on Fang Binxing At Wuhan
University," *Guardian* (20 May 2011) at web.archive.org/web/20200923200125/https://
www.theguardian.com/world/2011/may/20/china-great-firewall-creator-pelted-shoes;
Tania Branigan, "China's Great Firewall Not Secure Enough, Says Creator," *Guardian*
(18 February 2011) "running dog" at web.archive.org/web/20200813094051/https://
www.theguardian.com/world/2011/feb/18/china-great-firewall-not-secure-internet.

77. Allen-Ebrahimian, "Man Who Nailed Jello," quote "rules Weibo"; Griffiths, *Great
Firewall of China*, 275–83; Xinmei Shen, "The Story of China's Great Firewall,
the World's Most Sophisticated Censorship System," *South China Morning Post*
(7 November 2019) at web.archive.org/web/20201120224212/https://www.scmp
.com/abacus/who-what/what/article/3089836/story-chinas-great-firewall-worlds
-most-sophisticated; Lianrui Jia and Xiaofei Han, "Tracing Weibo (2009–2019): The
Commercial Dissolution of Public Communication and Changing Politics," *Internet
Histories* 4, no. 3 (2020): 304–32 at doi.org/10.1080/24701475.2020.1769894; Murong
Xuecun, "Scaling China's Great Firewall," *New York Times* (17 August 2015) "not free
online" at web.archive.org/web/20200218082926/https://www.nytimes.com/2015/08/18
/opinion/murong-xuecun-scaling-chinas-great-firewall.html.

78. For China's global internet connections, see web.archive.org/web/20201112033056
/https://global-internet-map-2018.telegeography.com/; its connection to Taiwan is
visible at web.archive.org/web/20201025013058/https://asia-pacific-map-2012
.telegeography.com/.

79. Griffiths, *Great Firewall of China*, 28, 113, 297, 299; Lily Hay Newman, "The Attack
on Global Privacy Leaves Few Places to Turn," *Wired* (4 August 2017) "Apple complied"
at web.archive.org/web/20201118223122/https://www.wired.com/story/china-russia
-vpn-crackdown/; Liza Lin and Yoko Kubota, "China's VPN Crackdown May Aid

Government Surveillance," *Wall Street Journal* (17 January 2018) at web.archive.org
/web/20201109041507/https://www.wsj.com/articles/chinas-vpn-crackdown-may-aid
-government-surveillance-1516189155; Anthony Spadafora, "China VPN Crackdown
'To Control Coronavirus Message,'" *Tech Radar* (19 February 2020) at web.archive.org
/web/20200702043140/https://www.techradar.com/news/china-cracks-down-on-vpns
-in-bid-to-control-coronavirus-message; Grady McGregor, "China Deploys a Favorite
Weapon in the Coronavirus Crisis: A Crackdown on VPNs," *Fortune* (25 February
2020) at web.archive.org/web/20201021194543/https://fortune.com/2020/02/25
/coronavirus-china-vpn/; Anthony Spadafora, "China Cracks Down on VPN Use
Following Coronavirus," *Tech Radar* (27 February 2020) "authorities throttle" at web
.archive.org/web/20201101024019/https://www.techradar.com/news/coronavirus-sees
-china-crack-down-on-vpns; Jacob Fromer and Owen Churchill, "Chinese App Called
Tuber Provides a Still-Censored Look over Beijing's 'Great Firewall,'" *South China
Morning Post* (10 October 2020) at web.archive.org/web/20201013033938/https://
www.scmp.com/news/china/article/3104949/chinese-app-called-tuber-provides-still
-censored-look-over-beijings; Xinmei Shen, "Chinese Browser That Helped Users
Bypass Great Firewall Disappears after Racking Up Millions of Downloads," *South
China Morning Post* (12 October 2020) at web.archive.org/web/20201014120631/www
.scmp.com/abacus/tech/article/3105106/chinese-browser-helped-users-bypass-great
-firewall-disappears-after.

80. Griffiths, *Great Firewall of China*, 131–58, 195–201; "China Has Turned Xinjiang into
a Police State Like No Other," *Economist* (31 May 2018) "police state" and "IJOP" at web
.archive.org/web/20200701231137if_/https://www.economist.com/briefing/2018/05/31
/china-has-turned-xinjiang-into-a-police-state-like-no-other; Danny O'Brien, "Massive
Database Leak Gives Us a Window into China's Digital Surveillance State," Electronic
Frontier Foundation (1 March 2019) at web.archive.org/web/20200701212003/https://
www.eff.org/deeplinks/2019/03/massive-database-leak-gives-us-window-chinas
-digital-surveillance-state; "Xinjiang and the World: The Persecution of the Uyghurs
Is a Crime against Humanity," *Economist* (17 October 2020) at web.archive.org/web
/20201015124606/www.economist.com/leaders/2020/10/17/the-persecution-of-the
-uyghurs-is-a-crime-against-humanity.

81. Griffiths, *Great Firewall of China*, 247–73, 285–305; Max Seddon, "Russia's Chief Internet
Censor Enlists China's Know-How," *Financial Times* (26 April 2016) at web.archive
.org/web/20201125142928/https://www.ft.com/content/08564d74–0bbf-11e6–9456
-444ab5211a2f; Steven Feldstein, "The Road to Digital Unfreedom: How Artificial
Intelligence Is Reshaping Repression," *Journal of Democracy* 30, no. 1 (2019): 40–52 at
doi.org/10.1353/jod.2019.0003; William Chalk, "China's Digital Imperialism: Shaping
the Global Internet," SupChina (2 July 2019) at web.archive.org/web/20201107231008

/https://supchina.com/2019/07/02/chinas-digital-imperialism-shaping-the-global
-internet/; Golnaz Esfandiari, "Iran to Work with China to Create National Internet
System," Radio Free Europe / Radio Liberty (4 September 2020) at web.archive.org
/web/20201126001700/https://www.rferl.org/amp/iran-china-national-internet
-system-censorship/30820857.html; Nick Bailey, "East African States Adopt China's
Playbook on Internet Censorship," Freedom House (24 October 2017) at web.archive
.org/web/20201113020733/https://freedomhouse.org/article/east-african-states-adopt
-chinas-playbook-internet-censorship.

82. Lindsay Gorman and Matt Schrader, "U.S. Firms Are Helping Build China's Orwellian
State," *Foreign Policy* (19 March 2019) at web.archive.org/web/20201013052310
/https://foreignpolicy.com/2019/03/19/962492-orwell-china-socialcredit-surveillance/;
Ryan Gallagher, "How U.S. Tech Giants Are Helping to Build China's Surveillance State,"
The Intercept (11 July 2019) "recruited Western companies" at web.archive.org/we
/20201002212210/https://theintercept.com/2019/07/11/china-surveillance-google-ibm
-semptian/; Paul Mozur and Don Clark, "China's Surveillance State Sucks Up Data: U.S.
Tech Is Key to Sorting It," *New York Times* (22 November 2020) at web.archive.org/we
/20201123194804/https://www.nytimes.com/2020/11/22/technology/china-intel
-nvidia-xinjiang.html.

83. McKenzie Funk, "How ICE Picks Its Targets in the Surveillance Age," *New York Times*
(2 October 2019) at web.archive.org/web/20201011154655/https://www.nytimes
.com/2019/10/02/magazine/ice-surveillance-deportation.html; Beth Kindig, "Palantir
IPO: Deep-Dive Analysis," *Forbes* (29 September 2020) at web.archive.org/web
/20201002151735/https://www.forbes.com/sites/bethkindig/2020/09/29/palantir-ipo
-deep-dive-analysis/; Angel Au-Yeung, "Palantir IPO Cements Billionaire Fortunes
for Cofounder Peter Thiel and CEO Alexander Karp," *Forbes* (30 September 2020)
at web.archive.org/web/20201009084120/https://www.forbes.com/sites/angelauyeung
/2020/09/30/palantir-ipo-cements-billionaire-fortunes-for-cofounder-peter-thiel-and
-ceo-alexander-karp/; Michael Steinberger, "Does Palantir See Too Much," *New York
Times* (21 October 2020) "dangerous" at web.archive.org/web/20201023055055
/https://www.nytimes.com/interactive/2020/10/21/magazine/palantir-alex-karp.html.

84. Japreet Grewal, "Internet Shutdowns in 2016" (Centre for Internet and Society, India,
2016) at web.archive.org/web/20170716081218/https://cis-india.org/internet
-governance/files/internet-shutdowns; see data at web.archive.org/web/20201117234241
/https://www.internetshutdowns.in/; Paul Schemm, "Ethiopia Shuts Down Social Media
to Keep from 'Distracting' Students," *Washington Post* (13 July 2016) at web.archive.org
/web/20201108094008/https://www.washingtonpost.com/news/worldviews/wp/2016
/07/13/ethiopia-shuts-down-social-media-to-keep-from-distracting-students/; Julia
Powles, "Google and Microsoft Have Made a Pact to Protect Surveillance Capitalism,"
Guardian (2 May 2016) at web.archive.org/web/20201108111409/https://www.theguardian

.com/technology/2016/may/02/google-microsoft-pact-antitrust-surveillance
-capitalism.

85. See the 2016 Foxconn company report (Hon Hai Precision Industry Co., Ltd) at web
.archive.org/web/20200503143623/http://www.foxconn.com/Files/annual_rpt_e/2016
_annual_rpt_e.pdf; Chang Liang-chih and Frances Huang, "Foxconn Plans to Increase
China Workforce to 1.3 Million," Focus Taiwan (19 August 2010) at web.archive.org
/web/20100822195546/http://focustaiwan.tw/ShowNews/WebNews_Detail.aspx?ID
=201008190012&Type=aECO; Jill Disis, "Foxconn Says Its Big iPhone Factories in
China Are Back to Normal," CNN Business (15 May 2020) at web.archive.org/web
/20200518173100/https://www.cnn.com/2020/05/15/tech/foxconn-china-factories
-intl-hnk/index.html.

86. Cheng Ting-Fang and Lauly Li, "How China's Chip Industry Defied the Coronavirus
Lockdown," *Nikkei Asian Review* (18 March 2020) at web.archive.org/web
/20201125191541/https://asia.nikkei.com/Spotlight/The-Big-Story/How-China
-s-chip-industry-defied-the-coronavirus-lockdown; "Wuhan East Lake Hi-Tech
Development Zone," Hubei Provincial People's Government (21 May 2013) at web
.archive.org/web/20180101152822/http://en.hubei.gov.cn/business/zones/201305
/t20130521_450067.shtml; Sidney Leng and Orange Wang, "China's Semiconductor
Drive Stalls in Wuhan, Exposing Gap in Hi-Tech Production Capabilities," *South China
Morning Post* (28 August 2020) at web.archive.org/web/20201126224029/https://www
.scmp.com/economy/china-economy/article/3099100/chinas-semiconductor-drive
-stalls-%20wuhan-exposing-gap-hi-tech

87. Schaller (2004), quote on 532 ("electrons").

88. Jess Macy Yu and Karl Plume, "Foxconn Reconsidering Plans to Make LCD Panels at
Wisconsin Plant," Reuters (29 January 2019) at web.archive.org/web/20201125185346
/https://www.reuters.com/article/us-foxconn-wisconsin-exclusive-idUSKCN1PO0FV;
Josh Dzieza, "Wisconsin Denies Foxconn Tax Subsidies after Contract Negotiations Fail,"
The Verge (12 October 2020) at web.archive.org/web/20201125213310/https://www
.theverge.com/2020/10/12/21512638/wisconsin-foxconn-tax-subsidies-lcd-factory-
rejected.

CHAPTER 11.
THE QUESTION OF TECHNOLOGY

1. Don DeLillo, *The Silence* (New York: Scribner, 2020), quote on 59 ("crushed our
technology"); Annie Sneed, "Environmental Thinker Bill McKibben Sounds
Warning on Technology," *Scientific American* (15 April 2019) at web.archive.org
/web/20201108001835/https://www.scientificamerican.com/article/environmental
-thinker-bill-mckibben-sounds-warning-on-technology/. *Contrast* Andrew Marantz,
"The Dark Side of Techno-Utopianism," *New Yorker* (23 September 2019) at web

.archive.org /web/20200616151400/https://www.newyorker.com/magazine/2019/09/30 /the-dark-side-of-techno-utopianism.

2. See Ruth Oldenziel, *Making Technology Masculine: Men, Women and Modern Machines in America, 1870–1945* (Amsterdam: Amsterdam University Press, 1999), chapter 1; Eric Schatzberg, *Technology: Critical History of a Concept* (Chicago: University of Chicago Press, 2018).

3. For shifting meanings of "technology," see Thomas J. Misa, Philip Brey, and Andrew Feenberg, eds., *Modernity and Technology* (Cambridge: MIT Press, 2003), 1–30; Thomas P. Hughes, *Human-Built World: How to Think about Technology and Culture* (Chicago: University of Chicago, 2004), 2–15, 175ff; and David E. Nye, *Technology Matters: Questions to Live with* (Cambridge: MIT Press, 2006), 1–15.

4. Keith Pavitt, "Innovation Processes," in *The Oxford Handbook of Innovation* (Oxford: Oxford University Press, 2005), ed. Jan Fagerberg, David C. Mowery, and Richard R. Nelson, 93.

5. Cheryl R. Ganz, *The 1933 Chicago World's Fair: A Century of Progress* (Champaign, Ill.: University of Illinois Press, 2008) at muse.jhu.edu/book/33490.

6. W. Patrick McCray, "From Lab to iPod: A Story of Discovery and Commercialization in the Post-Cold War Era," *Technology and Culture* 50 no. 1 (2009): 58–81, quotes on 58 ("iPod") and 60 ("research monies") at doi.org/10.1353/tech.0.0222; Matthew N. Eisler, " 'The Ennobling Unity of Science and Technology': Materials Sciences and Engineering, the Department of Energy, and the Nanotechnology Enigma," *Minerva* 51 (2013): 225–51 at doi.org/10.1007/s11024–013–9224-z; Cyrus C. M. Mody and Hyungsub Choi, "From Materials Science to Nanotechnology: Interdisciplinary Center Programs at Cornell University, 1960–2000," *Historical Studies in the Natural Sciences* 43 no. 2 (2013): 121–61 at jstor.org/stable/10.1525/hsns.2013.43.2.121; Cyrus C. M. Mody, *The Long Arm of Moore's Law: Microelectronics and American Science* (Cambridge: MIT Press, 2017), quote on 22 ("scientific landscape").

7. Richard A. Goldthwaite, *The Building of Renaissance Florence* (Baltimore: Johns Hopkins University Press, 1980), 425.

8. Nicholas Terpstra, *Abandoned Children of the Italian Renaissance: Orphan Care in Florence and Bologna* (Baltimore: Johns Hopkins University Press, 2005), 120–24; oatmeal gruel at web.archive.org/web/20201112023744/http://cookit.e2bn.org /historycookbook/121-gruel.html. Charles Dickens' *Oliver Twist* (1838) has the infamous plea for gruel, "Please, sir, I want some more" owing to rations of "three meals of thin [oatmeal] gruel a day, with an onion twice a week, and half a roll of Sundays" at www.gutenberg.org/ebooks/730.

9. See Amartya Sen, *Development as Freedom* (New York: Knopf, 1999); and Haider Khan, "Technology, Modernity and Development" in Misa et al., *Modernity and Technology* (Cambridge: MIT Press, 2003), 327–57.

10. Historical studies over a long span of time often are the best way for seeing clearly how technical choices entail cultural forms and social developments; see, for example, Otto Mayr, *Authority, Liberty, and Automatic Machinery in Early Modern Europe* (Baltimore: Johns Hopkins University Press, 1986); Cecil O. Smith Jr., "The Longest Run: Public Engineers and Planning in France," *American Historical Review* 95 (1990): 657–92; and Francesca Bray, *Technology and Gender: Fabrics of Power in Late Imperial China* (Berkeley: University of California Press, 1997). For "paths not taken" see David Noble, *Forces of Production: A Social History of Industrial Automation* (New York: A. A. Knopf, 1984); Ken Alder, *Engineering the Revolution: Arms and Enlightenment in France, 1763–1815* (Princeton: Princeton University Press, 1997); and Eric Schatzberg, *Wings of Wood, Wings of Metal: Culture and Technical Choice in American Airplane Materials, 1914–1945* (Princeton: Princeton University Press, 1999).

11. See Thomas J. Misa, *A Nation of Steel: The Making of Modern America, 1865–1925* (Baltimore: Johns Hopkins University Press, 1995), 262–82. Recent thinking about technology and economics is summarized by W. Brian Arthur, *The Nature of Technology* (New York: Free Press, 2009).

12. Gabrielle Hecht, *The Radiance of France: Nuclear Power and National Identity after World War II* (Cambridge: MIT Press, 1998), 88–102, 127–29, quote on 15.

13. For research on technology and Europe, see publications of the Tensions of Europe and Inventing Europe networks, including Thomas J. Misa and Johan Schot, "Technology and the Hidden Integration of Europe," *History and Technology* 21 no. 1 (2005): 1–19; Erik van der Vleuten and Arne Kaijser, eds., *Networking Europe: Transnational Infrastructures and the Shaping of Europe, 1850–2000* (Sagamore Beach, Mass.: Science History Publications, 2006); Mikael Hård and Thomas J. Misa., eds., *Urban Machinery: Inside Modern European Cities* (Cambridge: MIT Press, 2008); Ruth Oldenziel and Karin Zachmann, eds., *Cold War Kitchen: Americanization, Technology, and European Users* (Cambridge: MIT Press, 2009); Alexander Badenoch and Andreas Fickers, eds., *Materializing Europe: Transnational Infrastructures and the Project of Europe* (Houndmills, UK: Palgrave Macmillan, 2010). Palgrave Macmillan published a six-volume series entitled *Making Europe* (2013–2019) at www.palgrave.com/gp/series/14816 *or* www.makingeurope.eu.

14. Adam Max Cohen, *Technology and the Early Modern Self* (Houndmills, UK: Palgrave MacMillan, 2009), 137–207.

15. Lee Vinsel and Andrew L. Russell, *The Innovation Delusion: How Our Obsession with the New Has Disrupted the Work That Matters Most* (New York: Currency, 2020), quote on 7; see Ed Zitron's Twitter thread (January 2020) at web.archive.org /web/20200627000435/https://twitter.com/edzitron/status/1083476320808398849.

16. Winfried Baumgart, *Imperialism: The Idea and Reality of British and French Colonial Expansion, 1880–1914* (Oxford: Oxford University Press, 1982), quotes on 21.

17. A. P. Thornton, *The Imperial Idea and Its Enemies* (New York: St. Martin's Press, 1966), 54.

18. For the interplay between technology, nature, and responses of native indigenous peoples to imperialism, see Daniel Headrick's *Power over Peoples: Technology, Environments, and Western Imperialism, 1400 to the Present* (Princeton: Princeton University Press, 2010).

19. During the Renaissance, Alberti wrote extensively about how difficult and time consuming it was to become and remain a player in court culture, with one's mind always on how to please one's courtly patron. During the industrial era, there were economic and legal barriers to technology. Arkwright restricted access to his patented spinning technology through licensing agreements stipulating access only by large, one-thousand-spindle mills.

20. For technology dialogue, see Arnold Pacey, *Technology in World History* (Cambridge: MIT Press, 1990).

21. Henry Etzkowitz, "Solar versus Nuclear Energy: Autonomous or Dependent Technology?," *Social Problems* 31 (April 1984): 417–34.

22. Peter Harnetty, *Imperialism and Free Trade: Lancashire and India in the Mid-Nineteenth Century* (Vancouver: University of British Columbia Press, 1972), 6.

23. Karl Marx, *Das Kapital*, vol. 1, chapter 15, quotes see section 5 ("weapons against") and ("refractory hand of labour") at web.archive.org/web/20200619063637/www.marxists .org/archive/marx/works/1867-c1/ch15.htm.

24. Richard A. Goldthwaite, *The Building of Renaissance Florence* (Baltimore: Johns Hopkins University Press, 1994), 84.

25. Cris Shore, *Building Europe: The Cultural Politics of European Integration* (New York: Routledge, 2000).

26. Such oppositional agency is highlighted in Andrew Feenberg, *Questioning Technology* (New York: Routledge, 1999), chapter 5; for the progressive and democratic potentials of the internet, including expansion of the public sphere, see Feenberg, *Technosystem: The Social Life of Reason* (Cambridge: Harvard University Press, 2017), chapter 4.

27. Misa, *Nation of Steel*, 282; Åke Sandberg, Jesper Steen, Gunnar Broms, Arne Grip, Lars Sundström, and Peter Ullmark, *Technological Change and Co-determination in Sweden* (Philadelphia: Temple University Press, 1992).

28. Ramesh Srinivasan, *Whose Global Village? Rethinking How Technology Shapes Our World* (New York: NYU Press, 2017) quote on 18 ("strategically employ technologies") at jstor.org/stable/j.ctt1bj4qkd.5. On ICT4D, see Sarah Underwood, "Challenging Poverty," *Communications of the ACM* 51 no. 8 (August 2008): 15–17 at doi .org/10.1145/1378704.1378710; Dorothea Kleine, "The Value of Social Theories for Global Computing," *Communications of the ACM* 58 no. 9 (September 2015): 31–33 at doi.org/10.1145/2804246; Maletsabisa Molapo and Melissa Densmore, "How to Choose

a Mobile Phone for an ICT4D Project," In Proceedings of the Seventh International Conference on Information and Communication Technologies and Development (New York: Association for Computing Machinery, 2015) at doi.org/10.1145/2737856 .2737897.

29. Arthur P. J. Mol, "The Environmental Transformation of the Modern Order," in *Modernity and Technology*, Misa et al., 303–26.

30. Jack Snyder, "The Modernization Trap," *Journal of Democracy* 28 no. 2 (April 2017): 77–91, quote on 77 ("illiberal") and 78 ("selling out") at doi.org/10.1353/jod.2017.0026; Cade Metz, "When A.I. Falls in Love," *New York Times* (24 November 2020) ("great time") at web.archive.org/web/20201130175524/https://www.nytimes.com/2020/11/24 /science/artificial-intelligence-gpt3-writing-love.html.

31. Bharati Ray, "The Genesis of Railway Development in Hyderabad State: A Case Study in Nineteenth Century British Imperialism," *Indian Economic and Social History Review* 21 (1984): 45–69, quote on 54.

32. See Misa, "Retrieving Sociotechnical Change from Technological Determinism," in *Does Technology Drive History? The Dilemma of Technological Determinism* (Cambridge: MIT Press, 1994), ed. Merritt Roe Smith and Leo Marx, 115–41.

33. Mitchell F. Rice, "Information and Communication Technologies and the Global Digital Divide," *Comparative Technology Transfer and Society* 1, no. 1 (2003): 72–88 at https:// doi.org/10.1353/ctt.2003.0009; World Summit on the Information Society, "Digital Divide at a Glance" (2003–2005) at web.archive.org/web/20201106232353/https://www .itu.int/net/wsis/tunis/newsroom/stats/; 2019–20 World Bank data at web.archive.org /web/20201119020150/https://data.worldbank.org/indicator/IT.NET.USER.ZS.; Ramesh Srinivasan, *Whose Global Village? Rethinking How Technology Shapes Our World* (New York: NYU Press, 2017) quote on 1 ("neither global") ("excluded") at jstor.org/stable /j.ctt1bj4qkd.4.

34. ITU statistics by country 1990–2019 at web.archive.org/web/20201104013331/ https://www.itu.int/en/ITU-D/Statistics/Pages/stat/default.aspx. See Tony Romm, " 'It Shouldn't Take a Pandemic': Coronavirus Exposes Internet Inequality among U.S. Students as Schools Close Their Doors," *Washington Post* (16 March 2020) ("how challenging") at web.archive.org/web/20201005224154/https://www.washingtonpost .com/technology/2020/03/16/schools-internet-inequality-coronavirus/; Dan Levin, "In Rural 'Dead Zones,' School Comes on a Flash Drive," *New York Times* (13 November 2020) at web.archive.org/web/20201122161158/https://www.nytimes.com/2020/11/13 /us/wifi-dead-zones-schools.html.

35. See Ross Bassett, *The Technological Indian* (Cambridge, MA: Harvard University Press, 2016); Divy Thakkar, Nithya Sambasivan, Purva Kulkarni, Pratap Kalenahalli Sudarshan, and Kentaro Toyama, "Unexpected Entry and Exodus of Women in

Computing and HCI in India," CHI '18: Proceedings of the 2018 CHI Conference on Human Factors in Computing Systems (April 2018): 1–12 at doi.org/10.1145 /3173574.3173926. *Contrast* Greg Baumann, "Google Cracks Silicon Valley's Wall of Silence on Workforce Diversity, Confirming That It Is Largely White and Male," *Silicon Valley Business Journal* (28 May 2014) at web.archive.org/web/20160602122513/https:// www.bizjournals.com/sanjose/news/2014/05/28/google-largely-white-and-male-cracks -silicon.html.

36. Ronald Deibert, John Palfrey, Rafal Rohozinski, and Jonathan Zittrain, eds., *Access Controlled: The Shaping of Power, Rights, and Rule in Cyberspace* (Cambridge: MIT Press, 2010), quote on 524 ("heavily censored regions"). The 2010 Internet Censorship Report compiled data from Reporters sans Frontières (en.rsf.org) and the OpenNet Initiative (opennet.net) at web.archive.org/web/20181126111720/https://woorkup.com /internet-censorship-report/. Lara Farrar, "Cashing in on Internet Censorship," CNN (19 February 2010) "market is growing" at web.archive.org/web/20201109032624 /http://edition.cnn.com/2010/TECH/02/18/internet.censorship.business/index.html.

37. See Freedom House's 2020 Freedom on the Net report "dismal year" and "dramatic decline" at web.archive.org/web/20201116183120/https://freedomhouse.org/report /freedom-net/2020/pandemics-digital-shadow For quantitative measures, see Ram Sundara Raman, Prerana Shenoy, Katharina Kohls, and Roya Ensafi, "Censored Planet: An Internet-wide, Longitudinal Censorship Observatory," CCS '20: Proceedings of the 2020 ACM SIGSAC Conference on Computer and Communications Security (2020): 49–66 at dl.acm.org/doi/pdf/10.1145/3372297.3417883.

38. Michael Adas, *Machines as the Measure of Men* (Ithaca: Cornell University Press, 1989). See also David Arnold, *Science, Technology and Medicine in Colonial India* (Cambridge: Cambridge University Press, 2000); Michael Adas, *Dominance by Design: Technological Imperatives and America's Civilizing Mission* (Cambridge: Belknap Press, 2006); and Gabrielle Hecht's "Infrastructure and Power in the Global South" (2019) at web.archive .org/web/20201201181318/https://www.academia.edu/37728377/Infrastructure_and _Power_in_the_Global_South.

39. Mohandas K. Gandhi, *Indian Home Rule or Hind Swaraj* (Ahmedabad: Navajivan Publishing, 1938), quotes from preface and sections 8 Condition of India, 9 Railways, 12 Doctors, and 19 Machinery at web.archive.org/web/20200725161303/https://www .mkgandhi.org/hindswaraj/hindswaraj.htm. On Gandhi in modern India, see Larry Ceplair, "Mohandas K. Gandhi (1869–1948) and Jawaharlal Nehru (1889–1964)," in *Revolutionary Pairs* (Lexington: University Press of Kentucky, 2020), 98–135; and as a social reformer and dramatist, see Aditya Nigam, "Reading 'Hind Swaraj' Today," *Economic and Political Weekly* 44 no. 11 (2009): 41–47 at jstor.org/stable/40278613.

40. See Kim Vincente, *The Human Factor: Revolutionizing the Way People Live with Technology* (New York: Routledge, 2004); Andrew Feenberg, *Between Reason and*

Experience: Essays in Technology and Modernity (Cambridge: MIT Press, 2010); and the essays by Feenberg (on philosophy of technology), Haider Khan (on development policies) and Arthur Mol (on environmental thinking), in Misa et al., *Modernity and Technology*.

41. Howard Rheingold, *Virtual Reality* (New York: Simon & Schuster, 1991) quote on 349–50 ("force of its own"), describing the "titillating fantasy" of teledildonics.

ESSAY ON SOURCES

Listed here are those works that I have drawn on repeatedly. I have generally created text notes for direct quotes, numerical data, or specific points.

Scholarship on Renaissance technology once began with Bertrand Gille's classic *Engineers of the Renaissance* (Cambridge: MIT Press, 1966), but today the first stop, for Leonardo at least, is Paolo Galluzzi's "The Career of a Technologist," in his *Leonardo da Vinci: Engineer and Architect* (Montreal: Montreal Museum of Fine Arts, 1987). Vivid illustrations, stunning mechanical models, and informed text accompany Claudio Giorgione, ed., *Leonardo da Vinci: The Models Collection* (Milan: Museo Nazionale della Scienza e della Tecnica Leonardo da Vinci, 2009). The consequences of Renaissance-era perspective informs Eugene Ferguson's *Engineering and the Mind's Eye* (Cambridge: MIT Press, 1992). The 500th anniversary of Leonardo's death prompted major exhibits in Paris (Louvre), in Florence (Uffizi), and in Milan at the Sforza Castle, "Last Supper," and at the National Science and Technology Museum.

A neat synthesis of fine-arts scholarship is Peter Burke's *The European Renaissance: Centres and Peripheries* (Oxford: Blackwell, 1998), while Jerry Brotton's *The Renaissance Bazaar: From the Silk Road to Michelangelo* (Oxford: Oxford University Press, 2002) takes a global view. Court patronage is examined in Mary Hollingsworth's *Patronage in Renaissance Italy* (Baltimore: Johns Hopkins University Press, 1994) and David Mateer's *Courts, Patrons, and Poets* (New Haven: Yale University Press, 2000). On Alberti as a court-based scholar, architect, and engineer, see Anthony Grafton's *Leon Battista Alberti* (New York: Hill & Wang, 2000) and Joan Gadol's *Leon Battista Alberti* (Chicago: University of Chicago Press, 1969).

Pamela Long's *Openness, Secrecy, Authorship* (Baltimore: Johns Hopkins University Press, 2001) surveys technological knowledge since antiquity, illuminating authorship and patronage during the Renaissance, while late sixteenth-century Rome is the focus of her *Engineering the Eternal City* (Chicago: University of Chicago Press, 2018). Bert Hall's *Weapons and Warfare in Renaissance Europe* (Baltimore: Johns Hopkins University Press, 1997) contextualizes the "gunpowder revolution." For specific topics, see Richard A. Goldthwaite's *The Building of Renaissance Florence* (Baltimore: Johns Hopkins University Press, 1980), Luca Molà's *The Silk Industry of Renaissance Venice* (Baltimore: Johns Hopkins University Press, 2000), and W. Patrick McCray's *Glass-Making in Renaissance Venice* (Aldershot, England: Ashgate, 1999). For shipbuilding see Frederic Chapin Lane's classic *Venetian Ships and Shipbuilders of the Renaissance* (Baltimore: Johns Hopkins University Press, 1992; original 1934); and Robert C. Davis's *Shipbuilders of the Venetian Arsenal* (Baltimore: Johns Hopkins University Press, 1991).

For Gutenberg see Albert Kapr's definitive *Johann Gutenberg: The Man and His Invention* (Aldershot, England: Scolar Press, 1996), supplemented with Elizabeth L. Eisenstein's two-volume study *The Printing Press as an Agent of Change* (Cambridge: Cambridge University Press, 1979) or her one-volume illustrated *The Printing Revolution in Early Modern Europe* (Cambridge: Cambridge University Press, 1983). Critical of Eisenstein is Adrian Johns, *The Nature of the Book: Print and Knowledge in the Making* (Chicago: University of Chicago Press, 1998) and *Piracy: The Intellectual Property Wars from Gutenberg to Gates* (Chicago: University of Chicago Press, 2011).

The key source for Dutch commerce in the seventeenth century is Jonathan Israel's *Dutch Primacy in World Trade, 1585–1740* (Oxford: Clarendon Press, 1989), a masterful synthesis and goldmine of data. Israel provides a detailed view of Dutch culture and politics in *The Dutch Republic: Its Rise, Greatness, and Fall, 1477–1806* (Oxford: Clarendon Press, 1995). A portrait of Dutch society from rich to poor is Paul Zumthor, *Daily Life in Rembrandt's Holland* (Stanford: Stanford University Press, 1994).

Other crucial sources include Karel Davids and Jan Lucassen, eds., *A Miracle Mirrored: The Dutch Republic in European Perspective* (Cambridge: Cambridge University Press, 1995), Karel Davids and Leo Noordegraaf, eds., *The Dutch Economy in the Golden Age* (Amsterdam: Netherlands Economic History Archives, 1993); and Joel Mokyr, *Industrialization in the Low Countries, 1795–1850* (New Haven: Yale University Press, 1976). Compare Richard W. Unger, *Dutch Shipbuilding before 1800* (Assen: Van Gorcum, 1978) with Carla Rahn Phillips, *Six Galleons for the King of Spain: Imperial Defense in the Early Seventeenth Century* (Baltimore: Johns Hopkins University Press, 1986). For the Dutch tulip mania, enliven Peter M. Garber, *Famous First Bubbles: The Fundamentals of Early Manias* (Cambridge: MIT Press, 2000) with Mike Dash, *Tulipomania* (New York: Crown, 1999).

For useful entrees, see Maxine Berg and Pat Hudson, "Rehabilitating the Industrial Revolution," *Economic History Review* 45 (1992): 24–50; and Joel Mokyr's *The British Industrial Revolution: An Economic Perspective* (Boulder: Westview, 1993), pp. 1–131. For a wider view, see Robert C. Allen, *The British Industrial Revolution in Global Perspective* (Cambridge: Cambridge University Press, 2009) and Jeff Horn, Leonard N. Rosenband, and Merritt Roe Smith, eds., *Reconceptualizing the Industrial Revolution* (Cambridge: MIT Press, 2010). Ivy Pinchbeck, *Women Workers and the Industrial Revolution, 1750–1850* (London: Frank Cass, 1930) is an indispensable source.

On London, begin with M. Dorothy George's classic *London Life in the Eighteenth Century* (New York: Capricorn Books, 1965; original 1925). Valuable social histories include David R. Green, *From Artisans to Paupers: Economic Change and Poverty in London, 1790–1870* (London: Scolar Press, 1995); L. D. Schwarz, *London in the Age of Industrialisation: Entrepreneurs, Labour Force and Living Conditions, 1700–1850* (Cambridge: Cambridge University Press, 1992); and Roy Porter's entertaining *London: A Social History* (Cambridge: Harvard University Press, 1994).

The literature on Manchester is voluminous. See Sidney J. Chapman, *The Lancashire Cotton Industry* (Manchester: Manchester University Press, 1904); Martin Hewitt, *The Emergence of Stability in the Industrial City: Manchester, 1832–67* (Aldershot, England: Scolar Press, 1996); Geoffrey Timmons, *The Last Shift* (Manchester: Manchester University Press, 1993); and Steven Marcus, *Engels, Manchester, and the Working Class* (New York: Random House, 1974). For the city's business structure, see Roger Lloyd-Jones and M. J. Lewis, *Manchester and the Age of the Factory* (London: Croom Helm, 1988); a deft comparative study is Harold L. Platt, *Shock Cities: The Environmental Transformation and Reform of Manchester and Chicago* (Chicago: University of Chicago Press, 2005).

On Sheffield see David Hey, *The Fiery Blades of Hallamshire: Sheffield and Its Neighbors, 1660–1740* (Leicester: Leicester University Press, 1991); Sidney Pollard, *A History of Labour in Sheffield* (Liverpool: Liverpool University Press, 1959); Geoffrey Tweedale, *Steel City: Entrepreneurship, Strategy and Technology in Sheffield, 1743–1993* (Oxford: Clarendon Press, 1995); Tweedale, *Sheffield Steel and America: A Century of Commercial and Technological Interdependence, 1830–1930* (Cambridge: Cambridge University Press, 1987); and K. C. Barraclough, *Sheffield Steel* (Ashbourne, England: Moorland, 1976).

For "alternative paths" to industrialization, see Charles F. Sabel and Jonathan Zeitlin, eds., *World of Possibilities: Flexibility and Mass Production in Western Industrialization* (Cambridge: Cambridge University Press, 1997) and Philip Scranton, *Endless Novelty: Specialty Production and American Industrialization, 1865–1925* (Princeton: Princeton University Press, 1997). Cotton comes full circle—again, a driver of world history—in Sven Beckert's *Empire of Cotton: A Global History* (New York: Alfred A. Knopf, 2014).

For imperialism and technology, start with Daniel Headrick's *The Tools of Empire: Technology and European Imperialism in the Nineteenth Century* (Oxford: Oxford University Press, 1981); *The Tentacles of Progress: Technology Transfer in the Age of Imperialism, 1850–1940* (Oxford: Oxford University Press, 1988); and *The Invisible Weapon: Telecommunications and International Politics, 1851–1945* (Oxford: Oxford University Press, 1991). For economics see Lance E. Davis and Robert A. Huttenback, *Mammon and the Pursuit of Empire: The Economics of British Imperialism* (Cambridge: Cambridge University Press, 1988). A valuable synthesis is P. J. Cain and A. G. Hopkins, *British Imperialism, 1688–2000*, 2nd edition (Harlow, England: Longman, 2002).

On technology and British imperialism, see Daniel Thorner, *Investment in Empire: British Railway and Steam Shipping Enterprise in India, 1825–1849* (Philadelphia: University of Pennsylvania Press, 1950); Roy MacLeod and Deepak Kumar, eds., *Technology and the Raj: Western Technology and Technical Transfers to India, 1700–1947* (New Delhi: Sage, 1995); Zaheer Baber, *The Science of Empire: Scientific Knowledge, Civilization, and Colonial Rule in India* (Albany: SUNY Press, 1996); Michael Adas, *Machines as the Measure of Men* (Ithaca: Cornell University Press, 1989); Michael Adas, ed., *Technology and European Overseas*

Enterprise (Aldershot, England: Variorum, 1996). There are many valuable case studies in *Indian Economic and Social History Review*. For native Indian industry, see Tirthankar Roy, *Artisans and Industrialisation: Indian Weaving in the Twentieth Century* (New Delhi: Oxford University Press, 1993); and Roy, *Traditional Industry in the Economy of Colonial India* (Cambridge: Cambridge University Press, 1999).

On "railway imperialism," see Clarence B. Davis and Kenneth E. Wilburn, Jr., eds., *Railway Imperialism* (Boulder: Greenwood, 1991) and John H. Coatsworth, *Growth against Development: The Economic Impact of Railroads in Porfirian Mexico* (De Kalb: Northern Illinois Press, 1981). Detail on the Indian railways can be found in Ian J. Kerr's *Building the Railways of the Raj, 1850–1900* (New Delhi: Oxford University Press, 1995) and *Railways in Modern India* (New Delhi: Oxford University Press, 2001).

The section on German chemistry draws on John Joseph Beer, *The Emergence of the German Dye Industry* (Urbana: University of Illinois Press, 1959); Jeffrey Allan Johnson, *The Kaiser's Chemists: Science and Modernization in Imperial Germany* (Chapel Hill: University of North Carolina Press, 1990); Henk van den Belt and Arie Rip, "The Nelson-Winter-Dosi Model and Synthetic Dye Chemistry," in W. E. Bijker et al., eds., *The Social Construction of Technological Systems* (Cambridge: MIT Press, 1987), 135–58; Anthony S. Travis, *The Rainbow Makers: The Origins of the Synthetic Dyestuffs Industry in Western Europe* (Bethlehem: Lehigh University Press, 1993); L. F. Haber, *The Poisonous Cloud: Chemical Warfare in the First World War* (Oxford: Clarendon Press, 1986); and Peter Hayes, *Industry and Ideology: IG Farben in the Nazi Era* (Cambridge: Cambridge University Press, 1987).

Paul Israel's *Edison: An Inventive Life* (New York: Wiley, 1998) supersedes earlier popular biographies by Josephson, Conot, Clark, Harris, and Baldwin. Additional sources include Reese V. Jenkins et al., eds., *The Papers of Thomas A. Edison* (Baltimore: Johns Hopkins University Press, 1989 et seq.) online at edison.rutgers.edu/digital.htm; Paul Israel, *From Machine Shop to Industrial Laboratory: Telegraphy and the Changing Context of American Invention, 1830–1920* (Baltimore: Johns Hopkins University Press, 1992); Thomas P. Hughes, *Networks of Power: Electrification in Western Society, 1880–1930* (Baltimore: Johns Hopkins University Press, 1983); and Robert Friedel and Paul Israel, *Edison's Electric Light: Biography of an Invention* (New Brunswick: Rutgers University Press, 1986).

On corporate restructuring in the United States, see Louis Galambos and Joseph Pratt, *The Rise of the Corporate Commonwealth* (New York: Basic Books, 1988); W. Bernard Carlson, *Innovation as a Social Process: Elihu Thomson and the Rise of General Electric, 1870–1900* (Cambridge: Cambridge University Press, 1991); Alfred D. Chandler, Jr., *The Visible Hand: The Managerial Revolution in American Business* (Cambridge, Mass.: Belknap Press, 1977); and Naomi R. Lamoreaux, *The Great Merger Movement in American Business, 1895–1904* (Cambridge: Cambridge University Press, 1985).

Terry Smith, *Making the Modern: Industry, Art, and Design in America* (Chicago: University of Chicago Press, 1993) and Thomas P. Hughes, *American Genesis* (New York: Viking, 1989) examine modernism in art, architecture, and technology. The essential source on Italian Futurism is Pontus Hulten, ed., *Futurismo e Futurismi* (Milan: Bompiani, 1986), including an invaluable 200–page "dictionary" with biographies and manifestos. On de Stijl, see Carsten-Peter Warncke, *De Stijl, 1917–31* (Cologne: Taschen, 1998) and Joost Baljeu, *Theo van Doesburg* (London: Studio Vista, 1974). On the Bauhaus see Reyner Banham, *Theory and Design in the First Machine Age* (Cambridge: MIT Press, 1980) and John Willett, *The New Sobriety, 1917–1933: Art and Politics in the Weimar Period* (London: Thames & Hudson, 1978).

Sources on German household rationalization and the Frankfurt building program, in English, include Martina Heßler, "The Frankfurt Kitchen: The Model of Modernity and the 'Madness' of Traditional Users, 1926 to 1933," in Ruth Oldenziel and Karin Zachmann, eds., *Cold War Kitchen: Americanization, Technology, and European Users* (Cambridge: MIT Press, 2009), 163–84; Mary Nolan, *Visions of Modernity: American Business and the Modernization of Germany* (New York: Oxford University Press, 1994); and Debra Coleman, Elizabeth Danze, and Carol Henderson, eds., *Architecture and Feminism* (New York: Princeton Architectural Press, 1996).

On "modern" materials, see Pearce Davis, *The Development of the American Glass Industry* (Cambridge: Harvard University Press, 1949); Ken Fones-Wolf, *Glass Towns: Industry, Labor, and Political Economy in Appalachia, 1890–1930s* (Urbana: University of Illinois Press, 2007); Barbara L. Floyd, *The Glass City: Toledo and the Industry That Built It* (Ann Arbor: University of Michigan Press, 2015); and Thomas J. Misa, *A Nation of Steel* (Baltimore: Johns Hopkins University Press, 1995).

On the German military, see William Carr, *Arms, Autarky and Aggression: A Study in German Foreign Policy, 1933–1939* (New York: W. W. Norton, 1972); Alan Milward, *War, Economy, and Society, 1939–1945* (Berkeley: University of California Press, 1977); Mark Walker, *German National Socialism and the Quest for Nuclear Power, 1939–1949* (Cambridge: Cambridge University Press, 1989).

On the Manhattan Project see Richard G. Hewlett and Oscar E. Anderson, Jr., *The New World, 1939–1946*, vol. 1 of *A History of the United States Atomic Energy Commission* (University Park: Pennsylvania State University Press, 1962); Richard Rhodes, *The Making of the Atomic Bomb* (New York: Simon & Schuster, 1986); and Barton C. Hacker, *The Dragon's Tail: Radiation Safety in the Manhattan Project, 1942–1946* (Berkeley: University of California Press, 1987).

On postwar computing and electronics, see Stuart W. Leslie, *The Cold War and American Science* (New York: Columbia University Press, 1993); Kenneth Flamm, *Creating the Computer* (Washington, DC: Brookings Institution, 1988); Thomas J. Misa, "Military Needs, Commercial Realities, and the Development of the Transistor, 1948–1958," in *Military Enterprise and Technological Change*, ed. Merritt Roe Smith (Cambridge: MIT Press,

1985), 253–87; Arthur L. Norberg and Judy E. O'Neill, *Transforming Computer Technology: Information Processing for the Pentagon, 1962–1986* (Baltimore: Johns Hopkins University Press, 1996); Kent C. Raymond and Thomas M. Smith, *Project Whirlwind: The History of a Pioneer Computer* (Bedford: Digital Press, 1980); Paul N. Edwards, *The Closed World: Computers and the Politics of Discourse in Cold War America* (Cambridge: MIT Press, 1996).

Writings on globalization vary greatly in perspective. Contrast Thomas L. Friedman's upbeat view in *The Lexus and the Olive Tree* (New York: Farrar, Straus & Giroux, 1999) with Jerry Mander and Edward Goldsmith, eds., *The Case against the Global Economy* (San Francisco: Sierra Club Books, 1996). A wide-ranging collection is *Globalization and the Challenges of a New Century*, edited by Patrick O'Meara et al. (Bloomington: Indiana University Press, 2000). For a Marxist analysis, see Michael Hardt and Antonio Negri, *Empire* (Cambridge: Harvard University Press, 2000). The *Economist* magazine's pro-globalization perspective shapes John Micklethwait and Adrian Wooldridge, *A Future Perfect: The Essentials of Globalization* (New York: Crown Business, 2000).

On fax machines see Jonathan Coopersmith's *Faxed: The Rise and Fall of the Fax Machine* (Baltimore: Johns Hopkins University Press, 2016), and Susanne K. Schmidt and Raymund Werle, *Coordinating Technology: Studies in the International Standardization of Telecommunications* (Cambridge: MIT Press, 1998).

Sources on McDonald's include John F. Love, *McDonald's: Behind the Arches* (New York: Bantam, 1995; original 1986); James L. Watson, ed., *Golden Arches East: McDonald's in East Asia* (Stanford: Stanford University Press, 1997); Robin Leidner, *Fast Food, Fast Talk: Service Work and the Routinization of Everyday Life* (Berkeley: University of California Press, 1993); and John Vidal, *McLibel: Burger Culture on Trial* (New York: New Press, 1997). For insider views see Ray Kroc, *Grinding It Out: The Making of McDonald's* (Chicago: Henry Regnery, 1977); and George Cohon's lively *To Russia with Fries* (Toronto: McClelland & Stewart, 1999).

On the internet and World Wide Web, start with Stephen J. Lukasik, "Why the Arpanet Was Built," *IEEE Annals of the History of Computing* 33 no. 3 (2011): 4–20; Janet Abbate, *Inventing the Internet* (Cambridge: MIT Press, 1999); and Shane M. Greenstein, *How the Internet Became Commercial* (Princeton: Princeton University Press, 2017). On computer networks see Urs von Burg, *The Triumph of the Ethernet: Technological Communities and the Battle for the LAN Standard* (Stanford: Stanford University Press, 2001). On internet privatization, see Jay P. Kesan and Rajiv C. Shah, "Fool Us Once Shame on You—Fool Us Twice Shame on Us: What We Can Learn from the Privatizations of the Internet Backbone Network and the Domain Name System," *Washington University Law Quarterly* 79 (2001): 89–220.

Technology and systemic risk is a sprawling topic. A foundational work is Charles Perrow, *Normal Accidents: Living with High Risk Technologies* (Princeton: Princeton University Press,

1999). A key cultural history is David Nye, *When the Lights Went Out: A History of Blackouts in America* (Cambridge: MIT Press, 2010). Two critiques of energy consumption are Paul R. Josephson, *Motorized Obsessions: Life, Liberty, and the Small-Bore Engine* (Baltimore: Johns Hopkins University Press, 2007) and Thomas L. Friedman, *Hot, Flat, and Crowded* (New York: Farrar, Straus and Giroux, 2008). The commercial internet is surveyed in William Aspray and Paul Ceruzzi, eds., *The Internet and American Business* (Cambridge: MIT Press, 2008) and Shane Greenstein, *How the Internet Became Commercial* (Princeton: Princeton University Press, 2016). For global shipping, start with Marc Levinson's *The Box: How the Shipping Container Made the World Smaller and the World Economy Bigger* (Princeton: Princeton University Press, 2006).

Dominance of the digital engages recent history. Scholarly writings on Moore's Law include Robert Schaller, "Technological Innovation in the Semiconductor Industry: A Case Study of the International Technology Roadmap for Semiconductors (ITRS)" (PhD thesis, George Mason University, 2004), summary at doi.org/10.1109/PICMET.2001.951917; Ethan Mollick, "Establishing Moore's Law," *IEEE Annals of the History of Computing* 28 no. 3 (2006): 62–75; and Christophe Lécuyer, "Driving Semiconductor Innovation: Moore's Law at Fairchild and Intel," *Enterprise and Society* (forthcoming) at doi.org/10.1017/eso.2020.38. A valuable overview is Andrew Blum, *Tubes: A Journey to the Center of the Internet* (New York: Ecco, 2012). On changes in the internet, see David D. Clark, *Designing an Internet* (Cambridge: MIT Press, 2018); Evgeny Morozov, *The Net Delusion: The Dark Side of Internet Freedom* (New York: Public Affairs, 2011); and Laura DeNardis, *Protocol Politics: The Globalization of Internet Governance* (Cambridge: MIT Press, 2009). On China see James Griffiths, *The Great Firewall of China: How to Build and Control an Alternative Version of the Internet* (London: Zed, 2019). Compare Andrei Soldatov and Irina Borogan, *The Red Web: The Struggle between Russia's Digital Dictators and the New Online Revolutionaries* (New York: Public Affairs, 2015).

INDEX

Page numbers in **bold** refer to figures.

Appropriation (*continued*)
231–34, 253, 337. *See also* Technology-culture dynamics
Arab Spring (2011), 303–4, 313, 338
Architecture, modern, xiii, **151**, 161, 166–69, **167**, **169**, **173**, **177**, 331; and Bauhaus, 163–75; and CIAM, xiii, 175–79; and de Stijl, 163–70; and Italian Futurism, 159–63; and Museum of Modern Art, 159, 175, 179, 360n.30. *See also* Berlin; Chicago; Frankfurt; International Style; Modernism; New York; Urban planning
Arkwright, Richard, 57, 70–74, 394n.19
Arnold, Gen. Henry H. (Hap), 193
ARPANET, 235–41, 260, **298**, 299, 306, 308. *See also* Advanced Research Project Agency; Internet
Artificial intelligence, 210, 276, 294, 313, 316, 319, 335
ASCII. *See* American Standard Code for Information Interchange
ASM Lithography, 293, 295–96
Associated Press, 220
AT&T Corporation (American Telephone and Telegraph Company), 141, 143–45, 220, 242, 259, 286–87, 290–91, 297–99, 304–309, **305**. *See also* Bell Telephone Laboratories, Western Electric
Atom bomb: cost of, 195, 199, 321; Hanford (Washington) plutonium factory, 187, 190–93, **191**, 321; and Japan, 192–95, **194**; Los Alamos, 185, 192–93, 205–6, 209, 321; Metallurgical Laboratory, 186, 191–92; Oak Ridge uranium factory, **187**, 187–93, 196, 206, 321; Soviet Union, 198; test at Trinity, 192–93. *See also* Manhattan Project
Atomic Energy Commission (AEC), 196–97, 205–6, 209; Shippingport (Pennsylvania) reactor, 196–98, **197**, 201. *See also* United States Department of Energy
Auschwitz, 129. *See also* Germany, and National Socialism

Automata, 9, 31, 75–78, 331
Automated Radar Terminal System (ARTS), 262–63. *See also* Air traffic control
Automated Targeting System (ATS), 276. *See also* Shipping industry
Automobiles, 9, 25, 150, 159–60, 165, 214, 247, 252, 284; industry, 149, 154, 158–59, 163, 256–58, 289, 329. *See also* Car culture

Baidu (China), 312–13
Balla, Giacomo, 150, 160–61
Ballistic Missile Early Warning System (BMEWS), 203
Baltimore & Ohio Railroad, 112, 133
Baran, Paul, 210, 235–36
Barclay, Charles: porter brewing company, 63–65, 67
Bardeen, John, 200
Barlow, John Perry, 300, 301–2, 304–5, 311, 383n.49. *See also* Cyberculture
BASF (Badische Analin- & Soda-Fabrik), 124–29
Bauhaus, 151, 159, 163, 168–75
Bayer (Friedrich Bayer & Co.), 124, 126, 129
Beck, Ulrich, 279–80, 376n.70
Beckert, Sven, 91, 401
Beer: consumption, 44, 62–64, 64, 67, 215, 348n.18, 350n.13, 350n.19; production, 63–66, **64**, **65**. *See also* Alcoholic beverages
Beerhouse Act (1830), 67
Behrens, Peter, 168, 174
Beijing (China), 98, 231–32, 252, 312, 315
Belgian Congo, 119, 187. *See also* Africa; Congo; South Africa
Bell, Alexander Graham, 131, 143
Bell System Technical Journal, 259
Bell Telephone Laboratories, 189, 208, 297–98; and radar, 199–200; and transistor, 200–205. *See also* AT&T; Western Electric

Bengal Technical Institute, 106

Berlage, Hendrik Petrus, 165–66

Berlin, 124, 132, 133, 135, 184; GEHAG, 172–73; and modern architecture, **151**, 163–64, 168, 170, 172–74, 176

Berners-Lee, Tim, 242–44, 367n.35

Berry, Wendell, 319

Bessemer, Henry, 69, 80, 83, 89, 152–54

Besson, Jacques, 27–28

Biringuccio, Vannoccio, 2, 20, 26–27

Birmingham (England), 58, 60, 65–66, 69, 79

BITNET, 239

Blair, Tony, 280

Boccioni, Umberto, 150, 159–61

Bohr, Niels, 190, 195, 212

Bombay (Mumbai), 94, 96, 99–105, 108, 109, 332

Borgia family, 2, 10, 16

Bové, José, 226

Brahe, Tycho, 29

Bramah, Joseph, 69–70

Brand, Stewart, 300

Brattain, Walter, 200

Braudel, Fernand, 32, 58

Brazil, 32, 47–48, 52, 91, 216, 249, 310, 317–18

Bridgewater, Duke of, 72, 76

British Association for the Advancement of Science, 69

British East India Company, 41; activities in India, 92–98; dissolved in 1858, 104, 108; employees in London, 62, 94; and opium, 98–99; and railroad building, 108–9; and shipbuilding, 61–62

British South Africa Company, 116, 118

Brown, Lester, 258

Brunel, I. K., 62

Brunel, Marc, 70

Brunelleschi, Filippo, 3–5, 12–13, 345n.21

Burma, 95, 99, 102, **110**, 339

Burns, Mary, 78

Bush, George W., 274–75

Bush, Vannevar, 147, 206, 209; and OSRD, 186

Calcutta, 94–97, 99–105, 108, 109, 112

Canada, 218, 229–31, 249, 254, 308, 337; Calgary (Alberta), **255;** and patents, **127;** railroads in, 112–14, 120. *See also* West Edmonton Mall

Canadian Pacific Railway, 114

Canals: Bridgewater, 72, 76; Erie, 112; Florence, 10; Ganges, 99; Grand (China), 98; Grand Junction (Britain), 60; Milan, 9, 160; Panama, 119; St. Lawrence-Great Lakes, 33, 113; Suez, 96, 103, 119

Cape Town to Cairo Railway, **118**, **120**

Capitalism, 31, 40–41, 58, 79, 91; "gentlemanly," 59, 349n.4; and global economy, 40–48, 91, 213–19, 228–30, 245, 331–35; and imperialism, 94, 120; and industrial revolution 57–91; organized capitalism, 122, 140; "Pentagon capitalism," 180, 211. *See also* Dutch Republic; Global economy; Venture capital

Captain Crunch. *See* Draper, John

Car culture, 226, 254–58. *See also* Automobiles

Carlyle, Thomas, 77

Caro, Heinrich, 124–26, 149

CCITT (Comité Consultatif International Télégraphique et Téléphonique), 221–22, 226

Censorship, 23, 268, 311. *See also* Internet, censorship

Central Intelligence Agency (CIA), 268–69, 287, 316

Cerf, Vinton, 235, 237–39, 260. *See also* Internet

CERN (Conseil Européen pour la Recherche Nucléaire), 243

Dutch West India Company (WIC), 41, 47–48

E-85, 255–58, 370n.25. *See also* Ethanol
E-commerce, 216, 242–45, 251, 260, 297, 302, 312–14, 369n.9. *See also* Internet
Earth Policy Institute, 258
East Indian Railway Company, 108, 112
Eastman, George, 146
Eastman Kodak, 146, 187, 321
eBay, 244, 303
Eckert, J. Presper, 206
Ecological modernization, 334. *See also* Environmental movement
Economics: of empire, xii, 96, 113; of technology, 23, 285, 320–23. *See also* Capitalism; Global economy
Edison, Thomas, 130–44, 168, 320, 326; and AC electricity, 135, 138; and "bugs," 132; early fax, 220; and invention, 130–31, 139, 141, 149
Edison Electric Light Company, 131, 133, **134**, **136**
Edison General Electric, 137–38, **136**, 140

Egypt, 19, 92, 118, 154, 182, 282; and global economy, 219; and internet shutdowns, 303–6, 317, 339, 384n.56; and Suez Canal, 96. *See also* Arab Spring
Einstein, Albert, 186
Eisenhower, Dwight: and atom bomb, 192–93; Atoms for Peace, 196–98; and "military-industrial complex," 180, 211
Eisenstein, Elizabeth, 21–24, 399–400
Electrical engineering, 106, 141, 143–48, 196, 285
Electrical industry, xvi, 122–23, 129–47, **134**, **136**, 140–43; blackouts, 147, 249–50, 265–66, 325; brittle power, 250; high-voltage direct current (HVDC), 265, 325; Pearl Street Station, **134**, 135
Electricity grid, 148, 249–50, 261, 264–67,

270, 280, 325, 341; failure in India (2012), 249–50; incompatibility in Japan, xvi; Inter-Control Center Communications Protocol (ICCP), 266
Electronics. *See* Computers; Integrated circuit; Transistor
Elementarism, 169–70. *See also* Architecture, modern; Modernity, and art
Elsevier, Louis, 29
Email, xiv, 224, 238–40, 297, 335, 367n.32. *See also* Internet
Empire. *See* Imperialism
Energy crisis (1970s), 213, 254, 289
Energy Independence and Security Act (2007), 256
Energy policies, 254–58, 329–30
Energy Policy Act (2005), 256, 330
Energy systems, xvi-xvii, 53, 131–40, 247–51, 265–70, 278, 325
Engels, Friedrich, 77–79, 89, 122
Engineering, 2, 3, 4–5, 10, 15, 34, 59, 67–69, 122, 171, 190–91, 202–5, 284, 308, 320–22; in China, 24; education in Germany, 124; education in India, 105–6; education in U.S., 143–48; hydraulic, 10, 24; Leonardo and, 2–5, 10–15; as profession, 141, 143–44
Environment, xvi, 65, 89–90, 211, 214–15, 225, 230, 252–54, **255**, 279–80, **288**, 319, 323, 328–29
Environmental movement, 280, 334–36, 340; pollution, xvii, 60, 77–79, 86–87, 214, 247, 249, 279, 323, 329, 334, 377n.6
Ercker, Lazarus, 26, 27
Estonia, 267, 270
Ethanol, 255–58, 280, 371n.28. *See also* E-85
Ethernet, 235–38, 309, 404
Europe, 17–18, 21–26, 29, 30–38, 41–42, 47, 60, 91, 99–103, **103**, 106, 109, 133–35, 214–19, **216**, 222, 224–27,

personal computing, 238; and SABRE, 209; and SAGE, 208–10; System 360, 209–10, 222, 284

ICT4D. *See* Information and communication technologies for development

Idaho National Laboratory, 265–66

Imperialism, ix, xii, 75, 92–121, 214, 231–33, 324, 327–29, 335–36

Implementing Recommendations of the 9/11 Commission Act (2007), 276–77

India, 41, 45–48, 152; Bangalore, 101, 338; Chandigarh, 252; cotton industry, xii, 41, 46, 90–91, 99, 104–5, 108, 119, 328–29; and court technologies, 2, 30; electricity grid failure (2012), 249–50; Ganges River, 95–96, 99, **111**; and global economy, xiv, 91, 215, 228, 234, 246, 269–70, 339–40; Grand Trunk Road, 99; internet, 304, 307, 317–18, 338–39; Mutiny of 1857–58, xii, **93**, 98, 101–4, 112, 327; and opium, 96–99, 119; Public Works Department, 93, 99, 104–6; railroads, 104, 106–13, **110, 111**, 327; telegraph system, 99–103, **100**, **103**, 327. *See also* Bombay (Mumbai); Calcutta; Madras

Indian Ocean, 96, 103, 119

Indonesia: 41, 45–46, 53; and global economy, 214, 217, 219, 246

Industrial districts, 35, 52–53, 57, 59–60, 69, 83, 87–89, 323, 329

Industrial research, xiii, 122–25, 130, 141–45, 191, 320–22. *See also* Science-based industries

Industrial revolution, x–xii, 57–91, **71**, 214–15, 218, 279, 320, 331–32; in China, 213–14, 283, 317–18, 323, 329, 334, 339; "dark satanic mills," 69, 89; and living conditions, 60, 64, 77–79, 83–87; protoindustrial, 52, 347n.7; second (from 1870s), xii, 92, 104, 122–49, 159, 172

Industrial science, 125, 141–43, 149, 181, 321–22

Industrial society, xi, xiii, 57–58, 89, 122, 279

Information and communication technologies for development (ICT4D), 333–34

Information highway (infrastructure), 240, 244, 250, 260, 267, 270, 302–3

Information Processing Techniques Office (IPTO), 210, 235–38. *See also* Advanced Research Project Agency

Information systems, xv, 99, 130, 207–9, 219–25, 234–45, 247, 250, 258–61, 263–70, 278. *See also* Internet

Insull, Samuel, 137, 140

Integrated circuit, 203–5, 281–89; ASIC, 293–94. *See also* Moore's Law; Transistor

Intel, 201, 281–86, 289, 291, 294–97, 316, 377n.4. *See also* Fairchild Semiconductor; Integrated circuit; Silicon Valley

International Business Machines. *See* IBM

International Maritime Organization, 278

International Monetary Fund (IMF), 219, 323

International Style (architecture), 174–75, 360n.30. *See also* Architecture, modern

International Technology Roadmap for Semiconductors (ITRS), 292–94, 296

International Telecommunication Union (ITU), 221–22, 226

Internet: ARPANET, 235–41, 260, **298**, 299, 306, 308, 383n.45; cafés, **243**; censorship, 268, 311–16, 326, 329, 339; civilian origins, 205, 235–36, 238–41; domain-name system (DNS), 240–41, 314; email, 224, 235, 238–40, 297, 335, 367n.32; exchanges, 305, 309–12, 315; freedom, 244, 300–6, 311–17, 326, 339; Internet Protocol (TCP/IP), 235–41,

Nuclear-powered airplane, 181, 201, 361n.2

Oak Ridge (TN) uranium factory, **187**, 187–96, 206, 321. *See also* Atom bomb

Obama administration, 303, 311, 314. *See also* Internet freedom

Office of Naval Research (ONR), 206–7

Office of Scientific Research and Development (OSRD), 186. *See also* Atom bomb; Bush, Vannevar; Radar

Ohmae, Kenichi, 214

OPEC oil embargo (1973), 254, 256

Opium trade, xii, 46, 93–99, 119

Opium wars (1840–42, 1856–60), 93–99, 119, 315

Oppenheimer, J. Robert, 185, 211

Oppositional agency, x, xviii, 226–27, 280. *See also* Protest movements

Oracle (company), 284, 291, 297

Orange Free State, 115–16. *See also* South Africa

Orphans, 4, 323, 348n.18, 392n.8

Ottoman Empire, 2, 9, 37; and court technologies, 2, 30. *See also* Turkey

Oud, J. J. P., 166–69, **167**, 174, 178

Owen, Wilfred, 128

Pacey, Arnold, 2, 24, 328–29

Pacific Railroad Act (1862), 112–14

Packet switching, 210, 235–36. *See also* Internet

Pakistan, 94, 228, 269–70, 317, 338

Palantir Technologies, 316

Paris, xv, 16, 22, 60, 123, 133, 150, 159, 160, 163, 174, 182, 234, **250**

Patents and patent law: in Dutch Republic, 53; and Edison, 130–38; and electrical industry, 136–42; in England, 73, 394n.19; and facsimile, 220; in Germany, 126–27; statistics on, **127**; and transistors, 202

Pavitt, Keith, 127, 320

Pearl Street (power) Station, **134**, 135. *See also* Edison, Thomas

Peninsular & Oriental Holding, 273–74

Peninsular & Oriental Steam Navigation Company, 92, 108

Pentagon: "capitalism," xiv, 180; complex (1941–43), 186. *See also* Military-industrial complex

Perkin, William, 123–25

Perspective. *See* Geometrical perspective

Philippines, 41, 46, 251, 327, 339; global economy, 214, 219

Pinchbeck, Ivy, 74, 351n.27

Plantin, Christopher, 20–24, **22**, 33

Plutonium, 190, 198, 273, 324, 330; and atom bomb, 190–95, **191**

Poison gas (warfare), 127–28, 148, 182, 185, 324

Port Authority of New York and New Jersey, 248–49, 274

Porter industry, 62–70, **67, 69**. *See also* Beer

Ports (harbors), 38, **42, 61**, 105, 162, 181, **274**. *See also specific port cities*

Portugal, 34–35, 45–48, 214–15, 225, 249, 325

Postel, Jon, 308

Pouzin, Louis, 238

Predator drone aircraft, 269

Price-Anderson Act (1957), 330

Printing: as industry, 20–26, 59, 88; moveable type, ix, 2, 18–20, **22**, 30. *See also* Protestant Reformation

PRNET, 236–37

Project Lexington, 181

Protestant Reformation, 21, 33–34, 51, 54

Protest movements: anti-globalization, 226–27, 239–43, 314, 362; anti-industrialism, 80–83, 91, 317–18; anti-war, 224

Purdue University, 199

Quinine, xii, 123, 327

Stimson, Henry, 192–93, 195
Stone & Webster, 145, 187, 188
Strategic Air Command (SAC), 237, 299
Strategic Defense Initiative ("Star Wars"), 181, 199, 211, 238
Suburban development, 251–54, **255**, 294, 369n.14
Suez Canal, 96, **103**, 119
Swadeshi movement, 119–20, 340
Sweden, 34–35, 215, 229, 307, 325; and industrial revolution, 90; warship *Vasa*, 38–39
Sylvania, 199, 203, 294
Syria, 154, 268, 304; and internet, 317, 339
Systemic risk, xv-xvii, 119, 249–80, **250**, 328, 332, 336, 404. *See also* Risk society
Systems Development Corporation, 208
Systems management, 145, 208, 264, 266
Szilard, Leo, 186, 195

Taipei (Taiwan), 41, 214, 220, 231, 248–49, 251, 292–95, 315, 318
Taiwan Semiconductor Manufacturing Co. (TSMC), 284, 293–96, 318
Taut, Bruno, 150–51, **151**, 159, 167–69, 174
Taylor, Frederick W., 172, 176, 178
TCP/IP. *See* Internet
Technical education, 104–6, 120, 122, 328, 330. *See also* Engineering, education
Technological determinism, x–xi, xv, 283, 311
Technological fundamentalism, 150, 179, 214
Technology: definitions of, 122, 319–20; and economics, 23–24, 26, 213–14, 281–83, 320–23; philosophy of, 340, 397n.40; and power, 119, 317, 326–30; question of, x, xiii, xv, 319–41; and science, 122–51, 175, 320–22, 327
Technology-culture dynamics, xv, 179, 320–24, 335–40; disjunctions, 336–38; displacement, 121, 220,

329–31; divisions, 74, 336–38. *See also* Appropriation, cultural; Digital divide
Technology dialogue, 328–29, 338
Technology transfer, 24, 35, 90, 286, 291, 328. *See also* Technology dialogue
Technopolitics, 324
Telegraph, 130–33, 220–21, 327; in India, xii, 99–104, **100**, **103**; and war, 100–101, 182, 327, 354n.15
Telephone system, 131, 145, 201–5, 221–23, 239, 258–60, 267
Television, 216, 219, 230, 271, 302, 304, 318; standards, 222
Teller, Edward, 190, 198
Tencent Holdings (China), 297, 312–14
Tennessee Valley Authority (TVA), 148
Terrorism, 246, 248, 272–74, 308
Tesla, Nikola: and AC electricity, 138; and radio, 140
Texas Instruments, 200, 204, 286, 289, 292, 294, 380n.22
Thailand, 214, 316, 323
Thermonuclear weapons, 198, 211, 297–98. *See also* Atom bomb
Thomson, Elihu, 137–38, 143–44
Thomson-Houston Company, 136–40
Times Facsimile Corporation, 221
Tokyo, xvi, 60, 231–33, 241, 292
Transistor, xviii, 200–205, 212–17, 232. *See also* Bell Telephone Laboratories; Integrated circuits
Transvaal, 115–17, 335. *See also* South Africa
Trojan horse, 273. *See also* Cyberwarfare
Truman, Harry S, 192–95
Turkey, 152; and global economy, 219, 246, 282; internet, 316–17. *See also* Ottoman Empire
Twitter, xvii, 302–3, 313, 315, 337, 384n.55

Uniform resource locator (URL), xviii, 242. *See also* Internet

Union Carbide Corporation, 187, 321

United States, 99, 119–22, 218, 255–59, 265, 271, 308–10, 316, 329–30, 336; and industrial revolution, 90, 92, 127, 140, 180–98; railroads, 112–15, 120, 153. *See also* Second industrial revolution; *individual (federal-government) acts*

United States Air Force, 181, 196, 201, 204, 206–9, 299. *See also* Strategic Air Command

United States Army: Air Force, 193, 198; Signal Corps, 202–3. *See also* Manhattan Project

United States Coast Guard, 273, 275, 278–79

United States Customs and Border Protection, 275–78

United States Department of Energy, 265, 296, 322. *See also* Atomic Energy Commission; Los Alamos

United States Department of Homeland Security, 267, 273, 275–77

United States Federal Aviation Administration (FAA), 262–64; Advanced Automation System (1981–1994), 263; "next generation" ATC (2007–2025), 264

United States General Accounting Office / Government Accountability Office (GAO), 263, 276–77

United States Immigration and Customs Enforcement (ICE), 275, 316

United States Navy: and atomic energy, 196–97; *Nautilus*, 196; Office of Naval Research, 206–7; and Project Whirlwind, 206–7

United States Secure Border Initiative, 258, 275–76

United States Steel Corporation, 140

United States Transportation Security Administration (TSA), 275

United States Treasury Department, 189

University of Pennsylvania, 199, 206

University of Texas, 287–88

Upton, Francis, 132

Uranium, 184–85, 198, 295; and gun-type atom bomb, 193–95, 273; isotope separation, 187–191, **187**

Urban planning, 162, 175, 251–52. *See also* Architecture; Modernism

Urbino, 2, 10, 12, 15–16

Ure, Andrew, 331

USENET, 239

User heuristic, xvii

Uyghurs. *See* Xinjiang Uyghur Autonomous Region

Uzbekistan, 225, 339

Venezuela, 48, 254; internet, 304, 339

Venice, 2, 9, 21, 31–32, 36–37, 41, 51, 154; Arsenal of, 29–30

Venture capital, 283, 289–91, 303, 312, 316; Kleiner Perkins, 290; Sequoia Capital, 291, 313

Verrocchio, Andrea del, 3–4

Vietnam, 214, 225, 248, 316, 339; war, 199, 211, 213, 299

Virus (computer), 261, 264, 266, 302. *See also* Coronavirus; Cyberwarfare

Vodafone, 304, 307–8

Von Braun, Wernher, xiv, 196

Von Neumann, John, 206–7

Wagner, Martin, 172

Walloons, 34–35, 45

"War on terrorism," 246, 272–74

Watson, James, 231

Watson, Thomas J., Jr. (IBM), 209

Watt, James, 57, 66, 69, 72; and Albion Mill (London), 66, 69; and "horsepower," 66

Weibo (China), 312, 314

West Edmonton Mall (Alberta), 251

Western Electric, 203, 208, 221, 380n.22. *See also* AT&T